The
Symbolic
Earth

The Symbolic Earth

Discourse and Our Creation of the Environment

James G. Cantrill & Christine L. Oravec
Editors

THE UNIVERSITY PRESS OF KENTUCKY

Scholarly publisher for the Commonwealth,
serving Bellarmine College, Berea College, Centre
College of Kentucky, Eastern Kentucky University,
The Filson Club, Georgetown College, Kentucky
Historical Society, Kentucky State University,
Morehead State University, Murray State University,
Northern Kentucky University, Transylvania University,
University of Kentucky, University of Louisville,
and Western Kentucky University.

Editorial and Sales Offices: The University Press of Kentucky
663 South Limestone Street, Lexington, Kentucky 40508-4008

Library of Congress Cataloging-in-Publication Data

The symbolic earth : discourse and our creation of the environment /
    James G. Cantrill and Christine L. Oravec, editors.
        p.   cm.
    Includes bibliographical references and index.
    ISBN 0-8131-1973-1 (cloth : alk. paper). — ISBN 0-8131-0883-7
(paper : alk. paper)
        1. Environmentalism. 2. Environmental policy—United States.
    I. Cantrill, James G. (James Gerard), 1955-   . II. Oravec,
    Christine L. (Christine Lena), 1950-   .
    GE195.S97   1996
    363.7—dc20   96-5956

This book is printed on acid-free recycled paper meeting
the requirements of the American National Standard
for Permanence of Paper for Printed Library Materials.

Manufactured in the United States of America

Contents

Introduction   1
    James G. Cantrill & Christine L. Oravec

Part I: The Field and Context of Environmental Discourse

1   Tracking the Elusive Jeremiad: The Rhetorical Character of American Environmental Discourse   9
    John Opie & Norbert Elliot

2   Naturalizing Communication and Culture   38
    Donal Carbaugh

3   To Stand Outside Oneself: The Sublime in the Discourse of Natural Scenery   58
    Christine L. Oravec

4   Perceiving Environmental Discourse: The Cognitive Playground   76
    James G. Cantrill

5   Environmental Advocacy in the Corridors of Government   95
    Michael E. Kraft & Diana Wuertz

6   Retalking Environmental Discourses from a Feminist Perspective: The Radical Potential of Ecofeminism   123
    Connie Bullis

Part II: Case Studies in Environmental Communication

7   "What to Do with the Mountain People?": The Darker Side of the Successful Campaign to Establish the Great Smoky Mountains National Park   151
    Bruce J. Weaver

8   Plastics as a "Natural Resource": Perspective by Incongruity for an Industry in Crisis   176
    Patricia Paystrup

9   Valuation Analysis in Environmental Policy Making: How Economic Models Limit Possibilities for Environmental Advocacy   198
    Tarla Rai Peterson & Markus J. Peterson

10   Liberal and Pragmatic Trends in the Discourse of Green Consumerism   219
    M. Jimmie Killingsworth & Jacqueline S. Palmer

11   The Mass Media "Discover" the Environment: Influences on Environmental Reporting in the First Twenty Years   241
    David B. Sachsman

12   Media Frames and Environmental Discourse: The Case of "Focus: Logjam"   257
    Harold P. Schlechtweg

List of Contributors   278

Index   281

Introduction

James G. Cantrill & Christine L. Oravec

> The soundest fact may fail or prevail in the style of its telling: like that singular organic jewel of our seas, which grows brighter as one woman wears it and, worn by another, dulls and goes to dust. Facts are no more solid, coherent, round, and real than pearls are. But both are sensitive.
> —Ursula K. Le Guin, *The Left Hand of Darkness*

The past three decades have witnessed a phenomenal growth in concern for and understanding of our environment. Since Earth Day 1970, environmental education in particular has gained a secure foothold in curricula across the country, promising a depth and breadth of ecological literacy unheralded in previous generations. We have also witnessed a steady increase in media attention to the environment, which in turn has influenced the national agenda. And responding as it always has to the vocalized concerns of its constituents, the U.S. government has ratified a series of treaties between ourselves and the Earth. The Clean Air Act of 1963, the Wilderness Act of 1964, the Endangered Species Act of 1966, the National Environmental Policy Act of 1970, the Clean Water Act of 1972, the Toxic Substances Control Act of 1976, the National Energy Act of 1978, the creation of Superfund in 1980, and the more recent Montreal and Rio Accords are but political benchmarks in the greening of public consciousness.

Insofar as the notion of environmentalism seems to have infused our daily lives, one would expect to find evidence that we are now well on our way to restoring and protecting what we have discovered is an all-too-fragile planet. Thus, many have received a sobering slap in the face when confronted with the extent to which the United States and other nations still perpetuate slow ecocide. Despite successful programs and public posturing, species continue to die off, land is irreparably stripped of life-giving forest, bioaccumulative toxins rain down from the heavens, and more people vie for fewer resources every year. Indeed, one need not be an alarmist to sense that the glimmer at the end of the environmental tunnel, touted with pride by industry and politicians, often seems accompanied by an apocalyptic rattle of the rails.

Undoubtedly, there are as many accounts for the failure of environmental rhetoric and reasoning as there are academic disciplines. Scholars have blamed everything from the shortsightedness of civic concern to the plotting of economic oligarchies in the shadows and a spiritual vacuum in the human condition. Yet there remains at least one, visceral thread weaving between such reasons for our continued assault on the environment. It is something we all feel from time to time, something we all recognize when viewing a blighted landscape or listening to the exchanges of environmental advocates, something we far too seldom articulate when thinking about the mess we have made of paradise.

Of our environment, what we say is what we see.

The environment that we experience and affect is largely a product of how we have come to talk about the world. One woman's ecological nightmare is another man's wise use of resources only because we are symbol-using creatures. To be human is to talk of what influences humanity, and in communicating about the environment, we reify what we take to be real. For all intents and purposes, the planet is a captive of our language community; the environment, beyond its physical presence, *is* a social creation, and this fact is often lost in the hoopla of crisis and deliberation. Furthermore, the only hope we have of ever preserving our environment is collectively to understand and alter the fundamental ways we discuss what we continually re-create.

The chapters in this book represent a departure from traditional considerations of the relationship between humans and their environment. Instead of focusing on the tangible interrelatedness of nature or the cold, intransigent face of environmental policies and problems, the contributors to this volume foreground the constitutive and constructive role of communication in approaching environmental issues. And rather than preferencing any one perspective on environmental discourse or privileging one analytic approach over another, we have invited respected colleagues from a variety of fields to reflect on the relationship between communication and our environment. Consequently, this collection represents the intersection among the disciplines of history, anthropology, communication studies, psychology, aesthetics, literature, journalism, rhetoric, and political science. These varied approaches embody an attempt to better appreciate what it means to construct, manipulate, and sometimes irrevocably change the only environment we will ever call "home."

The first part of *The Symbolic Earth* provides a foundation for approaching the nature of environmentalism from a communication perspective. In the initial chapter, John Opie and Norbert Elliot present a historical overview of how environmental discourse in the United States has changed during the past two hundred years; this history, in turn, has altered the manner in which we now approach the environment. In chapter 2, Donal Carbaugh identifies the influences of cultural factors on environmental thought and advocacy by comparing domestic and foreign depictions of natural settings. This theme carries through to Christine Oravec's analysis of environmental aesthetics in chapter 3; notably, her contribution points to the conclusion that, in evoking feelings in response to literal or figurative imagery, representations of nature employed in advocacy speak as much to the fundamental values of society as they do to issues and arguments. James Cantrill's review of research addressing the link between environmental discourse and human cognition suggests the extent to which our mental representations of the environment are more a product of socially induced biases than the result of informative communication about the environment or a particular persuasive campaign. In chapter 5, Michael Kraft and Diana Wuertz explore the role of communication strategies in lobbying efforts directed at elected representatives by environmental activist groups. Finally, Connie Bullis analyzes the growing awareness that issues of gender may powerfully affect how advocates understand and contribute to environmental circumstances. Taken as a whole, these six chapters survey the general elements that ground an understanding of how environmental perceptions, problems, and controversies are socially constructed through discourse.

Since advocacy and communication take place in particular settings, and typically for specific purposes, the second part of the book focuses on the manner in which medium and intent interact in the construction of environmental discourse and, hence, what people consider to be reality. These general and specific examples of communication about human relations to a natural environment draw together many of the theoretical and practical issues explored in previous chapters. Bruce Weaver looks back into history to discuss how protracted human presence in a natural environment influences the construction of a public context warranting federal preservation of land. In chapter 8, Patricia Paystrup deals with the plastics industry's attempt to recast its image in the United States; her analysis reinforces the notion that environmental preconceptions can severely undercut the viability of even well-financed persuasive campaigns. The following pair of studies also highlights the economics of environmental discourse and advocacy. First, Tarla Rai Peterson and Markus Peterson critique the way in which regulatory agencies too often rely upon purely economic considerations in evaluating and addressing public environmental policy. Then, in another genre, Jimmie Killingsworth and Jacqueline Palmer examine green consumerism, establishing the effectiveness of playing upon the environmental sensitivities of Americans to influence market behaviors. Next, David Sachsman explores the beginnings of environmental journalism in the United States as he reviews a series of pathbreaking studies of the media's role in promoting environmental awareness. Finally, in chapter 12, Harold Schlechtweg probes the relationship between political ideology and news reporting in his timely analysis of the media's portrayal of a major environmental issue. Case studies such as these may illuminate both the potential and the actual usefulness of critically inspecting environmental problems and practices, with a eye toward appreciating the role of communication in prompting awareness and policy.

In exploring the range of ideas embodied in this collection of writings, in coming to grips with the reasons for the success or failure of environmental communication, the reader should begin to appreciate the basic thread weaving through the fabric of the work. Pointedly, many of our authors focus on the link between the national ethos of a people and their relationship to the environment. This collective environmental disposition, forged in the crucible of history and shaped upon the enduring anvils of science and economics, exhibits an odd and sometimes precarious malleability. For example, the contemporary trend away from stringent environmental regulations in the United States, most vocally championed by members of the 104th Congress ushered into office in 1994, is only possible because the general tensile strength of environmentalism is continually weakened by competing forces in the public sphere. This tension, which dictates the choices we make in communicating about and within a threatened environment, too often promotes the use of emotion and castigation as the most effective means for promoting policy.

Although an evocational, blame-based discourse may be a potent tool in constructing the environment to reach the goals of advocacy, those contributing to *The Symbolic Earth* also recognize that it is a mistake to eschew implementational, instrumentalist, or apocalyptic rhetoric. Because we are bound to an empiricist culture, people need "logical" reasons to support their beliefs about environmental policy. Therefore, various contributors note that members of the public prefer to have somewhere to retreat when they are challenged or when they come to suspect that sentimentality is at the heart of the matter. Few are willing to be viewed as driven merely by gut feeling.

Yet at its most extreme, empiricism communicated in the form of scientific and technological expertise has become a tangible threat to environmental integrity. As a number of contributors demonstrate, such seemingly neutral, verifiable measures of environmental value as cost-benefit analysis dominate the realm of the knowable in the most pervasive infrastructures of society, government science, and business. The rhetorical nature of such "empirical facts" as the dollar value of a whooping crane is ignored or blithely denied by these dominant infrastructures. Certainly, the distortions of the communicative field inherent in appeals to the emotions are obvious, if difficult to withstand; what is not so obvious is the pervasive denial of such distortions, the narrowing of alternatives, and the unambiguous thinking that pervades the empirical realm. Claims to value-free or value-neutral communication among scientists, business leaders, and government officials must be questioned as carefully as popular exploitation of public sentiment.

In any event, our authors clearly imply that solving the puzzle of environmental communication in a time of danger and uncertainty requires diligent attention to myriad factors, including the forces of cognition, context, and culture. At the very center of those concerns, the knot that holds all of these elements together is the conundrum of language. How does the language with which we describe our world to ourselves and others affect our relationship to the environment? Clearly, references to the power of sublime tropes, place-namings, condensational symbols, and the like suggest that something quite powerful is located in the grist of environmental discourse. As we turn with increasing sophistication to an examination of language, we must not assume that the terms of the debate are unidimensional constructs; different people facing different circumstances may construct the environment in radically different ways while using the very same words. Furthermore, language not only functions as a tool for our use; language also uses us. Our identities with respect to our surroundings are constituted in and by the language that describes them.

Just as Ursula Le Guin pointed out when introducing her classic tale of an anthropologist coming to grips with a faraway culture, so too we hope that this text will underscore the very social nature of environmental "facts." It is a point variously foregrounded in each of the chapters and an issue that strikes at a core dilemma faced by those wishing to understand or practice effective environmental advocacy. When we communicate about the world, should we stress what we know or what we feel? Aim toward wisdom or beauty? Strive to form or inform?

Either of these choices harbors dangerous trade-offs between what is pragmatic and what is right. If we try to compromise by balancing both empirical and visceral ways of knowing when arguing for one or another environmental policy, we risk producing a shallow, convoluted, and ineffective discourse. Yet approaching the turn of the century, we must make a decision about how we should talk about what we know and feel, and we must do so before the opportunity is preempted by a collapsing global ecosystem. Our species has long since lost the luxury of mulling over abstractions such as population "bombs" or Malthusian nightmares. For every child born this second in some corner of the distracted globe, the job of communicating our way out of environmental peril becomes that much more difficult. And in a short time, environmental discourse may fall by the way to be replaced by more draconian measures to maintain a human presence on the planet.

But it is still this century, and we still have the opportunity to study and practice effective environmental communication. If the contributions to this volume represent or support a substantial shift in the way we think about environment and discourse, women and men may yet be sensitive enough to foster a more environmentally benign and prosperous world. The "fact" is, we still have much work to do and an environment in which to accomplish a great deal.

Part I

The Field and Context
of Environmental Discourse

1
Tracking the Elusive Jeremiad: The Rhetorical Character of American Environmental Discourse

John Opie & Norbert Elliot

> I became aware of the old island here that flowered once for Dutch sailors' eyes—a fresh, green breast of the new world. Its vanished trees, the trees that had made way for Gatsby's house, had once pandered in whispers to the last and greatest of all human dreams; for a transitory enchanted moment man must have held his breath in the presence of this continent, compelled into an aesthetic contemplation he neither understood nor desired, face to face for the last time in history with something commensurate to his capacity for wonder.
>
> F. Scott Fitzgerald, *The Great Gatsby*

Fitzgerald knew that American ethos was inextricably bound to geography. So did the colonial Puritans before him, and so do we. Across time and circumstance, Americans remain preoccupied with the environment and its relationship to the national character. The more we know about how our writers use rhetoric to frame positions about nature, the more we stand to learn about how rhetorical response reflects our national character and our perceptions of the natural world. This study, therefore, explores the aims and strategies—those observables such as features, forms, means, and tactics (Crusius, 1989, p. 118)—of environmental discourse.

Our method here is essentially diachronic. We analyze selected documents that illuminate the ways in which Americans have used language to advance positions about the environment. Examples of environmental discourse are taken from the following texts: Samuel Danforth's sermon "A Brief Recognition of New England's Errand into the Wilderness" (1670); Jonathan Edwards's posthumously published *Personal Narrative* (1765); William Bartram's *Travels* (1791); Ralph Waldo Emerson's *Nature* (1836); Frederick Jackson Turner's "Significance of the Frontier in American History" (1893); John Muir's *Yosemite* (1912), which describes his battle for Hetch Hetchy; Aldo Leopold's "Land Ethic," from *A Sand County Almanac* (1949); Rachel Carson's "Fable for Tomorrow," from *Silent Spring* (1962); Bill McKibben's *End of Nature* (1989); Al Gore's "Ships in the Desert," from *Earth in the Balance* (1992); and the Environmental Impact Statement as mandated by the National Environmental Policy Act. Through diachronic analyses of the texts, we establish a framework that illustrates synchronic commonalties of the various rhetorical strategies of these writers.

We regard each of the documents we address as a kind of jeremiad. Perry Miller applies this term to those particular political sermons that took their text

9

from the Old Testament book of the prophet Jeremiah and, as such, were ritualistic "castigations of the people for having defaulted" on their bond with the Lord (1956, p. 6). In seventeenth-century New England, such a political sermon was regularly used as what Sacvan Bercovitch aptly calls a "state-of-the-covenant address" (1978, p. 4). These sermons were delivered at public occasions, most notably on election days. As a rhetorical genre, the jeremiad has left its mark—sometimes explicitly, sometimes implicitly—on each of the texts we analyze, through one of two strategic means.

On the one hand, some jeremiad writers use what may be termed evocative strategies to persuade the reader to act in a certain way by means of associations that are, if we recall Aristotle's term in the *Rhetoric,* pathetic. Such techniques can move an audience to fear, pity, or compassion. We find such strategies used in the rhetoric of sensation of Edwards, the transcendental orientation of Emerson, the epic allusions of Turner, the religious imagery of Muir, the classical references of Leopold, and the apocalyptic visions of Carson, McKibben, and Gore. On the other hand, jeremiad writers use other methodological, or implementational, strategies to persuade the audience to take action in a prescribed way through ordered means that are, if we again recall the *Rhetoric,* logical. Such strategies are evident in the Ramist structure of Danforth's sermon, the classificatory systems of Bartram, and the National Environmental Policy Act's rules for the preparation of Environmental Impact Statements. These two rhetorical strategies reveal disparate cognitive, or psychological, dispositions toward the environment. If we generalize, we might say that writers employing evocative strategies tend to perceive the world as wonderful in its immediacy and in need of our intuitive perception for its maintenance; writers employing implementational rhetoric tend to view the world as chaotic and in need of control.

If we set aside the dissimilar strategies used by American jeremiad writers, however, we see unity in their purpose and effort. First, regardless of its strategy, each of the texts we examine chastises an audience for its failures. Second, this chastisement is used to persuade the audience; the jeremiad is thus persuasive discourse that attempts, in James Kinneavy's definition, to elicit a specific emotion or action (1980, p. 211). Third, as Bercovitch (1978) demonstrates, the jeremiad is a ritual by which the American effort revitalizes itself; in other words, the jeremiad is a ceremonial strategy used by writers to advance various positions that Americans should take toward the natural world. The essence of these rhetorical patterns and what they reveal is the subject of our study. Fourth, the jeremiad is used to obviate potentially dissimilar views and, thus, provide a positive message of hope. As we will show, Danforth and Edwards incorporate the secular realm into the spiritual; Bartram attempts to contain the abundant new discoveries in the American landscape through the order of botanical classification; Emerson unifies national with individual interests; Turner demonstrates that the externality of westward expansion forged the internal American spirit; Muir embeds the contemplative qualities of nature into a progressive national political agenda; Leopold brings the biotic natural world into the philosophic study of ethics; Carson, McKibben, and Gore solicit personal responses to forestall a global apocalypse; and the National

Environmental Policy Act aligns the protection of nature with the legislative process. Thus, the ritual of the jeremiad becomes a positive one: consensus is achieved as the nation proceeds to action (Bercovitch, 1978, pp. 25–26).

Resolving Conflict and the Rise of Ramism

Rhetorical form is driven by cultural circumstance. Indeed, if we agree with Niklas Luhmann in his *Ecological Communication* (1989, p. 30), we come to see communication as a social operation so powerful that it changes mental states. To understand the jeremiad itself, we must therefore begin with those particular circumstances that necessitated its development.

One of the great, if largely forgotten, controversies of the eighteenth century was whether the discovery of America had been a terrible mistake. Many Europeans felt that humanity would have been better off, and happier, if the New World had never been discovered. America seemingly stood outside natural science as well as traditional theology. The wilderness was chaotic, a place far from the world the colonists knew. In their world, life achieved order because it was sacred. Yet the discovery could not be abandoned because the secular errand had already begun. The New World had gained an economic importance; that is, after all, why the Virginia Company licensed a group of merchants to plunder America for timber and furs. From the beginning, therefore, there was a need to obviate tension between apparently antithetical but coexistent points of view. If the venture into the New World was to succeed, then the secular and the sacred had to be fused. This effort at fusion was precisely the force that shaped the American jeremiads (Bercovitch, 1978, p. 29).

In Raymond Williams's famous phrase, the Puritans established the "dominant culture" in America (Bercovitch, 1978, p. xii). This culture was not at all hierarchical but rather a fluid, free enterprise structure. This sense of free enterprise provided a rich medium for the growth of the Puritans' dual mission: to be both practical and spiritual guides. As Bercovitch puts it, "In their church-state, theology was wedded to politics and politics to the progress of the kingdom of God" (p. xiv). The jeremiad, or sermon, preached by Puritan ministers was the vehicle used to urge the American colonists to lean into their mission.

Of course, the mission had a grim side, ably demonstrated by Perry Miller (1956). As John Winthrop wrote in 1630 while still on the *Arbella*, "If we shall deal falsely with our God in this work we have undertaken and so cause Him to withdraw His present help from us, we shall be made a story and a by-word through the world" ([1630] 1963, p. 199). But these Old Testament warnings also contained the promise of ultimate success. The Puritans, as Bercovitch argues, were a people chosen as "instruments of a sacred historical design. Their church-state was to be at once a model to the world of Reformed Christianity and a prefiguration of New Jerusalem to come" (1978, pp. 7–8). Hence, God's punishments were corrective; vengeance became a sign of love. Rather than condemning, the jeremiad castigated in order to correct. Correct religious behavior, devotion to the Puritan mission, would ensure secular success. This fusion of dissimilarities is at the heart of the jeremiad: its aim is to obviate differences between sacred and secular ends.

If the Puritans dared to inhabit such a wasteland, they as God's people would need powerful, deliberative guidance. The rhetorical training of those who would lead the communities in the wilderness thus stands at the center of American beginnings. Students were trained in the classical tradition of the Renaissance as well as the religious tradition of the Reformation. Fundamentally, the curriculum was Latin, its elegant rationality shaping the language of all the professions; Aristotle was either translated into Latin or studied in the original Greek, with that language's accompanying abstractions of ontology and metaphysics. Eventually Hebrew, the site-specific pragmatic language of the priests, prophets, and seers, was added.

Interpreting this classical curriculum, S. Michael Halloran (1982) finds that rhetorical training was the most important subject in the curriculum because it was the means by which all other subjects were communicated. The result of this rhetorical training was, as Halloran points out, the graduation of a student who relished the opportunity to declaim. Additionally, this classical training provided a sense of organization and structure that informed colonial writing on nature. For colonial orators such as Samuel Danforth, the world was rational and could, consequently, be conquered by reason. Graduating from Harvard or Yale, students were prepared to lead by the relentless application of method. The most methodological type of persuasive rhetoric, logic, produced predictable outcomes. This orderliness was a real comfort amid the wilderness of the external world.

In the context of this classical training, we must recognize the power of Ramist logic, which profoundly influenced the direction of Puritan thought and, thus, the genre of the jeremiad. Petrus Ramus (1515–72), a Renaissance rhetorician, "robbed the classical Rhetoric of *inventio* (discovery and evaluation of the material) and *dispositio* (classification and organization of the material) for the benefit of Logic" (Ong, 1958, p. 173). The fact that Ramus became a Protestant near the end of his life and was murdered in the Saint Bartholomew's Day Massacre helped to establish his popularity at Harvard. Elevated to the status of a saint in Puritan America—Cotton Mather called him "that Great Scholar and Blessed Martyr" (Morison, 1936, p. 188)—Ramus exerted a dominant influence at Harvard, and on America's clergy, until the first quarter of the eighteenth century. Perhaps the influence of Ramist logic can be seen most clearly in the clergy's communicative forms—including the jeremiad.

It is the break Ramus made with Aristotle that is most significant for our study. The classical rhetoric of Aristotle and Quintilian was rich in invention and arrangement of ideas. The thirty-nine topoi of Aristotle's *Rhetoric*—ways of analyzing, describing, and extending ideas—served as heuristics, or stimuli, that provided multiple ways of investigating a phenomenon. In addition, classical rhetoric offered various rhetorical techniques for the arrangement of thought after the heuristic process of invention. Ramus, however, bypassed the invention of ideas and the diversity of their presentation and focused only on method. Ramist method was based on Solon's Law, a technical procedure originally devised for plotting building ordinances in ancient Athens (Ong, 1958, p. 280; 1977, p. 190). The Ramist system itself was, therefore, based on a method of surveying the physical world. Accordingly, human thought became a concrete phenomenon, able to be manipulated like any physical object in nature. As a consequence of the absence of invention, Ramist discourse neglected the free exploration and development of ideas. Rigidity of approach is the consequence of Ramist rhetoric. "Ramist thought," Walter Ong writes, "was in effect the perfect closed system: for it contained 'method' as one of its parts, and method prescribed how the logic that contained it was to be organized and consequently how all thought was to be organized 'logically' as a collection of closed fields separated from one another" (1977, p. 191). Critical thought—including such aspects of rhetoric as the possibility of refutation, the limits of certainty, and the presence of multiple interpretations—was thus abandoned in favor of methodical thought.

It is not difficult to infer why Harvard-trained Puritan ministers would flock to Ramism. Methodical thought was the best way to deal with the ferocity and chaos of nature. It offered control over both nature and self. As Ong puts it, the result was a quantified, diagrammatic approach to physical reality and to the mind itself (1958, p. 306). Predictable outcomes were necessary for the conquest of both the external wilderness and the spiritual wilderness within.

The Election Jeremiad of Samuel Danforth

Samuel Danforth (1626–74) graduated from Harvard in 1643. A student of Cotton Mather's, Danforth was trained in the Ramist system of rhetoric. It is thus appropriate for us to analyze "A Brief Recognition of New England's Errand into the Wilderness"—an election sermon delivered in the form of a jeremiad on May 11, 1670, in Boston—for its adherence to method.

On important election days in Puritan New England, speakers had an opportunity to address the colony's governor, deputy governor, clergy, magistrates (who hoped to be reelected), and deputies. As A.W. Plumstead (1968) writes in his analysis of the context of these sermons, this was the only annual opportunity to speak to the full panoply of leaders of God's chosen people in New England.

Danforth took his text from Matthew 11.7: "What went ye out into the wilderness to see? A reed shaken with the wind?" ([1670] 1968, p. 57). After explicating the text according to other biblical sources and Greek etymology, Danforth set the doctrine in a way that recalled Winthrop's sermon of gloom forty years before: "Such as have sometime left their pleasant cities and habitations to enjoy the pure worship of God in a wilderness are apt in time to abate and cool their affection thereunto; but then the Lord calls upon them seriously and thoroughly to examine themselves, what it was that drew them into the wilderness, and to consider that it was not the expectation of ludicrous levity nor of courtly pomp and delicacy, but of the free and clear dispensation of the Gospel and the kingdom of God" (p. 61).

Danforth argued that his audience had a glorious but brief opportunity to construct a working balance between nature and civilization as they expanded across the fresh, uncorrupted continent. Unlike Plymouth's Pilgrims, who were running away, the Puritan settlers of Boston's Massachusetts Bay in 1631 came deliberately to take advantage of the untouched wilderness. Their stated mission was to construct, once and for all, the heavenly kingdom on earth using the pure biblical polity set forth in the New Testament. "The Bay Company," as Perry Miller wrote, "was an organized task force of Christians, executing a flank attack on the corruptions of Christendom . . . in a bare land, devoid of already established (and corrupt) institutions, empty of bishops and courtiers, where they could start *de novo,* and the eyes of the world were upon it" (1956, pp. 11–12). According to Danforth's jeremiad, a one-time opportunity was in the settlers' hands, and they were letting it escape their grasp. Their errand, Danforth proposed, was going badly. The job was becoming a colossal blunder. If the path was to be cleared, action had to be taken in a highly structured manner. In a world of harmony and consistency, the message of rigor had to be conveyed in a rigorous, albeit remorseless, rhetorical form.

Hence, more than the message of doom is significant. When Danforth began to speak in 1670, he presented his sermon in a rhetorical mode familiar to his audience, insofar as seventeenth-century Puritan sermons possessed a form sanctioned by the rhetoric of Peter Ramus. The sermon began with the identification of a specific biblical passage that would serve as the theme. An explication of the text in historical or philological terms followed. Then the orator announced the doctrine, the lesson to be drawn from the biblical text. Following the doctrine came the application to the listening audience of Massachusetts colonists. The sermon may thus be said to have four distinct sections: identification of biblical text, explication, doctrine, and application.

Significant here is the reliance of the jeremiad on Ramist method, correctly understood as a series of steps. The initial biblical quote from Matthew opens the sermon (step 1); the text is then rationally analyzed, historically and linguistically (step 2); the doctrine is formed from this analysis (step 3); and each application of the doctrine is analyzed for its use (step 4). The possibility of refutation, notation of the limits of certainty, the possibility of multiple interpretations—all are absent. Present are dogmatic propositions. Danforth's "Errand" sermon, as Ong says of the Ramist orientation, has "no acknowledged interchange with anything outside of itself" (1982, p. 134). There are, as Ong realizes, neither difficulties nor adversaries. The message is articulated through a narrow process. The characteristics of careful thought are replaced by what Ong has called "corpuscular epistemology" (1958) and "logocentrism" (1982), both terms meaning "a gross one-to-one correspondence between concept, word, and referent" (1982, p. 168).

There is good reason to spend so much time dwelling on the influence of the Ramist tradition on the jeremiad. First, the "Branches" (sections) and "Uses" (application) of the jeremiad initiated a continuous history of American jeremiads, or complaints, about the tragic fall of Americans from their mighty destiny. Americans have not acted rightly in their venture into the wilderness. Let the congregation be warned: individuals are in the wilderness for a specific purpose; should their affection for that purpose cool, the coal pits await. These messages, as we will demonstrate, continue in even the most recent American environmental discourse. Second, Ramist tradition carries a utilitarian orientation toward the environment; the natural world is merely the vehicle of Danforth's Puritan errand of anthropocentrism. Danforth has no aesthetic and contemplative reverie such as Fitzgerald would experience some two and a half centuries later. Plumstead writes, "Any joy the Puritans experienced came less from its [the wilderness's] intrinsic beauty than from thought of what they could do with it for God—how they could change it and make it into an 'enclosed garden' and a 'city on a hill'" (1968, p. 50). Ramist methodical thought found its purpose: promotion of the ethic of self-improvement. The presence of Ramism establishes the potentially ruthless orientation that governs the way Americans have been taught to deal with the environment.

What have we inherited from the jeremiad as expressed in 1670? In time, as Bercovitch (1978) has shown, jeremiads became secularized into high-minded critiques that carried immense moral weight against guilty Americans. Americans, the people without a history, who invented themselves, depended upon the speaker for guidance toward an uncertain future (Wills, 1978, pp. xix–xxi; Ritter, 1980, pp. 158–59). Leslie Fiedler observes that "we are forever feeling our pulses, because we feel, we *know,* that a little while ago it was in our power, new men in a new world (and even yet there is hope) to make all perfect" (quoted in Ritter, 1980, p. 164). David Noble (1965) adds that Americans must believe in the possibility that the nation might return to its covenant; if the people would repent and return to the old faith, history would be on their side.

So too endures the metaphor of a national saga embedded in the jeremiad. The jeremiad gained enduring value because its rhetorical form, though ritualistic, had a flexibility, according to Kurt Ritter, "to allow the Puritans [and succeeding generations] to redefine their identity once it became clear that their mission was not merely a temporary sojourn away from corrupt England, but a permanent errand into the wilderness" (1980, p. 157). This was the new history and destiny. The Puritans had the obligation to transform their new land from wilderness to a New Jerusalem, to turn from chaos to divine rationality. The environment was both profound disorder and the frighteningly all-powerful vehicle by which God punished his errant people. If nature, according to the Reverend Samuel Danforth, had the upper hand and the colonists were plagued with "earthquakes, dreadful thunders and lightnings, fearful burnings" (Plumstead, 1968, p. 74), then the people must recall the true meaning of the word *errand.* They are not on a little journey as inferiors; rather, they are on real business (P. Miller, 1956, p. 3). As Bercovitch writes, the wilderness "was a territory endowed with special symbolic import, like the wilderness through which the Israelites passed to the promised land" (1978, p. 15). Thus, the wilderness is the arena for both sacred and secular work.

It could have been merely ceremonial, but instead the jeremiad opened a long-standing American tradition of method and metaphor. Sometimes consciously, sometimes unconsciously, American writers have followed the rhetorical tradition of the jeremiad. In this tradition we are taught that we must make use of nature (itself the stage for the American experience) through a unique kind of dogmatic order (itself a product of remorseless method). We will see other influences on the American conception of nature, but the tradition of the jeremiad continues to endure, perhaps even to prevail.

Jonathan Edwards and the Rhetoric of Sensation

Perry Miller described the next phase in America's adventure with nature when he wrote that "the Great Awakening [of the 1740s] was the point at which the wilderness took over the task of defining the objectives of the Puritan errand." Miller wrote, "I am the more prepared to say this because Jonathan Edwards was a child of the wilderness as well as of Puritanism" (1956, p. 152). In the work of Jonathan Edwards (1703–58) we see the natural environment as a manifestation of the divinity of God. Thus, Edwards introduces into the tradition of the American jeremiad what we have termed the rhetoric of evocation. For Edwards, the will of God is communicated through the evocative power of language. Absent in Edwards is the Ramist methodology of Danforth; present is the use of language to incite a sense of divine power.

For many of us, Edwards is associated only with the notorious 1741 Enfield sermon. We interpret, as did the horrified orthodoxy of Boston, the screams of the congregation as evidence of the resurgence of chaos that would undermine hard-earned order. Edwards, stalking the wilds of western Massachusetts, heard the noises of revival as evidence of the human recovery of God's grace—goodness and justice and honesty—in the wilderness. For Edwards, the more venerable and raw the human spirit, the more capable it was of receiving God's grace. The wilderness, then, came to symbolize the unfettered spirit's potential for receiving grace, and it became the perfect setting in which that spirit should fulfill God's will. For Edwards, neither God nor his nature, nor a humanity filled with grace, could be contained in some Ramist conceptual box. Although a 1720 graduate of Yale, who thus received much the same education in logic as Danforth had at Harvard three-quarters of a century before, Edwards devoted himself to establishing connections between words and emotions.

Edwards explained the connection between human well-being and wilderness through the psychology of John Locke, as expressed in book 3 of the *Essay Concerning Human Understanding* (1690). From Locke, Edwards learned to read the ontological connection between the rashness of wilderness and the extremes of human sensibility. Following Locke, Edwards aimed to present neo-Platonic ideas, newly visible in the fresh wilderness, plain and naked through language—not simply the shadows on the cave wall, but white-hot reality. Disregarding the Ramist method of Danforth, Edwards redefined the world of ideas. Perry Miller writes that Edwards "so conceived [an idea] that it became the principle for organization and of perception not only for the intellectual man but for the passionate man, for the loving and desiring man, for the whole man. He conceded readily that a word can act as an emotional stimulus.... What he insisted upon was that by the words (used in place of a thing) an idea can be engendered in the mind, and that when the world is apprehended emotionally as well as intellectually, then the idea can be more readily and more accurately conceived" (1956, pp. 180–81). Thus, Miller described

Edwards's use of language as "sensational rhetoric" (p. 175). By living with the external wilderness, one could inhabit the kingdom of God. In his stunning *Personal Narrative,* written sometime after 1739 and published posthumously, Edwards described his first epiphany that the house of the Lord was in the virgin forest. After talking with his father about how the solitary mountains kindled his intimation of divinity, Edwards left his father's house for the pasture: "And as I was walking there, and looked up on the sky and clouds; there came into my mind a sweet sense of the glorious majesty and grace of God that I know not how to express. I seemed to see them both in a sweet conjunction, majesty, and meekness joined together. It was a sweet and gentle and holy majesty; and also a majestic meekness; an awful sweetness; a high, and great, and holy gentleness" ([1765] 1962, p. 60). Emerson would later infuse himself with the same sweet nature. For Edwards, such language itself fulfilled the goal of rhetoric: it created a disposition to the spiritual. The best aspect of the new American wilderness was that it revealed evidence of the unseen world, as much as it showed greenness in leaves, hardness in rocks, wetness in water, and edenic simplicity in Indians. For Edwards, the movement of the lily was a signifier of calmness, gentleness, and benevolence. Even those who are miserable love life, Edwards wrote in a notebook fragment, "because they cannot bear to lose sight of such a beautiful and lovely world" ([1726] 1980, p. 306).

Insofar as Enlightenment rationalism replaced theology in the late eighteenth century, Edwards had no theological heirs. But his rhetorical methods, as we will see, survive. His enduring contribution to the rhetoric of environmental discourse is summarized nicely by Miller: "So the word must be pressed and the rhetoric must strive for impression; it is a strength, not a weakness, of language that no matter how sensational it becomes, it has to depend upon something happening to the recipient outside and above its own mechanical impact" (P. Miller, 1956, p. 183). As a jeremiad writer, the Edwards of the Enfield sermon moved his listeners to fear, but his use of language as an evocative stimulus, seen explicitly in the *Personal Narrative,* also provided precedent for more positive evocative powers of language.

For Edwards, personal salvation was linked to public success (Bercovitch, 1978, p. 103), and language (the jeremiad) was the key to obviating the disparity between these concepts. Through its rejection of structured argument as the means of communicating divine insight, Edwards's jeremiad offered an alternative strategy by which the potentially conflicted private (sacred) world of the individual could be reconciled with the public (secular) world of the nation: people moved by language to work together in the wilderness to receive and fulfill God's covenant were bound to incur success.

Enlightenment Rationalism, Scientific Method, and Natural History

In the late eighteenth century, theology began to give way to botany. Among the many innovations left to us by the Age of Reason was the science of ecology (Worster, 1977). Two traditions established the place of nature. The first, embodied by Gilbert White (1720–93), was that of the parson-naturalist; the second, embodied by Carl Linnaeus (1707–78), was that of the anthropocentric classifier. Both traditions can be seen in the 1791 *Travels* of William Bartram (1739–1823).

In essence, Enlightenment rationalism treated nature as an insentient and inferior material world. A Manichaean dualism contrasted mindless nature with the infinitely superior world of human rationality and deistic spirituality. Humans, distanced from personal involvement with nature, could know—and perhaps control—the universal natural laws that govern nature. Nature was a mechanical automaton. Humanity could thus be freed from nature's grasp. As Donald Worster writes, "In its utilitarianism the Linnaean age of ecology strongly echoed the values of the Manchester and Birmingham industrialists and of the English agricultural reformers of the same period. . . . The ecology of the Linnaeans, then, dovetailed neatly with the needs of the new factory society" (1977, p. 53). Historically, this utilitarian orientation had gained momentum with the individualism of Renaissance humanism, had secured a method with Newtonian science, and now became dominant by means of the industrial revolution. Enlightenment savants concluded that the successful upward march of human civilization was to be defined by its progress toward autonomy. American history thus could be a narrative of the gradual conquest of the American wilderness, a demonstration of civilization's growing mastery of natural resources, a manifestation of built environments that celebrate their independence from nature.

According to this view, science was the supreme human activity, legitimate and authoritative because it was grounded in abstract universal principles that transcended any particular human context. Scientific method thus became, as Ramism had before it, the dominant rhetoric of Western culture (Halloran, 1982; C.R. Miller, 1978; Bazerman, 1988). Scientific objectivity prompted an even greater level of faith in method than the sermons of Danforth had. The application of scientific method gave one a key to the laws that God employed to govern the stars and planets. The New World thus became a great experimental laboratory that could be known through the process of cataloging all the phenomena of living nature. The imperial tradition—defined as the impulse to dominate the Earth (Worster, 1977, pp. 29–55)—left nothing untouched in its comprehensive inventories.

Americans believed that the wilderness had found its methodological master in the study of natural history. Virginia's Captain John Smith detailed the climate, the fertility of the soil, the shape of the terrain, the varieties of vegetation, and the different animals. The Swedish traveler Pehr Kalm, trained under Linnaeus himself, linked the distribution of plants with the difference in their habitats in Pennsylvania and New Jersey. The rules were to discover, describe, and classify according to Linnaeus by ordering nature into plant and animal kingdoms, each subdivided into phylum, class, order, family, genus, and species. Between 1748 and 1751 Kalm identified fifty previously unknown trees in northeastern America, while André Michaux and his son François André published volumes in which they sought to catalog all the trees of the New World.

There was significance in this impulse for classification and analysis. Natural history, for instance, suggested an environmental determinism: fevers, disease, debilitations, depended upon climate. America's best-known doctor, Benjamin Rush, attributed Philadelphia's yellow fever outbreak in the 1790s to exhalations from putrefied coffee on a wharf near Arch Street. Natural phenomena became interactive. The mission of science, therefore, was not only to reflect the world but also to improve it by understanding and anticipating these relationships. The self-help echo of Danforth is unmistakable. The human assignment in the new enlightened era was to convert the howling wilderness into a domesticated agricultural landscape. J. Hector St. John de Crevècoeur, a French intellectual who briefly enjoyed himself as a gentleman farmer in New York, wrote a brilliant set of letters about the assignment. As Benjamin Franklin had written in his 1743 prospectus for the new American Philosophical Society, "the first drudgery of settling new colonies . . . is now pretty well over," thus allowing attention to "all philosophical experiments that let light into the nature of things, tend to increase the power of man over matter, and multiply the conveniences or pleasure of life" (quoted in Boorstin, 1960, pp. 9–10).

The new fraternity of natural historians described in their journals the impact of environment upon plants, animals, and especially humans. John Bartram praised the diversity of plants, the remarkable soil, and the "clear and sweet water" of Pennsylvania and the Carolinas. His son William turned his own journey through the exotic southern United States into an adventure story about a wide-eyed wanderer through a Garden of Eden. William's famous *Travels,* published in 1791, elicited the praise of Coleridge and Wordsworth; he dazzled his readers with accounts in the rhetorical tradition of Gilbert White. For example, he wrote of the Alachua Savanna in Florida, "How the mind agitated and bewildered, at being thus, as it were, placed on the borders of a new world! On the first view of such an amazing display of the wisdom and power of the supreme author of nature, the mind for a moment seems suspended, and impressed with awe" ([1791] 1955, pp. 166–67). This is the kind of reaction Fitzgerald hoped for. Nevertheless, Bartram also exhibited the utilitarianism of Linnaeus. Here is an especially telling passage from the *Travels:* "Next day we passed over part of the great and beautiful Alachua Savanna, whose exuberant green meadows, with the fertile hills which immediately encircle it, would if peopled and cultivated after the manner of the civilized countries of Europe, without crowding or incommoding families, at a moderate estimation, accommodate in the happiest manner above one hundred thousand human inhabitants beside millions of domestic animals; and I make no doubt this place will at some future day be one of the most populous and delightful seats on earth" (p. 211).

It is worth noting the absence in Bartram's work of the emotive rhetoric of the jeremiad as an expression of doom; the optimism is evident in both passages quoted above. Yet both passages also display the language of sensation as expressed in Edwards's *Personal Narrative.* The description in the second passage, however, is not provided so that we feel the power of beauty and thus are drawn into an adoration of God. Rather, Bartram's rhetorical aim allows the reader to feel the power of secular promise and thus to become drawn into capitalist action. The jeremiadic ambiguities thus arise again: nature is at once a source of inspiration and a vehicle for productivity. The purpose of the errand into the wilderness (still both a sacred and a secular place) is to capture it in language and thereby to control it. Bartram's *Travels,* seen in this light, becomes a ritual (Bercovitch, 1978, pp. 24–26). In this rite, differences—the appreciation of nature versus its utilitarian use—are obviated, and a mode of consensus—the aesthetic response will encourage utilitarian use—is established.

What kind of rhetorical system was used in the eighteenth century to support such observations? After the bonds with England were broken, American universities shifted away from both the classical rhetoric of Aristotle and the methodical rhetoric of Ramus. As James Berlin observes, classical rhetoric in American colleges was associated with British colleges and a British way of life. New ideas continued to be imported, but now they came from Scottish, not English, colleges. Specifically, American universities embraced Scottish Common Sense Realism, a philosophy commensurate with American economic, religious, and aesthetic experience (Berlin, 1984, p. 6).

Two Scottish rhetoricians—George Campbell and Hugh Blair—wrote textbooks that became enormously popular in American universities. Blair's *Lectures on Rhetoric and Belles Lettres,* for example, published in 1783, went through 130 editions until the final one in 1911 (Berlin, 1984, p. 25). The rhetoric endorsed by Common Sense Realism was aligned with the rise of the classificatory schemes of the Enlightenment in two ways. First, the new rhetoric was inductive. It approached truth in a way readily adopted by a country interested in the practical benefits of science. Induction was naturally suited to descriptions of the cranes Bartram examined in the savannas of Florida. Second, the rhetoric of Common Sense Realism established forms for exposition, what have come to be known as the modes of discourse: narration, description, exposition, and argumentation (Connors, 1981). These modes allowed an ordering of experience that supported the classificatory schemes of the new science.

Yale adopted Blair's textbook in 1785, and Harvard followed in 1788, emphasizing the popularity of Common Sense Realism (Applebee, 1974, p. 9). The late eighteenth century was an age of observation, in which natural history became an emblem of the new American consciousness, including its utilitarian mode. The scientific method, fueled by the jeremiad and sustained by Scottish Common Sense Realism, uncovered nature's secrets and allowed the exploitation of the New World. Nature was a commodity, defined as a utilitarian tool in service to humanity. Bartram looked on the savannas and imagined suburbs.

The Transcendentalist Passion for Nature

Ralph Waldo Emerson (1803–82) regretted that no university had offered him a professorship in rhetoric (Berlin, 1984, p. 42). He was a skilled orator, and his voice resonates in essays such as *Nature* (1836) and *The American Scholar* (1837), each originally a public address. Emerson's celebration of nature—our modern ecosystems—stands in blatant opposition to the prescriptive response set by Campbell and Blair. With his intellectual forebear Jonathan Edwards, Emerson represents part of a minority report on America's environment. His rhetorical technique is therefore worth examining.

The new Transcendentalism protested against both Christian literalism and Enlightenment empiricism and sought to replace them by depicting the flow of a divine spirit from nature. According to these American Romantics, scientific rationalism could not adequately explain the passion and diversity of humanity, nor did it sufficiently plumb the depths and powers of nonhuman nature. Enlightenment scientists superficially labeled nature's features so that they could be used to human advantage. They saw enlightened humans as the potential masters of nature. The Romantics, on the other hand, were not interested in dissecting nature in order to control it. Rather, they wanted to embrace nature in a holistic, mystical union so as to be ennobled by it. They saw nature as the source of human enlightenment. For Emerson, as for other Romantics, rhetoric became a means of establishing and verifying the vital relationship between humanity and nature: "Words are signs of natural facts, particular natural facts are symbols of particular spiritual facts, [and] nature is the symbol of spirit" ([1836] 1983, p. 20).

The Romantics were fascinated with the unfettered, almost savage natural forces that first the Puritans and later the rationalists sought to control. Their rhetorical ambition was to liberate, not throttle, these primeval forces; metaphoric language was, of course, the perfect vehicle for expressing such forces. For Emerson, Berlin writes, "metaphoric language is the norm rather than the exceptional province of the philosophers" (1984, p. 48). Rhetoric requires neither the method of Ramus nor the vehicle of common sense. "The moment our discourse rises above the ground line of familiar facts, and is inflamed with passion or exalted by thought," Emerson wrote in *Nature*, "it clothes itself in images. . . . Hence good writing and brilliant discourse are perpetual allegories. This imagery is spontaneous. It is the blending of experience with the present action of the mind. It is proper creation. It is the working of the Original Cause through the instrument he has already made" ([1836] 1983, p. 23). Commenting on this passage, Berlin writes that the role of the orator is to unite the material and the ideal in metaphor: "Subject and object have meaning only in the creation of the unifying symbol" (1984, p. 48).

In his gentle jeremiad, *The American Scholar*, orated in 1837 before the Phi Beta Kappa Society in Cambridge, Emerson proposed that the American wilderness—nature still in its harmonious totality—could be the source of a new humanity peculiar to America. Human society is flawed; the source of renewal is nature. Nature is the seamless web of God, without beginning or end, the reflector of the infinite potential of humanity, which goes far beyond the capabilities of science. Recalling a noetic field captured in the rhetoric of Jonathan Edwards, Emerson believes the new Americans can break their chains. They must strive beyond mere language and science to apprehend enlightenment directly: "Man Thinking must not be subdued by his instruments. Books are for the scholar's idle times. When he can read God [in nature] directly, the hour is too precious to be wasted in other men's transcripts of their readings" ([1837] 1983, p. 58). Here was immanence with a vengeance, a rhetorical strategy leaning toward pantheism.

Jonathan Edwards, as we have seen, made his leap toward a blinding divinity in the physical universe. His unblinking look vaulted into the Romantic era and American Transcendentalism. Romantic Americans did not avert their faces from the divinity but looked wilderness straight in the eye. Everyone had immediate access to this pantheistic God, Emerson believed. We each become, in his infelicitous and consuming phrase, "a transparent eyeball" ([1836] 1983, p. 10). To enter the wilderness was to enter the divine presence. The United States had an advantage, Emerson knew: in its unspoiled terrain wild nature could still be experienced.

Emerson utilized the jeremiadic ambiguities. That is, he invited the individual to an endless process of incorporation through individuation. "Far from pressing the conflict between individual and society, Emerson obviated all conflict whatever by defining inward revolt and social revolution in identical terms, through the bipolar unities of the symbol of America" (Bercovitch, 1978, p. 184). For those following Emerson's jeremiad, the private vision of the individual was incorporated into the public mission of the nation. Thus, although the rhetorical strategies of the scientifically driven Bartram and the transcendentally inspired Emerson differ, we again see the unified aim of the American jeremiad.

Frederick Jackson Turner and the Rhetoric of History

Frederick Jackson Turner (1861–1932), the prophet of the westward movement, was an instructor in rhetoric before he began to teach history at the University of Wisconsin. He was honestly shocked when the director of the 1890 U.S. Census declared that the nation's frontier had come to an end. The frontier line between free land and settled land, which had moved westward for three hundred years, had been so overrun by civilization that it had effectively disappeared. Reflecting the anxiety that characterized the fin de siècle mood of Americans in the 1890s, Turner had doubts about the new forces of urbanization, capitalism, industry, and immigration from southern and eastern Europe. He had trouble connecting these new forces with the earlier golden age of exploration and settlement, and he was wary about the future of the individualism, democracy, and opportunity that he believed had come from the frontier experience.

Before we turn to an analysis of Turner's 1893 essay "The Significance of the Frontier in American History," it will be helpful first to identify those aspects of nineteenth-century post–Civil War rhetoric commonly employed in contemporary university textbooks. John Bascom, president of the University of Wisconsin during Turner's student days there, was a well-known textbook author (Curti & Carstensen, 1949, pp. 275–95). Bascom's *Philosophy of Rhetoric* (1866) and Adams Sherman Hill's *Principles of Rhetoric and Their Application* (1878) were important contemporary texts, both influenced by the Scottish Common Sense Realism— Ramist in approach—of George Campbell's *Philosophy of Rhetoric* (1776) and Hugh Blair's *Lectures on Rhetoric and Belles Lettres* (1783). So close were Bascom and Hill to the Scots, particularly in their lack of emphasis on the inventive power of rhetoric, that Berlin (1984) refers to them as American Imitators. Turner, however, was far more influenced by the rhetorical style and presentational strategies that Hill and Bascom advocated than he was by their Ramist methodology. Both Hill and Bascom advised orators to use presentational strategies drawn from Emerson, even though invention was to be limited, efficient, and logical.

In Turner's 1893 oratorical performance, he followed Bascom's advice to throw "the whole soul into a single current, setting outward effort." "The highest oratory," Bascom wrote in his textbook, "can only be called forth when the energies of the whole nature, with its fundamental forces, moral and religious, are to be aroused" (1866, p. 37). Turner declared independence from contemporary rhetoric with his famous 1893 presentation by not heeding Bascom and Hill's advice on the efficiency of rhetorical invention. Turner had become truly Emersonian by thinking, not merely speaking, in an inspired fashion. He abandoned the rhetorical strategies of academic logic in favor of evocative rhetoric.

Since Turner enjoyed a distinguished role as a student orator and as a young tutor in rhetoric to some two hundred students each semester, it should come as no surprise that his 1893 essay demonstrates great rhetorical sophistication. Because so little attention has been paid to these features of the essay—even though Turner first orated not only the 1893 paper but virtually all of his subsequent frontier papers over the next two decades—we will analyze some of its more interesting features (see Opie & Elliot, 1993).

The opening paragraph of the essay is designed, as Hill would have it, to give necessary information, engage attention, and win regard (Hill, 1878, p. 247). Turner begins by citing the 1890 bulletin from the superintendent of the Census. He quotes its finding: "At present the unsettled area has been so broken into isolated bodies of settlement that there can hardly be said to be a frontier line" ([1893] 1962, p. 1). Turner's key declaration is then conveyed in a periodic sentence, that grammatical structure especially designed for impact because, as Hill puts it, "the meaning is suspended till the end" (1878, p. 152). "This brief official statement," Turner says, "marks the closing of a great historic moment" ([1893] 1962, p. 1).

Turner goes on to employ metaphoric images of birth and growth that call to mind Charles Darwin's evolutionary theory: "Behind institutions, behind constitutional forms and modifications, lie the vital forces that call these organs into life and shape them to meet the changing conditions." American institutions, Turner continues, "adapt" themselves to changes; we have developed from the "primitive" ([1893] 1962, p. 2). During Turner's time, the theory of environmental determinism posited that geographic variations determined the nature of individuals. As Ray Billington, Turner's greatest successor, writes, to Turner this meant one thing: "The unique social and physical environment of the frontier endowed the pioneers with the distinguishing traits noted by travellers" (1973, p. 113). For Turner, the vital forces were the West. Just as Emerson had found metaphoric qualities in nature, so Turner found a symbol for democracy, individualism, and free enterprise in the West. He conveyed that proposition with a force that remains with us today.

But another kind of growth was also declared at the Chicago meeting of the American Historical Association, where Turner delivered his lecture. Twice in the first five paragraphs of the essay, Turner tells us that the "germ" theory of political development has been overemphasized ([1893] 1962, p. 2). There is more going on in the wilderness than "the development of Germanic germs," he slyly adds (p. 4). During his time at Johns Hopkins (1888–89), Turner had been exposed to the new scientific history then being imported from Germany. Influenced by linguistic methods, historians began looking at the Aryo-Teutonic residents of the Black Forest, who were believed to have developed democratic institutions; their descendants would later carry these germs to the New England town (Billington, 1973, p. 65). Sitting in the seminars at Johns Hopkins, Turner had listened to Herbert Baxter Adams, America's chief "germ" theorist. Now, in Chicago, Turner imagistically declared his intellectual independence: "The fact is that there is a new product that is American" ([1893] 1962, p. 4). Turner was "issuing a declaration of independence for American historiography" (Billington, 1973, p. 127). On this level, the essay can be understood as a jeremiad offered as a chastisement of imported methods of scholarship.

Turner ends the first section with a striking simile, recalled, no doubt, from his daily conversations with his friend, the geologist Charles R. Van Hise. "As successive terminal moraines result from successive glaciations, so each frontier leaves its traces behind it." Earth and stones, independence and democracy, each frontier leaves its glacial deposits behind it, and "when it becomes a settled area the region still partakes of the frontier characteristics." Turner concludes that "to study this is to study the really American part of our history" ([1893] 1962, p. 4). Throughout the essay, Turner conveys his propositions by means of epic descriptions. The frontier is crossed by successive "advance." We sit, as if at a parade, and watch the frontier recede. As we watch, we see the traders, the steamships, the railroads. From the early seventeenth century to the close of the nineteenth, civilization pushes forward. The United States "lies like a huge page in the history of society." "Line by line," Turner writes, "as we read this continental page from West to East we find the record of social evolution" (p. 11). We see Daniel Boone. Another wave rolls on.

Finally, Turner poses his most daring question. What were the influences of the frontier on the East and on the Old World? No longer is America a vehicle for the germs of Europe. Now the frontier is that which "promoted the formation of a composite nationality of the American people" ([1893] 1962, p. 22). It is the frontier that has promoted democracy in Europe. Turner concludes with what Hill called an argument from testimony, "the existence of the testimony being a sign of the truth of the matter testified to" (1878, p. 200). We hear the imagined voice of a representative from western Virginia who declared, "But, sir, it is not the increase of population in the West [which ought to be feared]. It is the energy which the mountain breeze and western habits impart to those emigrants" (Turner, [1893] 1962, p. 31).

In the last lines of the essay, Turner returns to the Homer of his freshman days at Wisconsin: "What the Mediterranean Sea was to the Greeks, breaking the bond of custom, offering new experiences, calling out new institutions and activities, that, and more, the ever retreating frontier has been to the United States directly, and to the nations of Europe more remotely. And now, four centuries from the discovery of America, at the end of a hundred years of life under the Constitution, the frontier has gone, and with its going has closed the first period of American history" ([1893] 1962, p. 38). The most important moment in American history—the moment that defines the very essence of American character—has passed. Whatever follows will be done in confinement. The environment now has limits.

In the jeremiadic tradition, Turner holds a significant place. First, with Danforth, he places a warning. As Billington points out, Turner believed that a major shift had taken place in the nation's psychology: "Never again would a stubborn environment help break the bonds of custom and summon mankind to accept its conditions. . . . Now Americans must learn to adjust their economy, their politics, their daily lives to live in a closed-space world" (1973, p. 128). Second, with Edwards and Emerson, Turner uses language evocatively; the metaphor comparing the Mediterranean Sea and the ever-retreating frontier is an excellent example of the influence of views of language found in the rhetorical textbooks of Hill and Bascom. Turner's descriptions of the advance of history and his reference to the Mediterranean Sea are both epic, not merely logical, rhetorical devices. Third, by using language in an evocative fashion, Turner implicitly abandons the reliance on logic advocated in both Ramist rhetoric and Scottish Common Sense Realism; in this, Turner is at one with Emerson. Fourth, and perhaps most significantly, Turner articulates a new aspect of the American myth, that of the dis- appearing frontier, the elusive mythopoetic West that embodied the essence of the American character even as it receded before us (Bercovitch, 1978, p. 164n). In this, Turner reinforces the potential of the jeremiad to obviate oppositional views and to renew the energy of the American experience: westward expansionism forged the American spirit.

From the appearance of Turner's 1893 essay until the present, the American consciousness has come to embrace the image of a receding West that holds, just out of our grasp, the power to forge the American character itself. In evoking that image of the wilderness, Turner articulated for the American people their desire to retain the pure West—a message as powerful today as Danforth's election day message was to his New England audience. The surety of this American desire was demonstrated in the first great American preservationist battle, the contest for Hetch Hetchy Valley.

John Muir Takes on the Efficient Conservationists

To understand the significance of the Hetch Hetchy struggle—a struggle for a Yosemite valley in which the rhetorical polarization of Ramist logic and Common Sense Realism versus the metaphoric and evocative power of language became manifest—we must return to nineteenth-century American Romanticism.

The Romantics ran directly against the strong American tradition of optimism and progress. The content of artist J.F. Cropsey's landscapes of Staten Island, for example, were described as follows: "The axe of civilization is busy with our old forests, and artisan ingenuity is fast sweeping away the relics of our national infancy. What were once the wild and picturesque haunts of the Red Man, and where the wild deer roamed in freedom, are becoming the abodes of commerce and the seats of manufactures" (quoted in P. Miller, 1956, p. 205). By the end of the nineteenth century, the jeremiadic question would become more sharply focused: What errand were Americans running in the wilderness now that the wilderness was disappearing? Perry Miller himself wondered. By 1900, he wrote, Americans found themselves running the Puritans' errand into the urban environment of New York, Detroit, and Gary (1956, p. 205). There thus arose a new environmental jeremiad about a tragic and inevitable conquest in which the purity of nature was pitted against the needs of the city.

In 1901 reality manifested Romantic fears. In that year the city of San Francisco put in a claim to use the Hetch Hetchy Valley in Yosemite National Park as the site for a municipal dam. To John Muir (1838–1914), the principal player in this drama, civilization was something to be tolerated, while true humanity belonged in the midst of wild nature. Like Thoreau, Muir left society to live deliberately. Yet escape was not possible, and at the end of his life, Muir found himself in the middle of a conservationist-preservationist debate over the fate of Hetch Hetchy that would demonstrate the ultimate force of rhetoric.

While a conversion experience can easily be exaggerated, it is true that Muir was transformed when, on an 1864 Canadian hike above the Great Lakes, he collapsed in tears at the sudden sight of a rare, exquisite wild orchid (*Calypso borealis*) in forlorn country (Fox, 1981, p. 45). Muir found himself merged with a pantheistic cosmos. God, or the supernatural, became identical with nature, or the living material world. Inspired with a sense of mission by the time he turned thirty, Muir abandoned conventional Christianity, found traditional scientific knowledge too narrow, and abhorred conventional consumer society. Science, he decided, atomized knowledge and split the living world into meaningless dead pieces.

Muir believed that people needed an empathic involvement in nature that was being lost in objective utilitarian science. Whereas the Puritans had feared the wilderness as the haunt of the devil, Muir feared the workings of machines, the abstract notions of Christianity, the social expectations of manly conquest, the arbitrary chronology of civilization, the dominion of man, and the commodification of nature (Cohen, 1984, pp. 25–26). He complained that modern science saw nature as matter in motion; this turned wilderness into a meaningless resource to be cut and sawed, eroded and paved. For Muir, nature—the most valuable of all entities—had a complete organic unity in which humans would find beauty, glory, spirituality, sacredness, and ultimate truth: "One is constantly reminded of the infinite lavishness and fertility of Nature—inexhaustible abundance amid what seems [in human terms] enormous waste" (quoted in Oelschlaeger, 1991, p. 189). As far as Muir was concerned, even the human soul, while real, did not set humans apart from the rest of the living world. Belief in human separateness only inspired arrogance, economic greed, and spoliation of humanity's own nest. People would only recover their sense of duty to the wilderness and get off their pedestals when they remembered that they existed as part of nature's community of life.

Muir hosted the aging Transcendentalist Ralph Waldo Emerson in May 1871 at Yosemite; the visit was an appropriate meeting of like souls. We should be wary, however, of the visit of the outdoors-minded President Theodore Roosevelt, who made a pilgrimage to Muir and Yosemite in 1903. In 1907 Muir would remind Roosevelt of that visit and urge the president to speak out against San Francisco's plans for the Hetch Hetchy Valley. Roosevelt, however, would empower Gifford Pinchot to dam the valley and fill it as a reservoir for San Francisco. On his visit with Muir, Emerson had called upon various members of the visiting party to recite poetry (F. Turner, 1985, p. 125); Roosevelt, apparently already committed to those things Muir despised, had just been spending a night of vacation in the Mariposa Grove.

The battle for Hetch Hetchy is well known and has been analyzed perceptively by Christine Oravec (1984). As Oravec sees it, the battle chiefly pitted conservationism against preservationism. The conservationists, endorsing utilitarian attitudes, argued for the public interest of the residents of San Francisco; the Hetch Hetchy Valley—with its high walls, narrow outlet, and flow of the snow-fed Tuolumne River—provided the ideal site for a municipal dam to provide water to the city 150 miles away. Conversely, preservationists argued that the aesthetic beauty of the forest served a larger national interest. The controversy between these two groups, Oravec concludes, shows how such ideas as "the public interest," "the national interest," and the "public" itself are "rhetorical notions shaped in response not only by the immediate context of the debate, but also by the legitimizing force of predominant social and political presumptions" (1984, p. 455). Oppositional positions, as we have seen, are fertile ground for the use of the jeremiad.

The battle for Hetch Hetchy evokes the oppositional, binary responses that persist today when Americans talk and think about the continent that they inhabit. This response may be seen in the rhetoric of the two primary antagonists, the preservationist Muir and the conservationist Pinchot. Muir, as a one-time college student who, like Turner, felt the genteel and literary influence of the University of Wisconsin's devotion to oratory and rhetoric, had read Emerson and Thoreau and Virgil's bucolic poems (Fleck, 1993). His rhetorical tendency was literary and reflective. Pinchot (1865–1948), chief forester in the Roosevelt Administration, was a disciple of Frederick W. Taylor's scientific management. His rhetorical tendency was scientific and referential. To demonstrate the rhetorical context for Hetch Hetchy, we provide samples of the language of these two men.

John Muir on Hetch Hetchy:

Hetch Hetchy Valley . . . is in danger of being dammed and made into a reservoir to help supply San Francisco with water and light, thus flooding it from wall to wall and burying its gardens and groves one or two hundred feet deep. The making of gardens and parks goes on with civilization all over the world, and they increase both in size and number as their value is recognized. . . . Nevertheless, like anything else worth while, from the very beginning, however well guarded, [the national parks] have always been subject to attack by despoiling gain-seekers and mischief-makers of every degree from Satan to Senators, eagerly trying to make everything immediately and selfishly commercial, with schemes disguised in smug-smiling philanthropy, industriously, sham-piously crying, "Conservation, conservation, panutilization," that man and beast may be fed and the dear Nation made great. Thus long ago a few enterprising merchants utilized the Jerusalem temple as a place of business instead of a place of prayer, changing money, buying and selling cattle and sheep and doves; and earlier still, the first forest reservation, including only one tree, was likewise despoiled.

These temple destroyers, devotees of raging commercialism, seem to have perfect contempt for Nature, and instead of lifting their eyes to the God of the Mountains, lift them to the Almighty Dollar.

Dam Hetch Hetchy! As well dam for water-tanks the people's cathedrals and churches, for no holier temple has ever been consecrated by the heart of man. [1912, pp. 255–57, 260–62]

Gifford Pinchot on conservation:

The first great fact about conservation is that it stands for development. There has been a fundamental misconception that conservation means nothing but the husbanding of resources for future generations. There could be no more serious mistake. Conservation does mean provision for the future, but it means also and first of all the recognition of the right of the present generation to the fullest necessary use of all the resources with which this country is so abundantly blessed. Conservation demands the welfare of this generation first, and afterward the welfare of the generations to follow.

The conservation of our natural resources is a question of primary importance on the economic side. It pays better to conserve our natural resources than to destroy them, and this is especially true when the national interest is considered. But the business reason, weighty and worthy though it be, is not the fundamental reason. In such matters, business is a poor master but a good servant. The law of self-preservation is higher than the law of business, and the duty of preserving the Nation is still higher than either. [(1910) 1971, pp. 40–44]

This analysis is from *The Fight for Conservation,* Pinchot's statement of progressivism. Later Pinchot would write of Hetch Hetchy, "As to my attitude regarding the proposed use of Hetch Hetchy by the city of San Francisco . . . I am fully persuaded that . . . the injury . . . by substituting a lake for the present swampy floor of the valley . . . is altogether unimportant compared with the benefits to be derived from its use as a reservoir" (quoted in Nash, 1982, p. 161).

How dissimilar the rhetorical strategies! In Muir, we find the ecological tradition of Gilbert White. Here is the rhetoric of sensation of Edwards's *Personal Narrative* and the metaphor of Emerson's *Nature.* Muir recalls the Garden of Eden in Genesis 2–3 and the expulsion of the money changers in Matthew 21. The passage evokes the temple in its religious imagery. In Pinchot, we find the scientific tradition of Linnaeus. Here is the Ramist logic of Danforth's election day sermon and the orderly yield of Common Sense Realism. Pinchot recalls the gospel of social Darwinism; his philosophy evokes the corporate world of management.

In their analysis of the rhetoric of American environmental politics, Jimmie Killingsworth and Jacqueline Palmer identify these two positions as representing the Enlightenment habit that placed science and mysticism at opposite ends of the spectrum. The vision of Muir regards nature as home; the vision of Pinchot regards nature as resource (Killingsworth & Palmer, 1992, p. 14). Killingsworth and Palmer identify the phenomena in the term *ecospeak* (p. 24). Ecospeak—a rhetorical form producing stark alignment, simplistic dualism, and numbing paralysis—emerges along bipolar lines. "On one side are the environmentalists, who seek long-term protection of endangered environments regardless of short-term economic costs. On the other side are the developmentalists, who seek short-term economic gain regardless of long-term environmental costs" (p. 9).

This polarization of thought has been with us from the beginning: the New England colonists were either with the Puritan errand or against it. Historically, such polarities are not easy to avoid in America, nor are they easy to subvert. The loss of Hetch Hetchy in 1913 to the conservationists embodies the nadir for the evocative tradition of Edwards, Emerson, Turner, and Muir in environmental discourse. The rhetorical powers of Ramism and Common Sense Realism were an overwhelming force. In addition, the new gospel of efficiency and systematization was becoming part of the national agenda; the influence of Pinchot on Congress in the twenty years after 1890 demonstrates the triumph of forest management professionals who would guarantee the American public future raw materials and profits (Marcus & Segal, 1989, pp. 201–2). In the face of such bureaucratic rationalism, Muir's prophetic voice sounds antiquated, perhaps even unrealistic. What the Puritans had most feared had happened: we had undertaken the errand and, going into the wilderness, had met ourselves as Satan coming out.

When the dam in Hetch Hetchy was actually built in 1923, many wondered if the advocacy strategies of Muir would ever regain their force. Would the status of evocative rhetoric regain even a minority position? Helpful in invigorating the language that Muir symbolized was the environmental ethic of Aldo Leopold.

Aldo Leopold and the Rhetoric of the Land Ethic

In 1935, Wisconsin wildlife manager Aldo Leopold (1887–1948) and his family acquired an abandoned 120–acre farm near Baraboo, Wisconsin. They began planting pine trees, as many as six thousand a year, together with other trees, shrubs, grasses, and flowers. They created a working laboratory to restore the environment to "aboriginal health." Leopold had watched the devastation caused by too many New Deal wasteland cleanup crews—the destruction of brush necessary for wildlife food and shelter, silted trout streams, and planted rows of identical trees that only reflected Pinchot-like "conservation technology" and not the natural makeup of the land. The intensive scientific management that he himself had learned at the Yale Forest School and had advocated in his 1933 book *Game Management* was part of the problem, he had realized by 1935, rather than the solution: scientific management saw nature as little more than a stockpile of raw materials to be shaped by human needs (Flader, 1974). Leopold realized that a different vision was needed.

Leopold took a remarkable step that modern environmentalism is still digesting: he turned the philosophies of Emerson and Muir into science. "Land," he wrote in an essay sometime in 1947 or 1948, "is not merely soil" ([1949] 1987,

p. 218). As a private landowner he personally devised the land-human connection that became the basis for the essay "The Land Ethic" in *A Sand County Almanac*. Land and humans, Leopold found, are yoked together in one history: "A land ethic changes the role of *Homo sapiens* from conqueror of the land-community to plain member and citizen of it. It implies respect for his fellow members, and also respect for the community as such" ([1949] 1987, p. 204). Leopold's achievement was to establish the bond between scientific insight and an ethical imperative, the litmus test of today's environmentalism.

Leopold introduces the evolution of ethics in a highly evocative passage indebted to the rhetoric of Emerson and Muir. He notes Odysseus's hanging of a dozen slave girls—"chattels"—because he suspected that they had misbehaved during his absence ([1949] 1987, pp. 201–2). Leopold, a graduate of Yale, evokes three responses by his use of this classical reference to book 22 of *The Odyssey*. First, he allows his audience to associate the ethical issues of which he was speaking with the values of the classical world itself. Second, as Roderick Nash points out, Leopold calls forth the Great Chain of Being, in which humanity held a position only midway between the lowliest creatures and divine beings. Hence, any attempt to put man at the center of the universe is implicitly absurd (Nash, 1982, p. 193). Third, Leopold illustrates that ethics evolve. Thus, a barbarous pattern that began in the classical world would not necessarily be carried into the strange New World. To complete the association, he writes that "individual thinkers since the days of Ezekiel and Isaiah have asserted that the despoliation of land is not only inexpedient but wrong. Society, however, has not yet affirmed their belief. I regard the present conservation movement as the embryo of such an affirmation" ([1949] 1987, p. 203). With the biblical allusion in place, Leopold has secured a sound space for the presentation of his proposition for a land ethic in which land is regarded as more than property with which we are free to do as we please.

For Leopold, environmental problems raise issues to fundamental moral questions about the quality of human society and its direction. How Americans behave toward their environment reveals their culture. This perception is the message of Leopold's jeremiad: the biotic natural world must be incorporated into the philosophic study of ethics. Landowners, he believes, must be participating citizens of the land community whose ethical duty is to restore the land to ecological integrity. Landowners are also rewarded with a powerful sense of belonging to something greater than themselves that continues after they are gone.

Thus, the tradition of the jeremiad is manifested in Leopold in his desire to invigorate his audience. Working to restore a tract of land transforms an ignorant, insensitive, and ordinary person into a sensitive, wise individual who recovers a sense of purpose. Human life is best understood in the context of nature, and not the other way around. Leopold's view reflects the continuous American tradition that recalls the Puritan sense of a mission and Edwards's vision of nature's beauty. Leopold renders beauty a criterion for the ethic of conservation and preservation. He enlarges beauty beyond the classic mountains, waterfalls, canyons, and lakes to include nonscenic swamps, dunes, prairies, and deserts. In the jeremiadic tradition, he obviates aesthetic polarities by absorbing them into a new vision of a natural aesthetic. The beauty of a swamp or a prairie is the historical or biological story it tells in both geologic time and human time. Following Danforth, Leopold thus wants those who went into the wilderness on the errand to examine themselves seriously and thoroughly, to find an ecological conscience.

Avoiding the Apocalypse: The Modern Jeremiad

Environmental jeremiads that persuade through evocation and those that persuade through prescribed methodology persist today, as a consequence of an American rhetorical tradition rooted, as we have observed, in the sermons of Danforth and Edwards. Rachel Carson, Bill McKibben, and Al Gore, for instance, all describe an environmental apocalypse. On the other hand, documents such as the Environmental Impact Statement (EIS) urge action through the systematic presentation of methodology.

Carson, McKibben, and Gore, like their predecessor Jonathan Edwards, wield language and manage to portray the fall of a wonderful, God-given world dependent upon our intuitive and inspired response for its survival. Carson's "Fable for Tomorrow" tells of a "town in the heart of America where all life seemed to live in harmony with its surroundings" (1962, p. 13). This edenic state falters when "a strange blight," "an evil spell" wrought by the people themselves, falls over the land. Similarly, in *The End of Nature,* McKibben describes the apocalypse as ongoing: "Our reassuring sense of a timeless future, which is drawn from that apparently bottomless well of the past, is a delusion" (1989, p. 5). Like Carson, McKibben uses the narrative as personally and graphically as possible: he reprimands us for putting nature away in the same way that some people "put their parents out of their lives and learn differently only when the day comes to bury them" (p. 70). McKibben's God, like Carson's, is waiting either for us to respond responsibly to the crisis or to compound original sin with terminal sin (p. 216).

The opening chapter of Al Gore's *Earth in the Balance* shares with Carson and McKibben's works a sense of human failure and destruction "on an almost biblical scale" (1992, p. 20). The book opens with Gore standing on the steel deck of a fishing ship, stranded in what was once the world's fourth largest inland sea, the Aral Sea. Now the sea is a desert because, ironically, an irrigation plan to feed the surrounding desert diverted the water that fed it. Like Carson and McKibben, Gore shocks us with what we have done, saddens us with our loss, shames us with our profligacy, and castigates us for defaulting on our bonds.

Coexistent with such sermons reminiscent of Edwards is a rhetoric in the Ramist tradition that systematically prescribes how we might use methodology to restore order. A locus of such rhetoric is found in the EIS, a report mandated as a prerequisite of any industrial changes having environmental impact. The organizational pattern for such documents is defined in part 1502 of the *Regulations for Implementing the Procedural Provisions of the National Environmental Policy Act* (U.S. Council on Environmental Quality, 1986). This pattern recalls, yet surpasses, the remorseless organizational rigor of Danforth's sermons. There must, first, be a statement of purpose and need for action—that is, a statement of the reason that an agency must perform an operation that may potentially damage the environment. This section of the report must be followed by a section that establishes potential alternatives to the environmentally dangerous action and the reasons these alternatives have not been adopted. An analysis of the impact of the action on the affected environment and an analysis of the direct and indirect effects of the action must then follow. The overt purpose of the EIS is therefore to "serve as an action-forcing device to insure that the policies and goals defined in the Act [the National Environmental Policy Act of 1969] are infused into the ongoing programs and actions of the Federal Government . . . [and to] provide full and fair discussion of significant environmental impacts" to the public (1986, p. 10). Thus, the EIS serves its jeremiadic purpose of obviation by incorporating the protection of nature into the legislative process.

In the EIS, we thus see the resurgence of the Ramist tradition of the colonial university, the Common Sense Realism of the nineteenth-century textbooks, the conservationist rhetoric of Pinchot, and the triumph of Progressivism against which Carson, McKibben, and Gore rebel. The EIS is a document in the tradition of Danforth, a rhetorical device that establishes safety through method, that ensures the success of the errand through organization. In the jeremiadic tradition, the EIS seeks to obviate differences by asking writers to pose alternatives and, gracelessly, to resolve them as a section of a document. Regardless of the particular circumstances that influence the writing of such a document, the methodological associations of command and control—thus producing safety through conformity—are a part of the collective consciousness.

If this picture is too bleak regarding the role of government in environmental discourse, there is perhaps hope. The attitudes of the former Office of Technology Assessment (OTA) toward the environment have shifted, as manifested in a landmark 1986 document, *Serious Reduction of Hazardous Waste: For Pollution Prevention and Industrial Efficiency*. Remediation of environmental waste is abandoned in favor of pollution prevention. The OTA approaches environmental problems in the spirit, if not in the language, of Leopold. Realizing that "manmade changes are of a different order than evolutionary changes, and have effects more comprehensive than is intended or foreseen," the writers of this government document take a holistic and ecological approach to waste: "Usually, hazardous industrial wastes are not destroyed by pollution control methods. Rather, they are put into the land, water, or air where they disperse and migrate. The result is that pollution control for one environmental medium can mean that waste is transferred to another medium" (U.S. Office of Technology Assessment, 1986, p. 218). Refreshing in its honesty, the document acknowledges that the management of environmental wastes is, after all, not the complete answer. The result is a modern jeremiad that, while attending to implementational strategies, nevertheless leads to an ethical imperative. The answer to environmental problems lies in technological designs that will reduce pollution within industrial plants themselves. Critical thought—including such aspects of rhetoric as the possibility of refutation, the limits of certainty, and the presence of multiple interpretations—thus returns as the OTA argues that we should dramatically shift our thinking away from the kind of anthropocentrism that allows only the most naive kind of technological fix. There exists at least the possibility that we may awaken from our dream of environmental domination.

In the end, perhaps we can be more hopeful than the Puritans who groaned about their errand into the wilderness. Perhaps a new Copernican revolution is at hand in which we will see an ecological approach to history, asserting that humans are not invincible masters of a passive material world. Whenever humans extract, reshape, and consume the biogeochemical environment around them, it will actively respond in beneficial ways, and also in ways that will not always be to our liking, including the creation of waste, pollution, ugliness, and possibly its own decline and death. Rather, ecological science shows that people have always lived inside environments characterized by dynamic interactions of energy, materials, and information among living things (including ourselves) and nonliving things. Simply on the grounds of our own selfish survival, we need to be watchful about the processes we start or invade. In each transfer of materials and energy, not only is some energy lost, but ecosystems change, creating new, different, and unexpected results.

Approaching the Rhetoric of American Environmental Discourse

We began this study with Danforth's 1670 election sermon and continued through the federal government's call for pollution prevention in 1986. This historical analysis allowed us to identify key environmental documents and to note some of the rhetorical conventions they employ.

Our vision of the origin of environmental discourse as jeremiad demonstrates the value of genre criticism as a sociopsychological category. As Charles Bazerman proposes, a genre offers a way of creating order in an ever-shifting world and is thus a rich source for inquiry (1988, pp. 6–7, 319–20). While the formal features of genres are fascinating to locate and analyze, important too are the ways that we can understand genre as a manifestation of the social and psychological aspects of culture. In closing, we will identify what we take to be three definitive characteristics of American environmental rhetoric that have emerged regardless of time and circumstance.

First, the aim of environmental discourse is persuasive. Environmental discourse is a rhetoric of advocacy. In the selections we have examined, the writers use their powers and the intrinsic qualities of the jeremiad to elicit from audiences a specific emotion, conviction, or action (Kinneavy, 1980, pp. 211–306). For Danforth and Edwards, the errand was to construct a heavenly kingdom on earth. For William Bartram, the errand was to classify the physical world and, thus, to understand and control it. For Emerson, the errand was to allow Americans to see themselves as having the potential for a transcendent and mystical union with nature. For Turner, the errand was to demonstrate that the receding West was the cauldron for the American character itself. For Muir, the errand was to preserve the wilderness. For Leopold, the errand was to invent a land ethic. For Carson,

McKibben, and Gore, the errand was to stop the scientific management of nature. And for the legislative architects of the EIS, the errand was, ostensibly, to ensure that the law was met and that the environment would be protected. Each took the mission seriously and tried to persuade the audience with a fervor that would have pleased Winthrop himself.

Second, environmental rhetoric is frequently either evocative or implementational in strategy, and each strategy has identifiable characteristics. In strategies of evocation, we find rich allusion to other texts, especially biblical and classical ones, as we saw in Leopold's use of Homer to explain the land ethic. Evocative strategies also extensively use the connotative and poetic aspects of language; here we recall Muir's poetic description of Hetch Hetchy. Present also in evocative rhetoric is careful attention to metaphor. Edwards, for example, posited that nature symbolized the designs of God; Emerson, that language was a sign of natural fact; Muir, that Hetch Hetchy was a temple. Even Gore's use of the Aral Sea as a symbol for the destruction of nature is highly metaphoric. Evocative rhetoric is close to the human lifeworld, celebrating the concrete and the tactile, as we saw in Muir's extended descriptions. Often, evocative strategies rely on narration and description, such as we saw in Carson's apocalyptic fable. And often, evocative strategies depend on the voice of the author, a technique we saw in McKibben's intensely personal response to global warming expressed in an analogy.

Conversely, environmental rhetoric that relies on method takes on different characteristics. While implementational environmental rhetoric is rich in the citation of other texts, it tends especially to draw on management and scientific studies. We saw such techniques in Pinchot's use of Taylorism and social Darwinism and in the requirement that data be used in the EIS. Implementational rhetoric also capitalizes on the denotative aspects of language noted as characteristic of logocentrism (Ong, 1982). Bartram, attempting to classify the natural contents of the New World, sought to make word and thing match in his botanical descriptions. There is thus in implementational strategies a subsequent denial of metaphor (as we saw in Pinchot's views on conservation), a reliance on systematic approaches (as in the sections of Danforth's sermon), and a need for the objectifications of science (as in the prescribed organizational pattern of the EIS). And just as the use of the first-person singular informs evocative strategies, so the depersonalized voice of scientific inquiry is the ideal of implementational strategies.

Readers of studies such as Walter Ong's *Orality and Literacy* (1982) will recall that cognitive (psychological) patterns are associated with each type of discourse. That is, the kind of thinking associated with evocative rhetoric is based in the immediate and fluid world of orality; the kind of thinking associated with implementational rhetoric is based in the objectified world of literacy. These patterns are everywhere in evidence in the two rhetorics. Yet as Ong suggests, we must be careful to avoid the stereotypes associated with the two cognitive patterns: evocative rhetoric is not a primitive rhetorical act, and implementational rhetoric is not a sophisticated counterpart. Turner, who held a Ph.D. from Johns Hopkins, was acutely aware that the implementational rhetoric of graduate school was inappropriate for the American Historical Association lecture presented at the

Columbian Exposition of 1893. Instead, Turner employed rhetorical strategies that were evocative to the audience of listeners, which included ordinary citizens as well as college professors. Turner—as would Al Gore—encouraged analysis by appealing to the mythic.

Again and again we see the values of the citizen played out against the values of science, as each advocate adopts characteristic rhetorical aims and strategies. Polarity is thus a third trait of American environmental rhetoric. As we have seen, implementational and evocative strategies tend to be understood as polar. Both strategies, however, are needed. Both manifest an orientation necessary for the preservation of the natural world in an alarmingly indifferent and powerfully technological society. The rhetoric of implementation is, after all, the rhetoric of modern science, and we must avoid disparaging references to its attempts at objectivity, however much those attempts have fallen short in the past. Similarly, the rhetoric of evocation embodies the voice of those who care about the natural world in another way; it is naive for scientists to disparage the vision of Emerson, Leopold, Carson, and McKibben, to classify it as mere literary discourse. As we have tried to demonstrate in this chapter, poets and scientists are equally valuable in the arena of American environmental advocacy. A favorable disposition toward the environment is brought about through the rhetoric of evocation. The manifestation of that disposition through policy is made possible by the rhetoric of implementation. Both are needed if we are to understand and preserve the environment as it embodies, to use Muir's phrase, the realm of possible insight.

Failure to recognize the legitimacy of either evocative or implementational rhetoric is, essentially, an insistence on polarizing civilization and nature, insofar as evocative rhetoric tends to speak on behalf of sacred nature while implementational rhetoric tends to speak for the secular pursuits of civilization. In truth, as we have come to understand, the species *Homo sapiens* is part of nature, and the healthy interaction of our species with the environment is so critical to its ultimate survival that the two forces cannot stand divided. We must enter the wilderness and find shelter there. As Danforth understood, we must perform our errand with a sacred mission to ensure secular success. As our species and nature cannot be divided, neither can the means by which we speak for them.

This is why the jeremiad will endure: it is the best rhetorical device for handling a most difficult subject—the representation of the American people in their environment. The jeremiad affords our culture the opportunity to rage with displeasure, to evoke the beauty of metaphor, to find safety in method, and to reconcile oppositions.

References

Applebee, A.N. (1974). *Tradition and reform in the teaching of English: A history.* Urbana, Ill.: National Council of Teachers of English.

Bartram, W. ([1791] 1955). *The travels of William Bartram.* M. Van Doren (Ed.). New York: Dover.

Bascom, J. (1866). *Philosophy of rhetoric.* Boston: Crosby & Ainsworth.

Bazerman, C. (1988). *Shaping written knowledge: The genre and activity of the experimental article in science.* Madison: University of Wisconsin Press.

Bercovitch, S. (1978). *The American jeremiad.* Madison: University of Wisconsin Press.

Berlin, J.A. (1984). *Writing instruction in nineteenth-century American colleges.* Carbondale: Southern Illinois University Press.

Billington, R.A. (1973). *Frederick Jackson Turner: Historian, scholar, teacher.* New York: Oxford University Press.

Blair, H. ([1783] 1965). *Lectures on rhetoric and belles lettres.* Harold F. Harding (Ed.). Carbondale: Southern Illinois University Press.

Boorstin, D.J. (1960). *The lost world of Thomas Jefferson.* Boston: Beacon Press.

Campbell, G. ([1776] 1963). *The philosophy of rhetoric.* L.F. Bitzer (Ed.). Carbondale: Southern Illinois University Press.

Carson, R. (1962). *Silent spring.* New York: Ballantine.

Cohen, M.P. (1984). *The pathless way: John Muir and American wilderness.* Madison: University of Wisconsin Press.

Connors, R.J. (1981). The rise and fall of the modes of discourse. *College Composition and Communication, 32,* pp. 444-55.

Crusius, T.W. (1989). *Discourse: A critique and synthesis of major theories.* New York: Modern Language Association.

Curti, M., & Carstensen, V. (1949). *The University of Wisconsin, 1848-1925: A history.* (Vol. 2). Madison: University of Wisconsin Press.

Danforth, S. ([1670] 1968). A brief recognition of New England's errand into the wilderness. In Plumstead, 1968, pp. 54-77.

Edwards, J. ([1726] 1980). The beauty of the world. In W.E. Anderson (Ed.), *The works of Jonathan Edwards* (Vol. 6, pp. 305-6). New Haven, Conn.: Yale University Press.

———. ([1765] 1962). *Personal narrative.* In C.H. Faust & T.H. Johnson (Eds.), *Jonathan Edwards: Representative selections, with introduction, bibliography, and notes* (pp. 57-72). New York: Hill & Wang.

Emerson, R.W. ([1836] 1983). Nature. In *Ralph Waldo Emerson: Essays and lectures* (pp. 7-19). New York: Library of America.

———. ([1837] 1983). The American scholar. In *Ralph Waldo Emerson: Essays and lectures* (pp. 53-71). New York: Library of America.

Fitzgerald, F.S. (1925). *The great Gatsby.* New York: Scribner's.

Flader, S. (1974). *Thinking like a mountain: Aldo Leopold and the evolution of an ecological attitude toward deer, wolves, and forests.* Columbia: University of Missouri Press.

Fleck, R.F. (1993). John Muir's transcendental imagery. In S.M. Miller (Ed.), *John Muir: Life and work* (pp. 136-51). Albuquerque: University of New Mexico Press.

Fox, S. (1981). *John Muir and his legacy: The American conservation movement.* New York: Little, Brown.

Gore, A. (1992). *Earth in the balance: Ecology and the human spirit.* Boston: Houghton Mifflin.

Halloran, S.M. (1982). Rhetoric in the American college curriculum: The decline of public discourse. *Pre/Text, 3,* pp. 245-69.

Hill, A.S. (1878). *The principles of rhetoric and their application.* New York: Harper & Bros.

Killingsworth, M.J., & Palmer, J.S. (1992). *Ecospeak: Rhetoric and environmental politics in America.* Carbondale: Southern Illinois University Press.

Kinneavy, J.L. (1980). *A theory of discourse: The aims of discourse.* New York: W.W. Norton.

Leopold, A. ([1949] 1987). *A Sand County almanac.* New York: Oxford University Press.

Locke, J. ([1690] 1959). *An essay concerning human understanding.* New York: Dover.

Luhmann, N. (1989). *Ecological communication.* J. Bednarz Jr. (Trans.). Chicago: University of Chicago Press.

Marcus, A.I., & Segal, H.P. (1989). *Technology in America: A brief history*. New York: Harcourt, Brace, Jovanovich.

McKibben, B. (1989). *The end of nature*. New York: Anchor.

Miller, C.R. (1978). Technology as a form of consciousness: A study of contemporary ethos. *Central States Speech Journal, 29,* pp. 228-36.

Miller, P. (1956). *Errand into the wilderness*. Cambridge, Mass.: Harvard University Press.

Morison, S.E. (1936). *Harvard College in the seventeenth century*. (Pt. 1). Cambridge, Mass.: Harvard University Press.

Muir, J. (1912). *The Yosemite*. New York: Century.

Nash, R. (1982). *Wilderness and the American mind*. 3rd ed. New Haven, Conn.: Yale University Press.

Noble, D. (1965). *Historians against history: The frontier thesis and the national covenant in American historical writing since 1830*. Minneapolis: University of Minnesota Press.

Oelschlaeger, M. (1991). *The idea of wilderness: From prehistory to the age of ecology*. New Haven, Conn.: Yale University Press.

Ong, W.J. (1958). *Ramus, method, and the decay of dialogue*. Cambridge, Mass.: Harvard University Press.

———. (1977). *Interfaces of the word: Studies in the evolution of consciousness and culture*. Ithaca, N.Y.: Cornell University Press.

———. (1982). *Orality and literacy: The technologizing of the word*. London: Methuen.

Opie, J., & Elliot, N. (1993). The significance of rhetoric in the Turner thesis. In J.G. Cantrill & M.J. Killingsworth (Eds.), *Proceedings of the conference on communication and our environment* (pp. 19-29). Marquette, Mich: Northern Michigan Printing Services.

Oravec, C.L. (1984). Conservation vs. preservationism: The "public interest" in the Hetch Hetchy controversy. *Quarterly Journal of Speech 70,* pp. 444-58.

Pinchot, G. ([1910] 1971). Wild land is wasted land. In J. Opie (Ed.), *Americans and environment: The controversy over ecology* (pp. 40-45). Lexington, Mass.: D.C. Heath.

Plumstead, A.W. (Ed.). (1968). *The wall and the garden: Selected Massachusetts election sermons, 1670-1775*. Minneapolis: University of Minnesota Press.

Ritter, K.W. (1980). American political rhetoric and the jeremiad tradition: Presidential nomination acceptance addresses, 1960-1976. *Central States Speech Journal, 31,* pp. 153-71.

Turner, F. (1985). *Rediscovering America: John Muir in his time and ours*. New York: Viking.

Turner, F.J. ([1893] 1962). The significance of the frontier in American history. In F.J. Turner (Ed.), *The frontier in American history* (pp. 1-38). New York: Holt.

U.S. Council on Environmental Quality. (1986). *Regulations for implementing the procedural provisions of the National Environmental Policy Act*. Washington, D.C.: GPO.

U.S. Office of Technology Assessment. (1986). *Serious reduction of hazardous waste: For pollution prevention and industrial efficiency*. Washington, D.C.: GPO.

Wills, G. (1978). *Inventing America: Jefferson's declaration of independence*. New York: Random House, Vintage Books.

Winthrop, J. ([1630] 1963). A model of christian charity. In P. Miller & T.H. Johnson (Eds.), *The Puritans: A sourcebook of their writings* (Vol. 1, pp. 195-99). New York: Harper & Row.

Worster, D. (1977). *Nature's economy: The roots of ecology*. San Francisco: Sierra Club Books.

2
Naturalizing Communication and Culture
Donal Carbaugh

The world is places.

—Gary Snyder, *The Practice of the Wild*

Communication occurs everywhere as part of natural contexts, physical spaces, and landscapes. Whether in riversides, mountain retreats, mountaintops, schoolrooms, courtrooms, living rooms, board rooms—each of these holds considerable force somewhere—or wherever, communication is radically "placed." In this sense, communication is always situated physically, in the particulars of place and time. Also, in turn, communication everywhere creates a sense of place, of the natural, of what is affirmed as emphatically and already there. Rather naturally, communication creates senses of (what is taken to be) sheer and utterly natural space. Communication can thus be conceived as radically and doubly "placed," as both located in places and as locating particular senses of those places.

By being within and by creating senses of places—from wilderness to Wall Street—communication helps cultivate particular ways of living as natural. Through everyday practices of communication, people everywhere cultivate ways of being placed with nature, in it, as it, ways of being within the natural realm. Taken together, then, and universally, communication occurs in places, cultivates intelligible senses of those places, and thus naturally guides natural ways of living within them.

This introductory and universal point also has a radically particular dimension: communication is not created the same in all natural places, and it does not create in all such places the same senses of—or relations with—the natural realm. Communication, therefore, as in and of natural worlds, is not only located in natural places; it also locates senses of natural spaces. Moreover, it localizes—that is, it creates—senses in socially distinctive ways, in particular cultural contexts, tilling specific tropes, fertilizing particular fields.

That communication occurs in natural space, that it also naturally creates senses of such places, that it guides sensible living in such places so conceived, and that it does so locally, thus variously, from place to place and people to people—these provide the starting points for the present essay.

If one starts here, by foregrounding communicative practices in natural and cultural space, with particular sensitivities to the natural senses of place being cultivated with communicative practices (and clearly there are other places to start), then questions arise about the links between "nature," "culture," and symbolic processes, and about the reflexive relationship between them. What is the relationship between specific symbolic practices and their natural environment? How, among people in specific contexts or communities, is nature (or place) symbolized? What expressive means are available for giving "nature" a voice? What meanings are associated with these expressive means? When are these used? By whom? What are the environmental, political, social, and interactional consequences of these expressive means, and the meanings that—in particular times and places—give voice to the natural?

These questions are not comprehensive, nor do they suggest simple answers, but perhaps they do suggest some initial probes, some paths to travel, so better to hear, and critically assess, what we (and others in other places) so often presume, a natural world. How we go about cultivating the natural, in time and place, the various features assumed and foregrounded (and forgotten), the various ways of natural living being nurtured (and negated)—all warrant our serious attention.

Natural and Cultural Dimensions of Communication

> There is an essential role for language studies [of the environment], for they are fundamental to exposing and then overcoming the presuppositions which entrench the distinction between nature and culture.
> —Max Oelschlaeger, "Wilderness, Civilization, and Language"

Because of certain well-worn features of our Euro-language, we are often caught in ideational duels between, for example, nature and culture, or between terms for contexts (e.g., wilderness, civilization, nature, homes, environment) and that which lives in those contexts (e.g., culture, plants, animals, humans). Further entrenching this picture of concentric entities is the view that language is a mere instrument for re-presenting what is already present in nature and culture, with its use involving a simple mapping of an objective something-out-there (in nature) or a something that is humanly common (in culture). Because of these well-worn linguistic ruts, we tend to speak and think about nature as an objective environment (sans culture), culture as a built environment (sans nature), and communication as simply a means of saying something about each.

These current cultivated tendencies make it easier, for a Western mind, to suppose that I am proposing to examine linguistic presentations of nature (or culture). And this would be correct. I am advocating this. This aspect of my proposal brings to the foreground the various ways human linguistic constructions shape meanings about natural space, and the consequences of these upon local and natural worlds. But I am also advocating a more basic point: that all systems of communication practices, as carriers of cultural meanings, and whether about "nature" or not, occur in natural spaces, naturally create ways of living in those places (bodies included), and thus are affected by and carry real physical consequences for those places. Whether one is speaking then about nature, or about cars, or families, or religion, or Disney, one's communication practices are a part of and consequential for nature's (and culture's) processes.[1]

I include cultural meaning systems as a constituent part of this general process and thus treat them likewise, for they are also part of natural space, influencing it and influenced by it. I do not assume that culture determines nature (although it does influence what is meant by *nature* and is thus consequential for nature). Nor do I assume that nature determines culture (although it does influence it). I do presume that these processes are related, that cultural meaning systems are part of and consequential for natural processes just as natural processes (broadly) give shape and form to all cultural systems. In other words, natural and cultural systems help shape each other and are radically consequential for each other. Keeping both in mind, I want to move in a particular direction and discuss how the social senses of each are being cultivated through particular practices of communication.

Communication is the basic social process in which natural and cultural senses are cultivated. Communication transforms raw space into a natural and cultural scene, into a place that is publicly meaningful in social terms. A condition and consequence of symbolizing activity, a process and outcome of communication, is the fashioning of places in humanly sensible, mutually intelligible, and actionable terms. Common senses of what is natural and cultural, then, are inextricably intertwined within human symbolic practices, with what is particularly intelligible about each of these processes deriving from local communicative practices.

From this naturalistic orientation, systems and practices of communication radically implicate cultural and natural processes and are thus consequential for both. But communication is not the totality of these processes. The order(s) of communication, while giving human expression to nature and culture, does not, and cannot, exhaust the natural or the cultural. To paraphrase Ralph Waldo Emerson and to redress current environmental difficulties, we must "know more from nature than we can at will communicate" (1987, p. 19). This Emersonian "knowing" productively points beyond the order of communication to ineffable resources that then might eventually seep into and enrich communication. This, I believe, is one reason we walk trails, climb peaks, or, for other reasons, visit sites of land use controversies, or landfills, oil spills, and so on. As a result of being a part of nature's spaces, we "know" more and thus might work to say something else, something better, about our places.

My fundamental starting point then is not simply that we talk about nature in distinctive ways, or that we talk about physical places in distinctive ways, or that different cultures represent things in distinctive ways. I take all of this to be true. But my fundamental point of entry into environmental issues and discourses about them is this: communication is the basic social process through which our natural ways and cultural meanings are being exercised socially. Further, whether this communication is explicitly about landscapes, lions, limousines, or whatever, in the process we implicate something of natural and cultural processes, with our communication being radically consequential for, if not the whole of, both the natural and the cultural. With regard to the main theme of this book, environmental communication is not just one type of communication that people sometimes produce (e.g., when they talk about "the environment"). As communication continuously and naturally (re-)creates places, it creatively integrates natural and cultural messages. At some level, these natural and cultural messages are being presumed and (re-)created as a condition for all systems of communication practices. Seen this way, environmental communication is the ever-present and multifaceted shadow of—natural and cultural—place in human symbolic action. It is being cast, contested, and cultivated in the communicative practices of all of our human communities.

In short, there is immanent in all systems of communication practices an environmental dimension, and it is being tailored and designed in locally distinctive ways. Whether talking about "the environment," "nature," or "culture," we are implicating each in our particular communicative practices, so we may as well erect some of our studies with these basic implications in view (see Ingold, 1991, 1992).

In the remainder of this chapter, I discuss some elements of this approach to environmental discourse and advocacy and apply some of its elements rather quickly to some social practices of communication. I focus on various cultural and natural dimensions of expression, with the general approach being a kind of naturalist's view of communication and culture. My belief is that the approach applies to symbolic practices generally, where "nature's objects" are explicitly discussed and where such is perhaps slightly less obviously explicit or pivotal. I begin by discussing one potent kind of communicative practice, place-naming, as a way of demonstrating the cultural and natural dimensions of communication. I will discuss, eventually, expressive forms in which "the natural" or "the environmental" dimension is a bit more hidden in the communicative practice.

Explicitly Radiating Nature and Culture

Naming places, and using such names in order to say various things, is a practice in all known languages, among all peoples. Through such a communication practice, people learn particular ways of identifying with their natural place, what specific spaces mean, vantage points from which to view these places or spaces, and ways of living (speaking, feeling) with them (Carbaugh, 1996, pp. 157–90).

The Western Apache of south central Arizona engage in one particularly powerful cultural practice of place-naming. During the course of some conversations, when wanting to comfort someone present, when speaking of absent parties who are close to those present, when wanting to do so with tact, and when traditional wisdom applies to serious errors in someone else's judgment, an Apache might say, "It happened at line of white rocks extends upward and out, at this very place!" followed by a pause of thirty to forty-five seconds, and then, "Truly. It happened at trail extends across a long red ridge with alder trees, at this very place!" followed by another pause (Basso, 1989, p. 105). When this kind of speaking is done successfully, an effect of smoothness, quiet, and softness is achieved. Keith Basso shows how such depictive language, for the Apache, accomplishes several outcomes: it yields very precise images of nature, symbolically positions persons in an exact natural space, and privileges one vantage point as optimal for viewing (thus looking forward) into that space, but it also evokes a communal history of entire sagas or tales that radiate from that very place (thus looking backward into time).

Basso describes the process of place-naming as "appropriating the landscape" (1989, p. 107), as involving an interpretation of the landscape, turning it into an expressive means, and using such means to achieve specific social ends. Basso's study superbly demonstrates how symbols of nature, while highlighting a depictive or imaging quality of language, can never be merely that. They also have other cultural and natural uses, from evoking historical wisdom through shared tales, to transforming worry about close others into hopefulness, to cultivating a rich relation with a natural world. Basso summarizes, "Such systems operate to place flexible constraints on how the physical environment can (and should) be known, how its occupants can (and should) be found to act, and how the doings of both can (and should) be discerned to affect each other" (p. 100).

Basso's study demonstrates how reflections on one communicative resource, place-naming, enables one to engage the various social and cultural uses of nature, with this communicative means having powerful semantic potential. Let me describe some similar patterns, based in other cultural worlds yet demonstrating what Basso found, that place-naming practices enable one to hear nature and culture anew.

Recently I had occasion to climb Mount Monadnock, on a wonderful ridge walk above the treeline in southern New Hampshire. Upon reaching the bald summit, I noticed some dates carved into the rock, "1834" and so on. Feeling irritated at this apparent necessity to write on the rock, I then scanned the wonderful panorama, 360 degrees, uninterrupted. Perched on the summit, again I noticed dates, and feeling annoyed, my mind turned backward, to an American past, which began "speaking" to me. This, indeed, was exactly the spot where Thoreau walked, exactly where his annoyance had been similarly aroused, because of the chink-chink of hammers on rock. Emerson also walked here, energized by this wonderful place. Yes, indeed, at this very place! After returning from the climb, I monitored uses of the place-name, Mount Monadnock, that I and my cohorts sometimes invoked. I began speaking and hearing in our words, not just a vivid physical picture, a looking outward into a space, but also the voices of ancestors dear to me, a looking backward into time. And further, upon invoking this and similar place-names, I felt specific moral precepts about nature, how it can and should be known, about us, how we can (and should) act within it, and about how the doings of both affect each other. In short, place-names (and these can include names of street corners as well as ridge tops) can provide powerful expressive means and meanings, and when invoked they do indeed ignite natural and cultural processes. There is great communicative work, a cultivation of nature and culture, getting done with place-names in particular and with communicative practices generally.

Like place-naming practices, other expressive forms or genres of communication explicitly identify "the natural." These forms are various and cannot be rigidly classified, since their parameters are by definition subject to cultural variability. The common link among these forms is a referential function of communication; that is, each form requires for its expressive power a particular relation between a word-phrase-image and a thing of the physical world. I wish therefore to cast a large net in order to include all communicative resources that are used to symbolize nature, thereby drawing attention to what might be called ethno-physical nomenclature, such as place-names and regional names (e.g., Takaki, 1984), local nomenclature of habitat, plants, and animals (e.g., Carbaugh, 1992, 1996), culturally loaded vocabulary for the body (e.g., White & Kirkpatrick, 1985), landscape paintings (e.g., Mulvey, 1983), landscape poetry, films, and so on. Considered most comprehensively, such phenomena would include the expressive forms that people use to render intelligible what Burke called "the sheerly natural" (1966, p. 373). Particular studies are suggested that focus on one phenomenon or another (e.g., place-naming or body-naming), or within a phenomenon (e.g., local nomenclature for habitat) a particularly telling instance (e.g., the spotted owl) (see Lange, 1992).

When communicative forms such as these are used, local and natural meanings are being radiated, and when social interaction carries forth unproblematically, a coherent statement about nature and culture is achieved. Coherence is "what participants hear (though generally they fail to notice it) when their work is going well" (Basso, 1989, p. 107), such as the meanings aroused when a natural resource, such as Mount Monadnock, is invoked. Three general functions of communication are foregrounded when nature is explicitly symbolized through such forms. These include, but are not necessarily limited to, the depictive, the cultural, and the social.

With regard to the depictive (cf. Ellis, 1983, pp. 229–31), some symbolic constructions function partially to portray (but never merely to represent) some aspect of the physical world, and that portrayal is always partial and selective. To create a vision with words or images is always to do so in one way rather than others. To mention Mount Monadnock, the timber wolf, or the human body is to focus a view, highlighting one image over possible others. Saying that such images are never merely re-presentational is an effort to highlight an ontological belief: communicative practices as natural and cultural phenomena are symbolic, and as such, they select from, locally design, and create particular senses of "the natural," suggesting particular configurations of attitudes and actions toward the natural and cultural, rather than possible others. Emerson's dictum perhaps says it best: "We know more from nature than we can at will communicate" (1987, p. 19). But communicate in and about nature we do, and in so doing we depict some things rather than others, cultivate some senses of that world, and of being with that world, while muting and deflecting others (the others including the "more" to which Emerson refers). This is the depictive function of communicative practices that is ever-present in all languages but is especially aroused when certain emotionally charged, sometimes physical, items are being explicitly symbolized, and being symbolized with.

I use the cultural function as a way of building upon and extending the depictive, thus placing it within a larger symbolic context of meaning-making (Carbaugh, 1989, 1990, 1993; Philipsen, 1987). The cultural suggests that "radiating" from the use of symbolic practices is a larger, historically grounded, multi-voiced semantic system of shared sentiments about what is and what ought to be (Carbaugh, 1988). This symbolic system and its parts, in my view, are always essentially contestable, and they suggest questions: What must be believed about nature, the person, social activities, relations, and emotion for this saying to be efficacious? What unspoken consensus must be present, for this imaging practice to have its local force? Thus, while the depictive function draws attention to a specific relation of an item with time and space, the cultural highlights the larger symbolic system, an ethos, of which the image is one particular part. Interpreting "cultural" messages thus may lead investigations in various directions (the particulars of which need to be discovered in each case), including discourses about religion (as in Emerson), about science (as in biology), about business (as with the continual oil spills), and so on, with each invoking shared premises about nature, what it is and should be, about persons, what they are and should or should not be, and about social life, the modes of action and relations that should or should not be (Carbaugh, 1990).

I have already invoked the social function. The social invites questions about the actual, interactional contexts of communication, about the places in which coherent ideational (and interactional) work is getting done. One might ask: How is it that the use of natural symbols not only depicts physical worlds but also positions people in specific human activities, creating identities (of present and nonpresent others), social relations (especially political allies and alliances), and patterns of action (structuring ways of living with nature)? In short, the social draws attention to exact social scenes where people act, how they act and cast shadows of place there (e.g., are they interpersonally, institutionally based), how they are related to each other (e.g., equal to unequal, close to distant), and the modes of action they cultivate together (e.g., cooperative, competitive).

To summarize the argument so far: Communication occurs in natural and cultural space, creates senses of that space, is consequential for that space, but varies by people, place, and time. Particular communicative forms are used that symbolize nature and culture, like place, animal, plant, and body names; these are used and interpreted culturally in order to accomplish multiple purposes. These purposes include, but are not necessarily limited to, the depictive, the cultural, and the social.

A Conceptual Framework Grounded in Place

Suggested above is an approach to and functional elements of a naturalistic study of environmental communication. Such a study would address the general problem of the relationship between communication, nature, and culture, would do so in full view of the natural and cultural dimensions and devices of communication, and would suggest responding with an anguished study of symbol use, anguished because of a constant attentiveness not only to communicative symbols, especially the use of words and images, but moreover to both the natural world consequences of those symbolic expressions and the cultural processes being cultivated in their use. Concepts such as context, symbol, code, discourse, and culture should help elaborate such a view. I discuss them here briefly, then apply them to two different sites of environmental communication.

With regard to *context,* suggested is the anchoring of studies both in rich descriptions of specific physical places and in descriptions of actual communication practices being used by particular people in these places. Where, generally and specifically, is this natural communicative practice being used? By and for whom? The physical setting, scene, participants (speaker, immediate and potential audiences), and topics of discussion provide important contextual information with regard to the natural processes, as well as the socially expressive means and the meanings associated with them (Hymes, 1972).

The concept *symbol* is fundamental. It is the basic material of expression, a strong toehold in situated communicative practices. As exemplified above with Apache sayings and the Mount Monadnock example, each such means affords a partial view (or hearing), a reflection, selection, accentuation, and deflection of reality (Burke, 1966, p. 45). In this sense, natural symbols, at least on some occasions, are potent expressive means, consisting of basic words and/or phrases and/or images, terms and/or tropes (metaphor, synecdoche, metonymy, etc.) that amplify sense in some directions, while muting others. The use derives from Kenneth Burke (1966) and is developed by Clifford Geertz (1973, esp. p. 89), with reviews and demonstrations appearing elsewhere (Carbaugh, 1988, 1996).

Symbols, though, are significant only within natural environments and larger systems of practices, within physical places and the clusters of symbols, contrastive agons, and mediating terms used there. To interpret these natural systems, the concept of *code* is useful. What symbols, along with their discursive meanings, cluster together in this place, for these purposes? The concept of code suggests interpreting any given symbol as part of a larger natural and symbolic system, pointing to comparisons (e.g., *eagle* versus *bear* as suggesting a coding of wildlife), contrasts (e.g., *eagle* versus *bear* as a coding of international conflict, the United States versus the Soviet Union), agonistic relations, and perhaps even their mediation by an epitomizing symbol (e.g., *negotiation* as a solution to international disputes). Such an analysis unveils beliefs of existence and values in the things that are naturally "said," what they suggest both about nature's ways and the ways of living with it. The term *coding* is useful here in order to move from assessments of structural relations (as with code) to modes of action (or how codes get practiced, constrained, and/or transformed in social scenes over time). This view of symbolic practices thus provides access to a worded world, just as the above focus on context provides access to the physical place in which it plays a consequential role. Both are essential for a balanced view, a view with a double allegiance to natural and cultural processes, bringing into communication both nature's ways and ways of living with it (Carbaugh, 1992).

Of course, any image from nature, or a symbolic expression of it (e.g., the term "Rocky Mountains"), might play a role in various places and codes of life. Living in some natural and cultural contexts suggests some ways of coding and cultivating nature more than others. Identifying a possible range of coding practices associated with such places would help give some sense to the multivocal, poly-phonous quality of these practices in nature and community, tracing the multi-faceted role of the item within and across the contexts of social life (Bakhtin, 1986).

Any item or symbol, and the natural and cultural codes of which it plays (a) part(s), may thus be arranged into larger units or *discourses*, or systems of symbols and codes. These can be defined variously, as, for example, along topical, actional, and/or affiliative lines, that is, by content, by the force of the action, and/or by social alliances and separations. For example, *eagle, winter wren*, and *raven* can play a role in a coding of bird life, which, when taken together with habitat (e.g., sugar maple, lodgepole pine) and topography (e.g., Mount Monadnock, Lake Sunapee), constitute a (partial) community and discourse of wilderness, defined topically. Other discursive communities, defined actionally, suggest ways in which expressive means are used to celebrate, antagonize, revolutionize, persuade, and so on. Defined affiliatively, one draws attention to the aspects of communities and codes that unite members (e.g., an epideictic discourse), separate one group from others (e.g., an oppositional discourse), stratify into constituencies, and perhaps arrange each subgroup within an overall hierarchy (e.g., a positional dis-course), and so on. The point here is that the interpretation may lead in various directions, each to be discovered in situ, given the local design of such systems. In this sense, critically exploring a community of symbolic practices in their natural and cultural scenes, through concepts of codes and discourses, provides less by way of what will be found, more by way of looking and listening. It is a sensitizing more than a definitive conceptual approach, a general way to ask about the mean-ingful use of, for example, an image in context, more than a "thing" to posit in advance and therefore "find."

This leads eventually to a view of *culture* as a system of symbols, codes of ex-pression, and the grand and supersensible discourses it creatively implicates. It is a system molded within and to context, a somewhat coherent set of practices that are consequential for nature, with its primary toehold in highly situated, socially con-stituted, mutually acted, and individually applied communicative practices.

Given these dimensions, functions, phenomena, and framework, we might then ask: What is the nature of the communal conversation in which "the envi-ronment" is expressed (conceived and evaluated)? Let us turn now to two brief demonstrations of environmental communication in which a keen sense of place (natural and cultural) is being created. The main objective of the demonstrations is to treat natural place as part of cultural communicative practices. The objective is analogous to the recent efforts in environmental advocacy to move from treatments of single entities (e.g., species, acts) to communities (e.g., ecosystems, places).

Green Roots in Finland

In midwinter, a few weeks after arriving in Finland, I took my five-year-old twin sons to a public swimming pool.[2] After being delightfully impressed by the quality of the public facility, we walked into the locker room area for men and proceeded to undress and put on our swimsuits. Because of our excitement to get into the water, we did not notice others around us as we rushed to the pool. After swimming, however, we came back into the locker room, and being less in a hurry, we found the showers arranged not in private stalls but in an open row. As we undressed by a shower, I looked by "our" shower for a place to hang our wet suits. Not finding any, I realized that if I wanted to hang up our wet suits, I needed to walk across the rather open locker room to the pegs set aside in one common place for the suits. As I walked to the pegs, I noticed that several men were standing upright, naked, arms across their chests, talking. Others were chatting similarly, with a foot propped up on a bench. Still others, *au naturel,* walked around to sinks, brushed hair, washed, went to sauna, and so on.

As my sons and I showered and went to sauna and then showered again, we noticed the extent to which the Finnish body was being used somewhat differently in public, at least differently from what we were accustomed to seeing (and doing) in the United States. We were part of a scene in which a rather matter-of-fact naturalness was being foregrounded. The delightful taken-for-granted quality of it all was striking, as one so often finds when confronting a different cultural world. Reflections were created in two directions as we got dressed. One set concerned our new environment in Finland, which, in turn, brought to mind our more familiar ways in the United States. I realized that I was "naturally" turning away from the view of others to hide myself from them, a gesture my Finnish contemporaries did not share.

Similarly, a few weeks later, while attending a wonderful "smoke" sauna where women and men used the same facility but at different times, I saw a large window overlooking a striking lake and forest scene. On the periphery was a hole in the lake ice where men and women could cool their bodies after the hot sauna. That this part of the scene was within public view, that some men and women proceeded in view to plunge through the cold water, and that this was all rather common fare for Finns was evident. Further, upon walking the public spaces of Finland, Americans at least are struck by some of the poster art, sometimes in the form of advertisements, that displays images of all parts of the body subtly, naturally. One poster presented men, two young and one middle-aged, nude, upright, in full profile, on a lakeshore, examining the scene, apparently ready to take a plunge. Another image showed a young woman, hands folded across her chest, but thereby accentuating, rather than hiding, her nipples.

This use of the body is a form of cultural expression and is, I think, tied to important Finnish codes. My point, of course, is not that all Finns "exhibit themselves"—as Americans have put it—in locker rooms or public saunas, or that they always display frontal nudity in their artwork. My point is that in Finland, one can notice such things being done as a part of routine life, and no Finn takes particular notice of such things. Finnish cultural meanings are invoked, some of which implicate cultural themes of naturalness, simplicity, and strength. As one Finn put it: "The body that requires no elaboration communicates strength."

Naturalness here has something to do with a public matter-of-factness, acknowledging the intrinsic quality of things as they are, and their limits, and is thus a part of a larger theme of modesty. Naturalness in this sense is opposed to artificialness as something made, civilized, borrowed, or seeking to become other than it is. Being natural is in this sense an affirmation of an unspoiled, sometimes holy (related especially to the sauna ritual), wild, even rural sense of being. Naturalness, in this way, implicitly de-emphasizes the unnatural, more civilized accoutrements of refined urban living, social stratifications, class distinctions, or luxurious "things." Foregrounded and valued is the simple matter-of-fact, limited, natural being, against its social corruption. As a Finnish woman said of being nude in sauna with various people: "You are a human being, plain and simple, no pretense."

Simplicity elaborates these meanings as it emphasizes and morally affirms a minimalist and noble attitude of getting by naturally, simply, with common sense. An air of elegance and plainness of style (notable also in Finnish design) is even apparent as one uses no more than is necessary to get the job done. A sense of strength integrates the natural and the simple and resonates with a rather agrarian past (and present). One ably inhabits rather harsh conditions (e.g., the hot, steamy sauna or the cold winter water and air), preferably with the simple and natural means that are available (e.g., the body).

This coding of the body as natural, simple, strong, is presumed and expressed in some public scenes and is part of many discourses of Finnish culture. Perhaps this coding is also most forcefully evident, albeit through a different symbol, as Finns describe and inhabit the summer cottage, an institution of Finnish life, a place set aside, ideally, for cultivating the simple, natural, hearty aspects of Finnish character.

Finland, a country of about 4.5 million people, has about 350,000 summer cottages, second homes used for summer holidays (Julkunen & Kuusamo, 1991, p. 217). Estimates are that two of three Finns have access to a summer cottage, and during June and July cities are quite empty as Finns migrate to their cottages. The image of the summer home is tied intimately to the landscape: the ideal cottage is perched on a slight hill or rise, in the forest, next to a lake. That the image is tied to the landscape, and the landscape to Finnish character, is a deep historical taproot of Finnish ways. A Finnish historian commented: "Finland's nationalism and identity were not based on history. . .[but on] a romantic infatuation with the landscape" (Klinge, 1992, p. 67).

In fact, the 1800s, the crystallizing era of Finnish nationalism, were replete with landscape art, poetry, and literature in which Finnish character is tied closely to the land, forests, and lakes. With even a short exposure to Finland, one sees and hears the prominence of "nature" to Finns. In the early 1900s, as Finns acquired material wealth through industrialization, many purchased summer cottages in order periodically to escape the communal living arrangements in urban industrial life. Returning to the forest, to "nature," one could cultivate the deeper, historically grounded values of simplicity, naturalness, and heartiness.

In the present, one can hear sprinkled into Finnish speech the place of the summer cottage and landscape. One woman was describing Finnish history and making the point that Finns have "always" been in the area now known as Finland by linking the people to their land and linking that land to the summer cottage:

We Finns, she said, "are grounded in our rocks, forests, and lakes. Y'know, we have the oldest rocks in the world here." A few seconds later, when discussing her summer cottage, she said: "Under each corner of the house are rocks stacked on other big rocks, and there the house sits." The Finns and their land are steadfast as rock, strong and enduring, in their natural home, with this strong and natural life tied to simple and modest living in a potent place, the summer cottage. Through these words and images one can hear a national and natural character, a landscape and locale, a simple and situated place for the hearty.

Another woman described her summer cottage, saying: "It's important for Finns to have the summer cottage, a place to be with nature." A man said of his: "It's on a hill, overlooking the lake, about two acres' worth." He smiled broadly. "And of course the sauna, an old sauna by the lake, with the trees and water around, sitting in there quietly, peacefully, with the birds singing. Nothing beats it." A woman noted similarly: "We Finns live very close to nature. . . .we go to the sauna for peace, quietness, refreshment mentally. We expect to come out feeling better." A recent five-part video by Finns on "how to become a Finn" contains much footage of the summer cottage, lakes, lakeshores, and water.

What I attempt to draw attention to here is the use of the body, the summer cottage, and the sauna as symbols of expression and to suggest that part of what is getting expressed with these symbols is a deep root and code of Finnish character. The expressions are grounded through uses of a particular landscape (the rocks, forests, lakes), situated with the summer cottage (and sauna), with the body used sometimes similarly. These images and forms of action create physical sites through which a simple, natural, hearty life is lived. Through these Finnish symbols and meanings, that are both of and about a place, potent images of life are conducted (with the body) and portrayed (with the summer cottage symbols). Through symbolic acts and expressions such as these, one is guided to living in place in a simple, natural, and hearty way, living an environment both naturally and culturally.

With these localizing expressions and codes, then, one can begin feeling, hearing, and seeing Finnish links between body and place, between patterns of living and locations, between being in and being as an environment. As the body can become a natural, simple site of unadorned being, so one can inhabit a natural site, the summer cottage (or sauna) in which natural energies can be restored, simple living conducted, a heartiness of soul nurtured. As the favored physical places of the society are celebrated, so too is the body conceived, as a site of and for being in place. That the attitudes of naturalness and simplicity, as well as modesty and limits, are associated with these routine practices, and that these attitudes provide some cultural bases for living, should be better understood. Environmental discourses, conceived culturally, penetrate routine living, and as they do so, they create ways of inhabiting, being with, and being part of nature's places.

A recent study of forest management in Finland, under the interesting title "The Forest and the Finns," begins a special section by saying: "The forests and trees have not only provided material sustenance but at the same time many beliefs have been linked with their use" (Reunala, 1989, p. 51). The essay goes on to describe the Finnish folk image of a "world tree" that supports the firmament of life, a belief that results in the preservation in timbered areas of special "memorial trees" as life-sustaining sites of both material and moral good fortune. Such a practice is yet another example of life conducted, as it were, close to the ground, a simple practice of naturalistic thinking, a source for the preservation of hearty living. It also anchors cultural reactions against clear-cutting and other forest management practices that exploit this place and its people.[3]

The cumulative effect of these symbols, codes, and discourses for Finns is, I believe, a pronounced minimizing of the distance between some everyday communicative practices and, as Finns put it, "nature's" processes. In other words, when they are lived, the practices create very little space between these symbolic forms and the physical environment. Just as the sauna historically was used as a sacred place for birthing and for cleansing the dead, so today the sauna, the summer cottage, and some forms of bodily conduct help keep Finnish life rather close to nature's ways. That Finnish communication cultivates such a link, and that the Finnish policies for living draw upon such links, should not go unnoticed. For through such symbols, forms, and codes, a naturalness, a simple strength, is being nurtured, in and of its space, of body and place.

Green Writing

We have focused our attention on communication patterns that cover visual images, routine conduct, and spoken comments by Finns. These patterns were identified, then interpreted, as depictive and cultural codes that established a particular—that is, a Finnish—relationship with natural space. Now we turn our attention to communication patterns that pervade a print medium by exploring one written text in detail. My objective here is simply to suggest how some features of a single text can be treated as richly depictive, cultural communicative resources.

This focal text is a journalistic article that appeared in the magazine *Wilderness,* a quarterly created for members of the Wilderness Society but also available to the general public. The magazine states the objectives of the society: "The Wilderness Society, founded in 1935, is a non-profit membership organization devoted to preserving wilderness and wildlife, protecting America's prime forests, parks, rivers, deserts, and shore lands, and fostering an American land ethic." The particular issue from which our text was taken is entitled "Saving the Wildlands of New England—A Puzzle of Possibilities." Our text, "Whose Woods These Are" by Norman Boucher, is the first in a series of three articles on the subject. The glossy medium and the society's stated objectives thus "place" this text squarely, but not exclusively, with a "green" audience, since the society generally advocates "preserving wilderness and wildlife." I will focus on the first part of the article:

> At last, the woods. After everything I've heard, I was afraid they'd appear different this year. I was afraid that this spring the hobblebush would fail to flower, that the winter wren, having read the newspapers, would choose to hide in the Georgia mountains and skip the tiresome journey north. For months I have been reading

dozens of reports and articles about these northern New England forests. For weeks I have been collecting opinions about what they are and what they're likely to become. In offices, motel rooms, restaurants, airplanes, and pickup trucks I have heard them sized up in such disparate images that I began to distrust my own memory, accumulated over two decades of happy scrambling at every season of the year. Finally, to escape this gloomy fog of confusion, I filled a pack and drove to the Percy Peaks, two isolated mountains in [the remote Nash Stream area of] northern New Hampshire so close to the Quebec border that the locals say those who get lost there come out speaking French. I was seeking one of the dimmest, most poorly marked trails I could recall. I needed the illumination of wildness. . . .I did not know two years ago that even my modest wilderness ambitions were being squeezed. I didn't realize what would soon happen to the woods of this valley and would make them a notorious landscape, one whose fate could affect the future of much of the region's wildest places. The time of innocence was already ending; the idea of New England wilderness by any definition was inching closer to absurdity. Already, as I lingered watching ravens drift unconcerned above the Percy Peaks, Wall Street had come calling on Nash Stream. [Boucher, 1989, p. 18]

Even a casual reading of this passage leads one to a very particular sense of place. Created, on the one hand, is a sense of the Percy Peaks area, a remote area in northern New Hampshire, pristine, untouched, ravens quietly soaring, but threatened! By what? Wall Street and the attendant gloom, confusion, and the disparate images that go with such a place. But of course there is more than this. Ways of living are being asserted: some are being nurtured, preserved; others are being criticized. There is clearly more here than "mountains" and "motel rooms." Yet how do we hear and see in this snippet of text, through these depictive symbols of nature, cultural messages at work? To begin, we might simply inquire about structural relations among symbols, with one preliminary eye toward a discourse of "wildness" and the other focused toward "Wall Street."

A discourse of wildness is signified here with several symbols. The use of the terms "the woods," "wildness," "New England wilderness," and "Nash Stream," as a "notorious landscape," all invoke a kind of place with which readers are assumed to be familiar, of which they can ably create images, and for which they would fight. But how, specifically, is this place symbolized, for present purposes? What specific objects are linguistically painted onto this cultural canvas? With what meanings?

Within this discourse of wildness at least three specific codes are activated explicitly. One concerns precise, valued images of habitat, such as "the woods," "northern New England forests," and "the hobblebush." Another creates valued images of wildlife, the bird life of "the winter wren" and "ravens." Thus we are offered some details of place through images of forests, plants, and birds. These codes of habitat and wildlife evoke common meanings—beliefs and values—about preserved lands, as being more or less pure, filled with free spaces for flight (by birds and people), set aside for their own value, creating, because of their pristine qualities, possibilities of human "illumination." A third code builds on images of a larger regional topography—"the Georgia mountains," "the Percy Peaks," "New Hampshire," "Quebec" and the related "speaking French"—thus quickly saying something important about "wildness" as radically contextual (in a remote region of New Hampshire) but embracing something beyond state (Georgia to New Hampshire), national (United States to Canada), and linguistic (English to French) boundaries. Thus, the code of topography "speaks" of places conceived generally as different, but it does so by stressing their—from this point of view—often hidden and muted interdependencies. The discourse of wildness, so built, speaks of habitat, wildlife, and topography, evoking common meanings of purity, freedom, and interdependencies and affording valued opportunities for "illumination."

Placed against this discourse, and developed in a characteristically polemical fashion, is what might be called a discourse of development. While not elaborated to the same degree as the discourse of wildness, suggested are elements of its own habitat, "offices, motel rooms, restaurants, airplanes, and pickup trucks," its own form of life, "disparate images," which result in a "gloomy fog of confusion," all—at this point—somewhat weakly associated with "Wall Street."

A third discourse identifies a relationship between wildness and development. Playing "wildness" against "development" creates both a fear of change ("afraid [the woods would] appear different") and an anxiety over movement from the "innocence" of wildness to the "absurdity" of development. In short, the old alarm is sounded: "wildness" is succumbing to "development," invoking a call to arms for allies and readers to be readied for battle. Yet again, the one seemingly uncontrollable mode of action (development) needs to be combated by another (preservation). So far as this goes, and at this level, the discourse reveals a familiar tune or plot line.

But what is familiar in it? And is this discourse preserving a discursive system that itself needs to be developed? Do the contents, propositions, oppositions, and morals of this discursive system (preservation over development) constrain its ability to transform environmental issues (e.g., the development of preservation discourse)? Are the traditional expressive means of nature operating to preserve the very problems they seek to solve?

My response is, well, yes and no. As for the affirmative, the primary depictive messages involve images of wildness as static, innocent, tranquil, and above all "natural." The "other" place is Wall Street and is characterized by toxic, cancerous growth, greed, jaded activity, and above all, in the sense of a learned and manufactured form of life, "culture." At this level, the depicted images speak cultural messages: they exploit and reproduce a fundamental opposition in American culture, nature versus culture (and the related polemic of theism/atheism). Nature here is God-made, given and pure. Culture is man-made, and a fall from grace. Such oppositions as these run very deep and are, if not literally present in the words of this text, hearable in its symbolic meanings. It is precisely these kinds of semantic structurings, among other things, that a cultural interpretation can expose: that is, a deeper hearing of such patterns that are typically presumed, unquestioned, and constitutive parts of the deeper meanings of the text.

Some elaborations of this theme should demonstrate the point further. Several possible cultural meanings may be brought to this and similarly structured texts, including—building on the above—oppositional themes of religious discourse (e.g., the saved environmentalist versus the sinner developer), of purity versus pollution, or spirituality versus immorality. Related are discourses that relate such cultural places to forms of action, making wilderness a place for rejuvenation and illumination, the other a place for exhaustion, confusion, and exploitation. A similar contrast basic to many American lives is the separation, and felt tension, between a sense of play and of work, which when combined with its Puritan roots yields a work hard, play hard syndrome. In our present text, this cultural theme becomes very interesting as the one group's playground is the site of the other's work. With regard to ownership, there are explicit overtones, in *wildness*, of the public good for the many, whereas *development* connotes the private interests of the one or the few. The latter triumph of private interests over the public good sounds themes of exploitation, pollution, immorality: the intrinsic corrupt(ible) nature of "cultural" institutions is immediately asserted and presumed. The Wild West confronts the Establishment East, and so on. All such symbolic oppositions raise a fundamental question of what it means to be a cultural actor, of what it means to be in natural places, suggesting responses in different moral discourses: one responds with meanings of purity, enlightenment, a union with nature, the other with meanings of corporate identity, exploitation, and multiple uses of nature.

The point here is not to give a comprehensive listing of discourses that come to bear on this communal conversation but rather to suggest some that may repay deeper analysis and interpretation. Indeed, part of the necessary work in cultural interpretation is the discovery of the most powerful discourses whose crisscrossing continua create the complexity of communication in cultural "texts" like this one. With regard to this written snippet, perhaps the above commentary suggests some leading candidates.

Concerning the social functions of this text, two faces are presented and agonistically related, with the agon resolved by praising the one. The face presented is that of the environmental advocate who strives against tremendous odds and almost overwhelming fiscal resources to preserve and protect wilderness and wildlife. The face attributed to the other is that of the greedy developer, the immoral profit mongerer, out to satisfy insatiable personal needs at considerable costs to others. In case the reader is unconvinced or unfamiliar with the latter type, the next paragraph in the article tells of Sir James Goldsmith, a "flamboyant" owner of a pulp and paper business, Diamond International Corporation. Goldsmith is described as an uneducated "crank" who "commuted between a wife in Paris and a mistress in London, fathering children with each." The story eventually gets quite complicated, with "the government" mediating between these two faces-groups, all in the name of what locals in northern New England call the working forest (Boucher, 1989). The struggle between the groups, as far as the discourse goes, tells of two peoples, one doing immoral work in the other's playground, the other recklessly playing around with the hard workings of locals. So constructed, the relationships portrayed are ones of strain and stress, and the primary mode of action is competition. The resolution offered is clear: Praise the preservationist, and damn the developer.

A full-blown analysis of the intricacies involved in this discursive production, as a cultural creation, would highlight the subtleties and depth of the communication. It would enable the reader to hold up for scrutiny the particular symbols, codes, and discourses being typically used but not typically scrutinized. This is a basic task of cultural study: to render scrutable that which is typically inscrutable. Once this is done, once the basic cultural discourses and oppositional faces are exposed, one is placed better to identify and assess the senses of place of which the discourses and faces are a part, as well as the discursive constructions being cultivated.

In the process, we can understand better how we reproduce our own well-worn ways and how we can create anew. In this regard, note how the term *working forest,* used by locals in the northern New England woods, stands at the borders of the two discourses. Themes of each are brought to bear and creatively played out. It is telling that the people who live closest to the issues have devised an expressive means able to embrace and express oppositions that for others, living elsewhere, are nearly impossible to integrate.

A similar dynamic, another effort at cultural creation or cultural integration, moves beyond these often impenetrable oppositions and concerns the general perspective being suggested here itself. How does one identify and redress old, tired tangles like those of nature versus culture? How does one design discourses able to transform old problems, in newly productive ways? The naturalistic approach advanced here is one such effort, an effort to hear (sense, see, feel) natural and cultural process in communication practices, to "hear" in the cultural the creation of nature, and to "see" in nature a culture at work. The approach seeks practical, as well as theoretical, goals.

In a single text like the *Wilderness* article, there lurk large natural and cultural meanings. From the familiar symbols of expression, to codes, to discourses, to competing faces, to tired ideological tangles—all become woven into rather routine practices of environmental communication. As we unveil the inner workings of these, we can, one hopes, become less habitual in their use and more able to enrich our knowledge about various places, of nature and culture.

Environmental Communication as the Creation of Natural and Cultural Place

By inquiring at the nexus of natural space and communicative processes, by arguing for a multifunctionally based interpretive and critical framework, and by briefly suggesting paths of inquiry around socially grounded texts, I hope to have shown some of the promise in a naturalistic approach to environmental communication. Surely there is much more to be said and much work that needs to be done. Of special importance is the grappling with highly particular, socially situated, symbolically constructed images in place. Specific case studies that trace the patterned use and interpretation of nature in communication and community are essential. This chapter suggests some movements in that direction, with Basso's 1989 study being exemplary. Such studies would enable a comparative assessment regarding available means for conceiving of, and evaluating, natural space, local meaning systems, and the attendant attitudes that these cultivate, and constrain (for some examples, see Bird, 1987; Bird-David, 1990; Callicott & Ames, 1988; Cox, 1973; Glacken, 1967; Hastrup, 1989; Ingold, in press; Myers, 1986; Rolston, 1987; Swagerty, 1984; Willis, 1990). Also, such studies should lead us to see our own taken-for-granted ways anew and to reflect upon them, freeing us from entangled webs we have helped to weave.

In the process, we shall be able better to respond to fundamental questions about communication, culture, and nature: How is natural space conventionally symbolized? What do these symbolic processes enable, and constrain, as situated social living? Are we reproducing and reconstituting troubles we seek to remedy? Or are we changing for the better the natural conditions, of nature and culture, in which we speak?

Notes

1. I want to include as "nature" and "natural space" all possible physical places and thus, following Roderick Nash, to include "a spectrum of conditions or environments ranging from the purely wild on the one end to the purely civilized on the other—from the primeval to the paved. This idea of a scale between two poles is useful because it implies the notion of shading or blending. Wilderness and civilization become antipodal influences which combine in varying proportions to determine the character of an area. In the middle portions of the spectrum is the rural or pastoral environment (the ploughed) that represents a balance of the focus of nature and man. As one moves toward the wilderness pole from this midpoint, the human influence appears less frequently. In this part of the scale civilization exists as an outpost in the wilderness, as on a frontier. On the other side of the rural range, the degree to which humanity affects nature increases. Finally, close to the pole of civilization, the natural setting that the wild and rural conditions share gives way to the purely synthetic condition that exists in a metropolis" (Nash, 1982, p. 6). Further, in this chapter I am particularly interested in the ways the dimension proposed here (or something like it) is conceived and expressed, related to actual natural worlds, and the ways different points on this spectrum are creatively invoked in the communication practices of distinctive human communities.

2. This report from Finland is based on fieldwork conducted during November 1992 and January–August 1993. The report is very preliminary and intended only as a demonstration of some of the possibilities of the approach advanced above. In the process, I am not trying to advocate that readers become Finns. I am trying simply to demonstrate that we can explore corners of life—typically untended—in which environmental-natural attitudes can be found. Such a study stands at the nexus of culture, nature, and communication, of natural and social studies, and can help provide an integrative view into environmental discourses. Offered in turn is a grounding of advocacy in the grass roots of local living.

3. These themes are portrayed visually and profoundly in the 1893 painting *Burning the Forest Clearing (The Wage Slaves)*, by the Finnish landscape artist Eero Jarnefelt. The images provided critical commentary on both the unfair labor practices of "slaves" (especially through the image of a young girl) and the "slash and burn" agricultural method at a time when the advantages and disadvantages of the method were being publicly debated (see Valkonen, 1992, 72–88).

References

Bakhtin, M.M. (1986). *Speech genres and other late essays.* V. McGee (Trans.), C. Emerson & M. Holquist (Eds.). Austin: University of Texas Press.

Basso, K. (1989). "Speaking with names": Language and landscape among the Western Apache. *Cultural Anthropology, 3,* pp. 99-130.

Bird, E.A. (1987). The social construction of nature: Theoretical approaches to the history of environmental problems. *Environmental Review, 11,* pp. 255-66.

Bird-David, N. (1990). The giving environment: Another perspective on the economic system of gatherer-hunters. *Current Anthropology, 31,* pp. 189-96.

Boucher, N. (1989). Whose woods these are. *Wilderness 53,* 186, pp. 18-41.

Burke, K. (1966). *Language as symbolic action: Essays on life, literature, and method.* Berkeley: University of California Press.

Callicott, J.B., & Ames, R.T. (Eds.). (1988). *Environmental philosophy: The nature of nature in Asian traditions of thought.* Albany: State University of New York Press.

Carbaugh, D. (1988). *Talking American.* Norwood, N.J.: Ablex.

———. (1989). Fifty terms for talk: A cross-cultural study. *International and Intercultural Communication Annual, 13,* pp. 93-120.

———. (1990). Toward a perspective on cultural communication and intercultural contact. *Semiotica, 80,* pp. 15-35.

———. (1992). "The mountain" and "the project": Dueling depictions of a natural environment. In C.L. Oravec & J.G. Cantrill (Eds.), *The conference on the discourse of environmental advocacy* (pp. 360-76). Salt Lake City: University of Utah Humanities Center.

———. (1993). "Soul" vs. "self": Soviet and American cultures in conversation. *Quarterly Journal of Speech, 79,* pp. 182-200.

———. (1996). *Situating selves: The communication of social identity in American scenes.* Albany: State University of New York Press.

Cox, B. (Ed.). (1973). *Cultural ecology: Readings on the Canadian Indians and Eskimos.* Toronto: McClelland & Stewart.

Ellis, D. (1983). Language, coherence, and textuality. In R. Craig & K. Tracy (Ed.), *Conversational coherence* (pp. 222-42). Beverly Hills, Calif.: Sage.

Emerson, R.W. (1987). *Emerson on transcendentalism.* E. Ericson (Ed.). New York: Unger.

Geertz, C. (1973). *The interpretation of cultures.* New York: Basic Books.

Glacken, C. (1967). *Traces on the Rhodian shore: Nature and culture in Western thought from ancient times to the end of the eighteenth century.* Berkeley: University of California Press.

Hastrup, K. (1989). Nature as historical space. *Folk, 31,* pp. 5-20.

Hymes, D. (1972). Models of the interaction of language and social life. In J. Gumperz & D. Hymes (Eds.), *Directions in sociolinguistics: The ethnography of communication* (pp. 35-71). New York: Holt.

Ingold, T. (1991). Against the motion. In T. Ingold (Ed.), *Human worlds are culturally constructed* (pp. 12-17). Manchester, Eng.: Group for Debates in Anthropological Theory, University of Manchester.

———. (1992). Culture and the perception of the environment. In E. Croll & D. Parkin (Eds.), *Bush base, forest farm: Culture, environment and development* (pp. 39-56). London: Routledge.

———. In press. Hunting and gathering as ways of perceiving the environment. In K. Fukui & R.F. Ellen (Eds.), *Beyond nature and culture.* Oxford: Berg.

Julkunen, J., & Kuusamo, A. (1991). Life in Finnish forests: Semiotics of a summer house. *Semiotica, 87,* 3-4, pp. 217-24.

Klinge, M. (1992). *Let us be Finns: Essays on history.* Helsinki: Otava.

Lange, J.I. (1992). Case study in American conflict: "Old-growth" and the spotted owl. In C.L. Oravec & J.G. Cantrill (Eds.), *The conference on the discourse of environmental advocacy* (pp. 183-94). Salt Lake City: University of Utah Humanities Center.

Mulvey, C. (1983). *Anglo-American landscapes.* Cambridge: Cambridge University Press.

Myers, F. (1986). *Pintupi country, Pintupi self.* Washington, D.C.: Smithsonian Press.

Nash, R. (1982). *Wilderness and the American mind.* 3rd ed. New Haven, Conn.: Yale University Press.

Oelschlaeger, M. (1992). Wilderness, civilization, and language. In M. Oelschaeger (Ed.), *The wilderness condition: Essays on environment and civilization* (pp. 271-308). San Francisco: Sierra Club Books.

Philipsen, G. (1987). The prospect for cultural communication. In L. Kincaid (Ed.), *Communication theory: Eastern and Western perspectives* (pp. 245-54). New York: Academic Press.

Reunala, A. (1989). The forest and the Finns. In M. Engman & D. Kirby (Eds.), *Finland: People, nation, state* (pp. 38-56). Bloomington: Indiana University Press.

Rolston, H. (1987). Engineers, butterflies, worldviews. *Environmental Professional, 9,* pp. 295-301.

Swagerty, W. (1984). *Scholars and the Indian experience.* Bloomington: Indiana University Press.

Takaki, M. (1984). Regional names in Kalinga: Certain social dimensions of place names. In E. Tooker (Ed.), *Naming systems* (pp. 55-77). Washington, D.C.: American Ethnological Society.

Valkonen, M. (1992). *Finnish art: Over the centuries.* Helsinki: Otava.

White, G., & Kirkpatrick, J. (1985). *Person, self, and experience: Exploring Pacific ethnopsychologies.* Berkeley: University of California Press.

Willis, R. (Ed.). (1990). *Signifying animals: Human meaning in the natural world.* London: Unwin Hyman.

3
To Stand Outside Oneself: The Sublime in the Discourse of Natural Scenery

Christine L. Oravec

The twentieth century has been labeled the age of anxiety, the technological age, and even the postmodern era. A lesser-known, but perhaps more specific, label for the century that has produced such forms of communication as generative grammar, electronic mass media, and nonrepresentational art might be the age of discourse. In this century, the term *discourse* has shed its traditional and exclusive association with continuous verbal or written prose. Instead, the term has expanded to include visual signals, nonverbal gestures, and such discontinuous fragments of signification as advertisements and product logos—in fact, all types and forms of symbolic communication.[1] The communicative force of these signs, however, still depends upon basic conventions of belief and understanding, that is, commonly held sets of connotations that provide their users with a convincing, though artificial, impression of what is real. The persuasiveness of these conventions rests, not in the accuracy of their reflection of reality, but in their convincing reproduction of what is taken for reality.

One of the most important discursive conventions for the study of environmental communication has been "the sublime." The literary and artistic convention known as the sublime dominated European and American aesthetic thought from the late eighteenth century to the early twentieth. As a convention, it prescribed a set form of language and pictorial elements for describing nature. It also encouraged a specific pattern of responses to nature that influenced the ways we look at and alter natural scenery. Consequently, it became useful for fulfilling rhetorical purposes of all kinds. For more than two centuries, artists and writers used the sublime to evoke emotional responses toward nature, to confirm aesthetic or ethical beliefs about nature, and to call attention to particular landscapes for settlement, tourism, or preservation. Indeed, we still employ conventions of the sublime today, in our written discourse as well as in our pictorial representations of nature, to mold and shape our responses to our environment.

In this chapter I wish to argue that sublime discourse, whether it is verbal or visual in its form and medium, is an integral part of the way we perceive nature, act with reference to it, and construct its relationship to ourselves. I will be discussing the historical convention of the sublime in some detail, but describing the convention is less important than what people do with the sublime and what it does to people. First, I will briefly explore the sublime as a perceptual screen through which we see nature. Then I will discuss the conventions of the sublime in visual representation and in written discourse, primarily during the nineteenth century in America. Finally, I will examine two contemporary controversies about the sublime that illustrate why the topic is of interest in today's environmental debates.

The Sublime as a Perceptual Screen

My first assumption is that the sublime convention acts as a screen, or a projection of human preferences upon the natural scene. Even as nature provides the pretext for the sublime convention, we view nature through the conventions of sublimity. This assumption is similar to the theme of a recent collection on Western art articulated by Jules David Prown: "[Our purpose] is to consider the processes whereby a scrim of myth has come to veil our view of the past, misleading by pleasing. The art of the West often purported to depict its subjects realistically, but perhaps it depicted more accurately the needs, values, and aspirations of its viewing audience" (1992, p. xi).

The products of Western artists, transparently accurate as they may have appeared, were carefully composed to conform to their viewers' aesthetic and social preconceptions of the beautiful, unique, and awe-inspiring. But as calculating as this sounds, the activity was by no means a completely conscious one. The artists were products of their own time, and in most cases they assumed that their era had as unmediated a view of nature as any other. Therefore the study of the convention of the sublime is a study of collective understanding or, if you will, preunderstanding, the "subconscious expressions articulated by choices made in the process of depiction not only in regard to inclusion and exclusion but as regards points of view and placement" (Prown, 1992, p. xiv).

Thomas Moran was one of those Western artists who established cultural preunderstandings about nature. According to Joni Louise Kinsey, "Moran's most evocative works were not simply pictures of their respective subjects; for many Americans the works became the places themselves in a sense. After seeing his paintings, visitors to the West often were unable to view the sites without preconceptions; the images forever changed perceptions of the land and its meaning for the culture that inhabited it" (1992, pp. 11–12). Indeed, many easterners who had never seen the Grand Canyon of the Yellowstone were first educated about its features by Moran's 1872 painting of it. *The Grand Canyon of the Yellowstone* was the first painting ever purchased by the U.S. government, having been acquired soon after the passage of the act making Yellowstone a national park. Thus its status as an accurate recording of the scene has been preserved for all time, not by verifying the painting by comparing it with its original subject, as one might verify a documentary photograph, but by keeping its subject from changing at all (Kinsey, 1992, p. 43). Even today, Moran's paintings of colorful geysers at Yellowstone and his views of Arizona's Grand Canyon from the South Rim have dictated the way we see those landscapes. His renditions have led us to prefer, or at least to recognize, a certain standpoint, a certain angle, even a certain time of day, for viewing these "natural" wonders.

Everyone who has been a tourist has experienced this self-reflexive effect of scenic representation. We travel to see the sights that we have become familiar with in pictures or descriptions, and we are less than satisfied if our preconceptions are confirmed. Oddly enough, however, it is often the scenery, not the representation, that dissatisfies us. Malcolm Andrews described this phenomenon: "There is a peculiar circularity in the tourist's experience. He values the kind of scenery which has been aesthetically validated in paintings, postcards and advertisements; he appraises it with the word 'picturesque'; and then he takes a picture of it to confirm its pictorial value" (1989, p. vii). It is as if the very words *sublime* or *picturesque* confer upon a portion of scenery a codified concept that is in turn indexed by a selected arbitrary sign, such as a photograph. This sign functions arbitrarily and abstractly with respect to the concept. As a result, since the scene has become signified, or reduced to an arbitrary measure, any remaining deviation from that pure reduction, any rough edges or uncontrolled variation, is "other," and therefore lesser, than its ideally perfect representation. Viewers become disappointed, not in the representation, but in the irreducible elements of the original scene.

Such an outcome, though paradoxical, is very important, since we often take the further step of changing nature radically to suit our preconceived notions of sublimity. As James Mullen suggested in 1874 to his readers in *Anthony's Photographic Bulletin*: "Let me advise you then to always have with you on your photographic trips, a spade and a good axe; the latter particularly will often be found 'a friend in need,' when it is desirable to cut a small tree or remove a branch that would otherwise obscure some important point of your view" (quoted in Masteller, 1981, p. 61). Apparently Eadweard Muybridge, an early photographer of Yosemite, who was also famous for his stop-action sequences of running horses and human figures, did just that.

> He has waited several days in a neighborhood to get the proper conditions of atmosphere for some of his views; he has cut down trees by the score that interfered with the cameras from the best point of sight; he had himself lowered by ropes down precipices to establish his instruments in places where the full beauty of the object to be photographed could be transferred to the negative; he has gone to points where his packers refused to follow him, and he has carried the apparatus himself rather than to forego the picture on which he has set his mind. [quoted in Robertson, 1984, p. 82]

In Yosemite Valley, artists and photographers cleared away undergrowth that interfered with their compositions, and early supervisors of the valley were all too willing to assist in the task.[2] For example, Galen Clark, the manager of Yosemite before it became a national park, ensured good views from the valley floor by using ax and fire to eliminate tall brush and trees. Since 1905, when management of the vegetation by the federal government began, the flora on the valley floor has grown up considerably, making the walls of the valley seem less high and obscuring the view even more than in Clark's day (Robertson, 1984, p. 94).

So far, I have illustrated how the sublime convention modified human perceptions of nature through visual examples. But writers also used sublime language in conventional ways. For example, Horatio Parsons, author of an 1836 guidebook to Niagara Falls, used the conventions of the sublime response when he wrote, "The effect of the Falls upon the beholder is most awfully sublime and utterly indescribable. The sublime, arising from obscurity, is here experienced in its greatest force. The eye, unable to discover the bottom of the Falls, or even to penetrate the mist that seems to hang as a veil over the amazing and terrific scene, gives place to the imagination, and the mind is instinctively elevated and filled with majestic dread. Here is 'All that expands the spirit, yet appals'" (quoted in McKinsey, 1985, pp. 158–59). Note how he described the sequence of emotions expected of his readers, guided their point of view, and addressed an implied second person. In fact, he was schooling his readers in the kind of responses expected of someone who truly appreciated the scenery of Niagara Falls. Parsons was obviously a writer interested in controlling his readers' reactions, but he was far from the only writer to do so. Authors influential in the early conservation movement, from Clarence King to John Muir to John Wesley Powell, used these same sublime formulas when they wrote for popular consumption.

Thus, whether represented visually or through verbal language, sublimity as a convention functioned actively to screen, or project, human expectations and desires upon the landscape. Although, as the very notion of a convention itself suggests, the sublime could not be identified directly with nature itself, its habitual association with what is "natural," and therefore "true" or "real," had consequences that affected the natural world through its impact upon readers and viewers.

The Conventions of the Sublime

Throughout history, the convention of the sublime has designated a formula, a code, or a set of rules for interpreting the meaning of what might otherwise be an arbitrary, random confluence of natural elements. During the Roman Empire, sublimity was a rhetorical characteristic primarily describing the public speaking of an imperial court. Its conventions coincided with those of the epideictic or ceremonial oration, in which the emperor served as the subject of elaborate, often incongruous comparisons and the audience functioned as evaluators of the speaker's skill at flattery. Later, epideictic oratory in the sublime mode enlivened many of the social and political institutions of Western civilization, such as the legislature, the bar, and the theater. The comparisons that orators used in these settings often evoked the sublimity of natural scenery. In describing a sublime scene, for example, a British traveler made explicit the connection between grand scenery and parliamentary oratory: "To give an Idea of the astonishing Variety which here crowds upon your mind is impossible and it may be well said to be the real sublime and beautiful conveyed in the Language of Nature infinitely more strong than the united Eloquence of Pitt, Fox and Burke" (quoted in McKinsey, 1985, p. 31). Despite its utility in the legislature and at the bar, however, the sublime epideictic speech as such remained one step removed from practical concerns by focusing upon the theatricality of the product rather than the determination of an outcome (see Longinus, 1985; Aristotle, 1941).

Interestingly, modern Western Europeans adapted the rhetorical conventions of the sublime not only to oratorical settings but also to historical landscape painting as a parallel form of ceremonial civic expression. The most influential early models were calm pastoral landscapes by Claude Lorrain and somewhat more tumultuous views by Salvator Rosa. The setting of the early sublime landscape contained foregrounds, middle grounds, and backgrounds, comparable to upstage, middle stage, and downstage in a modern theater. The foreground bordered the edge of the frame, prosceniumlike, and was represented by a large tree or a side of a cliff, called a coulisse. The middle ground was illuminated and often contained a human or animal figure, a river, or a ruin to lead the viewer's eye into the scene. Finally, the background of the landscape led to the horizon, framed by the objects in the foreground. The terminology was theatrical: the setting was populated by "scenery," "side-skips," and "screens" (Andrews, 1989, p. 29). In all, the effect signified benign domestication, with human beings inhabiting a friendly, harmonious domain.

Soon, however, this benign quality began to change. With the advent of the Romantic period, European pastoral painting began to include representations of true wilderness, particularly along the borders of the scene. This new, less classical type of landscape came to be known in England and America as the picturesque. American artists, in particular, employed picturesque conventions to adapt the visual language of the pastoral to somewhat rougher natural conditions than those common in Europe. Thomas Cole, for example, in his painting *The Oxbow* (1836), portrayed the settlements of the Connecticut Valley as surrounded by outcroppings of natural wilderness. Half of the painting was conventionally pastoral, with light, harmonious, settled nature in the middle ground; and the other half was portrayed as wilderness, with dark, uninhabited nature in the foreground. A line of blue hills reaching out the horizon filled the background. Furthermore, the standard picturesque conventions were stretched somewhat to accommodate this North American scene. The horizon line was raised so that the viewer saw the scene from above, and the scene itself was not as carefully framed on three sides as in most pastoral views.[3]

After about 1860 the mountain scenery west of the Mississippi River became better known to the general public, and artists attempted to apply European picturesque conventions to what proved to be a landscape filled with astonishing contrasts. These conventions, however, were not sufficiently flexible for describing the wildness of the scenery. American artists ultimately developed their own style, one that can be called the American sublime. A literary example here might help illustrate the relationship between the picturesque and the American sublime. In his book *The Mountains of California,* John Muir described a scene near Mount Lyell. He wrote that it was "in a high degree picturesque, and in its main features so regular and evenly balanced as almost to appear conventional— one somber cluster of snow-laden peaks . . . the whole surging free into the sky from the head of a magnificent valley, whose lofty walls are beveled away on both sides so as to embrace it all without admitting anything not strictly belonging to it. . . . Down through the midst, the . . . river was seen pouring from its crystal fountains... then sweeping on through the smooth meadowy levels of the valley" (Muir, [1894] 1977, pp. 49–50). Muir accurately labeled this scene picturesque, because the compositional elements of the picturesque were all there. Specifically, the foreground, represented by the angled walls of the glacial valley, neatly framed the middle ground, with its swollen river, which led the viewers' eye into a background terminating in towering peaks.

Yet Muir originally chose this picturesque view of Mount Lyell not to please himself but to accommodate some landscape artists whom he guided into the Sierras in 1872. Up until his discovery, they had been complaining to him: "All this is huge and sublime, but we see nothing yet at all available for effective pictures. ... Here [the] ... foregrounds, middle-grounds, and backgrounds [are] all alike" (Muir, [1894] 1977, p. 52). So he quite consciously searched until he found a suitable view. But as Muir knew, and the artists correctly guessed, the example of picturesqueness he found was not typical of the Sierras. Most of that mountain chain violated picturesque conventions by blurring the relationship between the various grounds and providing aspects that went far beyond any possible picture frame. This blurring, exaggeration, and violation of the conventions of the picturesque was the very essence of the American sublime.

To accommodate the wildness of such scenes, the representational style of artists and writers evolved from the realistic to the deliberately exaggerated, that is, from what Muir called "artistic bits capable of being made into warm, sympathetic, lovable pictures with appreciable humanity in them" to an emphasis upon great widths, depths, and contrasts ([1894] 1977, p. 49). This emphasis on the large and grand resulted in the depiction of gigantic objects, odd perspectives, and unusual atmospheric conditions and the blurring of the distinctions between elements of the picture. At its most extreme, the sublime depicted wildness as almost complete representational abstraction. Such landscapes as Moran's painting *The Chasm of the Colorado* (1873–74) dissolved into mere indications of gorges and cliffs, without geologic or scenic order and with no evidence of human scale or influence (Kinsey, 1992, pl. 2).

In fact, the sublime had always signified a general lack of verisimilitude or realism, because of its distortion of the facts of the scene and its use of hyperbole for effect. This exaggerated quality challenged the credibility of American explorers reporting on Western discoveries. When Jim Bridger or the early Yellowstone parties came back with outrageous descriptions or images of the West gained under difficult conditions, few people believed them. Later, even as the veracity of these reports was confirmed, artists such as Albert Bierstadt still received criticism for the imaginative liberties they took with their subjects (Kinsey, 1992, pp. 45–46; Anderson, 1992, pp. 10–15). Despite its association with "natural" referents, the accuracy or realism of sublime representation has always been in doubt. As I shall discuss below, the uneasy alliance in the sublime between accuracy in documentation and monumentality of expression evokes the most fundamental problems of representation itself.

Further attempts to accommodate the picturesque to wilder, more sublime scenes produced artistic innovations: changes in the size and shape of the canvas, resulting, for example, in the extreme viewpoints of monumental or upright can-vases, such as views from under a waterfall; multiple views within the picture frame, as in panoramas; and violations of the frame, as when mountains or rivers ran off the canvas's margin. Text and pictures often bore a reciprocal relationship to each other, particularly in more popularized publications, such as Muir's "coffee-table" book, *Picturesque California* (1888). For instance, a lithograph of a waterfall might be set within the text of a description of a waterfall, the text bor-dering the picture on both sides and running the length of the page. As for the text itself, the exaggerated character of the verbal description competed with the stylization of the lithographic art: "Amid the varied foams and the groundmists of the mountain streams that are ever rising from a thousand water-falls, there is an affluence and variety of rainbows scarce at all known to the careworn visitor from the lowlands. Both day and night, winter and summer, this divine light may be seen wherever water is falling in spray and foam, a silent interpreter of the heart-peace of Nature, amid the wildest displays of her power" (Muir, [1888] 1976, p. 87). When efforts to portray sublime scenes in written language contended with visual representations, the result sometimes was overblown and inexact. The total effect, however, seemed no more exaggerated than what the outstanding scenery of the West could bear.

Writers of sublime descriptions also attempted to create an American sub-lime by violating descriptive and narrative conventions. They featured descriptions of the natural scene that contained such linguistic signifiers as special adjectives (*wonderful, stupendous, dreadful, profound*), oxymorons connoting both fear and exaltation (*dread magnificence, awful grandeur*), metaphors, personification, cata-logs of features, broken diction, exclamations, shifts into present tense, second-person address, and the constant use of the word *sublime* as a shorthand term for a specific emotional response. These devices, all techniques of amplifica-tion or development, were the linguistic markers most associated with the sublime, according to Longinus, and they were also included in Aristotle's original concep-tion of the epideictic speech (McKinsey, 1985, chap. 2; Longinus, 1985, p. 71 [11.1]; Aristotle, 1941, p. 1358 [1368a27]). The result was a kind of epideictic oratory in praise of natural scenery, the very background and source of human endeavor.

To summarize this description of the conventions of the sublime, one might describe a continuum between the pastoralized picturesque, which is more "real-ist" or narrative in style, and the wilderness sublime, particularly the American va-riety, which is more semiotic or "abstract." This continuum corresponds roughly with the amount of human presence depicted in the scene. Yet the presence or ab-sence of human figures does not automatically indicate whether a depiction is sublime. Much has to do with the relationship of any human figures to the scene and to the reader or viewer.

In earlier versions of the pastoralized landscape, figures were placed within the frame simply to give it a human reference. In this capacity they were labeled staffage and seen as just another component of the sublime convention. Increas-ingly, however, and often in American views, human figures were portrayed observing the scene and registering their emotional reaction to it. This figuring forth of responses to the scene heightened the effect and provided a model for the viewer's own behavior. Raymond J. O'Brien writes: "Sublimity is universally enhanced by diminutive human figures depicted as shepherds, farmers, Indians, city types, or other 'idlers.' Invariably, they have their backs turned to the foreground, a technique that leads the viewer to identify with the figures and imparts a quality of isolation and spaciousness to the landscape in which people—dwarfed by nature—appear dominated by a powerfully compelling (perhaps frightening) environment" (1981, p. 174). Over time, social strata indicating stages or levels of civilization were successively inscribed on the landscape. In order of increasing cultural sophistication, the strata ranged from Indians to mountaineers, pioneers, settlers, farmers, industry, and urban dwellers. Correspondingly, the landscape underwent a transformation from wilderness to grazing land, farms, mines, and finally towns and cities. The inscription of stages of human development upon the landscape implied that the progressive settling of the West was inevitable, as natural as the forces of nature that were being tamed.

But the imagery of human existence became more problematic as the nineteenth century wore on. As the living landscape was being altered by human industry and settlement, some views of wilderness, such as those by Bierstadt and Moran, became more and more idealized, more and more abstract, and increasingly without human presence. This form of art seemed to deny or suppress the hard fact that the civilizing process was rapidly turning the original natural scene into urban landscape. Yet it did keep a kind of idealized, utopian conception alive that could serve as a model for preserving what little remained. In other, more popularized versions, the sublime convention was openly used for the purposes of promotion, settlement, and development of the land. But whether overtly or covertly, sublime representation always functioned to position the viewer with respect to specific social, political, or ethical issues (see Cronon, 1992, pp. 80–81).[4]

In order to accomplish this positioning, observers pictured in the landscape often functioned as stand-ins for the viewer, educating the viewer and directing attention much as a narrator unfolds a story. One particularly influential convention may have been derived from the art of Caspar Friedrich, a German Romantic painter. His paintings portrayed lone figures, often dressed in fashionable urban garments, elevated above the landscape and facing outward toward it, presumably awed and challenged by its sublimity (Koerner, 1990, fig. 77, *Wanderer above the Sea of Fog* [ca. 1818], p. 15). Indeed, Longinus's original definition of sublime rhetoric recognized the self-reflexive nature of such positioning, the feeling of both being in the scene and also being outside of it, viewing it (and oneself) from a higher or more distant perspective: "What is beyond nature drives the audience not to persuasion, but to ecstasy. . . . what is wonderful has a capability and force which, unable to be fought, take a position high over every member of the audience" (1985, p. 9 [1.3]). The word *ecstasy,* used here, literally means "a standing outside of" oneself (Longinus, 1985, p. 10). David Wyatt further specifies this sense of self-conscious duality: "There is something spectatorial about any encounter with the natural world, a tension between figure and ground that can throw us back on an awareness of ourselves as separate, human, perceiving. This is the burden of Thoreau's famous passage on 'thinking' in 'Solitude': 'With thinking we may be beside ourselves in a sane sense'" (1986, p. 16).

In sublime literature, narration provided the vehicle by which human activity entered the description of the scene. This narration frequently included a narrator with which the reader could identify, a first-person or eyewitness account, and a shift to second person to emphasize the readers' experience. Narration also revealed point of view—what one person could see from a viewpoint—thereby linking the literary and visual sublime. A characteristic example of second-person narration with point of view can be found in Muir's *Picturesque California.* Here Muir is describing the Sierra Mountains as if from a panoramic outlook:

> All along your course thus far. . . the landscapes are open and expansive. On your left the purple plains of Mono repose dreamy and warm. On your right and in front, the near Alps [of the Sierras] spring keenly into the thin sky with more and more impressive sublimity. But these larger views are at length lost. Rugged spurs and moraines and huge projecting buttresses begin to shut you in, until arriving on the summit of the dividing ridge between the head waters of Rush Creek and the northmost tributaries of the San Joaquin, a picture of pure wildness is disclosed far surpassing every other you have yet seen. [(1888) 1976, pp. 15, 18]

Such narrative conventions were often used by Muir, as well as others, to convey a sense of the grandeur and timelessness of the Western landscape and to involve readers in the construction of the scene.[5]

Furthermore, involving the viewer or reader in the production of the sublime scene inevitably led to the eliciting of particular attitudes and actions in those viewers of a moral, religious, or political nature. Outstanding scenery inevitably told a story, and of course every story had a lesson to be learned by those who aspired to an exalted moral or political character. Sublime scenery, then, was an important persuasive force. To illustrate the rhetorical power of sublimity, one might again turn to Longinus's *On the Sublime.* Only part of this treatise examined literary descriptions of natural scenery. True to its rhetorical foundations, it primarily discussed how the author or orator might influence the character and behavior of the great through praising their various qualities. By suggesting that great actions were desirable and by making leaders think that greatness came from themselves, the orator could tactfully induce them—or any audience, really—to pursue great actions (Longinus, 1985, pp. 104–5 [17.1]). To involve the reader or viewer in the observation or production of great scenery automatically implicated them in greatness.

By the modern European period, sublimity in the natural landscape was firmly linked to sublime morality. To be "sublime" was to display outstanding moral or religious qualities, and the phrase was applied to people as easily as to landscapes. Immanuel Kant, in his *Critique of Judgment,* drew the definitive link between the natural landscape and human moral aspiration. "In fact," he wrote, "a feeling for the sublime in nature cannot well be thought without combining therewith a mental disposition which is akin to the moral" (Kant, 1951, sec. 29, p. 109).

John Ruskin, the influential British art critic, also emphasized the moral elements of the sublime, and his conception determined its grand and expansive qualities for nineteenth-century America (Goetzmann, 1966, pp. 224–25, 377, 449).

Therefore, to be sublime was to be morally outstanding in some way, and the sublime was thoroughly moralized in America almost from the start. Americans were used to listening to sublime oratory emanating from pulpits or political podiums and extolling God, heroism, patriotism, and noble character. Muir himself wrote: "The mountains are fountains of men as well as of rivers, of glaciers, of fertile soil. The great poets, philosophers, prophets, able men whose thoughts and deeds have moved the world, have come down from the mountains—mountain-dwellers who have grown strong there with the forest trees in Nature's work-shops" (quoted in Teale, 1954, p. 321). This association of moral rectitude with natural scenery had powerful practical consequences. For example, in 1873 a mountain in Colorado was discovered upon which a white cross, etched in snow, remained visible for much of the year. What more sublime expression of God's handiwork than a mountain that bore the symbol of Christ's sacrifice? What mountain could be more deserving of preservation? Moran painted the Mountain of the Holy Cross, and the painting became one of the most frequently reproduced images of a particular mountain ever made. Interestingly, however, when the cross was partially obscured by a landslide, the mountain lost some of its moral connotations, and the effort to attain park status for it was abandoned.[6]

The set of conventions that brought together visual, narrative, personal, and moral connotations in one package was the sequence of emotions one might label the sublime response. This was a conventionalized series of three reactions familiar to anyone who encountered the sublime, as automatic as clapping one's hands after a performance. It had its origins in Longinus but was more a product of eighteenth-century associationist psychology and its attendant aesthetic theory, with later influences from Kant and German Romanticism.[7] The three stages were, first, apprehension, in which the individual subject encounters an object larger and greater than the self; second, awe, oppression, or even depression—in some versions, fear or potential fear—in which the individual recognizes the relative greatness of the object and the relative weakness or limits of the self; and, third, exaltation, in which the individual is conceptually or psychically enlarged as the greatness of the object is realized and the individual identifies with that greatness.

Emphasis could vary within these three parts. Sometimes the fear was a terror, or "terrific," as in the European gothic sublime employed in the romances of the early nineteenth century (Mary Shelley's *Frankenstein* is an example). Sometimes there was hardly any fear, as in the moral sublime and much of the American nationalistic sublime. Americans seldom employed the gothic sublime involving fear and terror except early in American history, when fear was a very real part of the encounter with nature.[8] Later, descriptions of American scenery generally followed the three-part sequence but downplayed the fear in exchange for awe. So Fanny Kemble wrote of her visit to the Hudson River in 1832: "I was filled with awe. The beauty and wild sublimity of what I beheld almost seemed to crush my facilities. . . . I felt as though I had been carried into the immediate presence of God" (quoted in O'Brien, 1981, p. 177). And Thomas Cole, in his famous "Essay on American Scenery" (1835), exclaimed: "And Niagara! that wonder of the world!—where the sublime and beautiful are bound together in an indissoluble chain. In gazing on it we feel as though a great void had been filled in our minds— our conceptions expand—we become a part of what we behold!" (quoted in McKinsey, 1985, pp. 211–12). Americans, then as now, preferred the positive, powerful, and uplifting quality of the sublime to the kind that reminded them of their limitations and their mortality. As we shall see, this positive quality of the sublime can work to instill hope and energy in environmental causes, but it also can encourage a dangerous dissociation between ourselves and the forces of destruction.

It may be oversimplifying to suggest that today's typical tourist to Niagara Falls might reproduce the reaction of Thomas Cole. But it is true that tourists wish to enter the sublime scene and become part of it in order to experience its expansive and enlightening effects. In this popular appeal lies sublimity's greatest advantage, and its greatest drawback, as a contribution to environmentalism. For instance, the preservation of natural scenery might just depend upon modern tourists' response to sublimity. But that very preservation might also depend upon tourists' being content to enjoy the scenery vicariously, through pictorial or literary representation. As a result, environmentalists argue that at some point people must recognize the difference between the sublime response and loving the scenery to death.

Because of the complexity of the sublime response, its relationship to environmental concerns is not easily determined. Therefore, without proposing specific answers, I shall sketch two general areas in which paradoxes of sublimity intersect with contemporary environmental issues.

The Sublime and Environmental Issues

The height of the sublime in the United States occurred in the last half of the nineteenth century, as the unique landscapes of the West became well known and the frontier became history, then nostalgia. The development of new communication technologies such as the motion picture, the telephone, the phonograph, and mass-produced photography focused public attention away from the oil paintings, lithographs, and verbal discourse that up until that time had always conveyed the sublime. But the sublime did not disappear; in fact, new expressions of sublimity were developed in the twentieth century through such media as art photography, fictional and documentary motion pictures, and advertising, both print and visual. Sublimity has remained a touchstone or grounding for our public conception of nature and, through nature, of the environment.

Along with new means of expression have come new definitions of the sublime. No longer considered simply an aesthetic device or a quality of moral character, the sublime has taken on several contemporary guises. I will discuss two. First, in concert with the twentieth century's obsession with language, the sublime is seen as a rhetorical trope or figure of discourse, a conventional linguistic device that represents or encodes our understanding of the natural world. And second, the sublime may be seen as a political tendency or an impulse toward policy, whether progressive or conservative. These two issues are hardly separate, but I will examine some questions in each area to show how the sublime functions as a contemporary concern.

First is the issue of the sublime as a rhetorical trope. Tropes are linguistic formulas, such as metaphors or analogies, that give an added dimension of meaning and artistry to words. Because successful tropes are repeated, they often become conventionalized to the point of cliché: for example, "Her words were as sublime as a mountain range." By calling attention to their own artificiality, however, tropes indicate that they cannot be reduced to their referents. That is, the sublimity of *words* and the sublimity of *a mountain range* are two different things, and using one adjective to refer to both is, to a degree, deceptive. Nevertheless, by employing the linguistic marker *as,* this trope demonstrably constructs a comparison between the two kinds of sublimity and invites us to recognize their similarities. In this way, tropes overtly display the way that language confers meaning upon a variegated and often fragmented world.

In everyday usage, however, a contradictory result can emerge. The more conventional that tropes and figures become within a language community, the more they can seem entirely "naturalized." In other words, conventionalized tropes can become so familiar that they appear to denote their referents directly and transparently, without any distracting interference stemming from the use of discourse as a medium. For *the sublime* to be thoroughly naturalized, for example, it would have to be as unobtrusive as the reference to the part of the body termed the hand in the phrase "she handed him the key." Of course, the sublime never attained such conventionality as to become part of daily language use in quite this way. Nonetheless, it *was* highly conventionalized, particularly in the nineteenth century, to the point at which a mere reference to sublimity in public discourse could evoke any and all of the connotations discussed earlier.

Thus, one would expect issues of conventionality and referentiality to arise when discussing the relationship of the sublime to the environment. The central question might be phrased this way: Does a sublime exist that is more or less unconventionalized, an authentic experience of nature against which other sublime discourses can be measured? Or is the sublime a practice of language use having no referential basis? With respect to these issues, the sublime functions as any other trope or figure, that is, as a piece of language, although it is uniquely identified with the nonlinguistic world. That is, because it is associated with what is "natural," the sublime connotes an authenticity and originality that is part of its very meaning; yet like rhetoric itself, it has a long-standing reputation for exaggeration and even falsehood.

We may approach the issues of referentiality and authenticity via a historical thesis. Elizabeth McKinsey, in her book *Niagara Falls: Icon of the American Sublime* (1985), argues that an "eclipse of the sublime" occurred after 1860. Before that time there existed an "authentic" sublime marked by "strong and genuine responses" that were experienced and communicated by the first explorers in the American wilderness (p. 157). The key to identifying the "authentic" was an original experience with nature—a discovery, if only a personal one. Later in the nineteenth century, she contends, repetition or familiarity with this "genuine" sublime produced "the rehearsed response, the preconceived idea, the accepted meaning," that is, "sentimentalization," comprising clichés and conventions. This sentimentalization occurred mostly in the popular arts, not the high arts, which attempted to reach for more "transcendent" meanings (pp. 130–31).

McKinsey forebodes grave consequences for this sentimentalization of the sublime. When we expect nature to conform to the sublime convention in the form of "picture spots" and "scenic views" and it does not, then nature itself is subject to the deflation of expectations, trivialization, and denigration. McKinsey supports her prediction with a historical argument. As the scenery changed with industrialization, the sublime depiction of it did not reflect these changes. This continuing sentimentalization and nostalgia, she says, encouraged visitors to forget the negative impact of civilization on the wilderness (1985, p. 164). McKinsey writes: "As men and women came more and more to deny and control the dark side of [nature], they ironically unleashed the negative aspect of their relation to nature in exploitation and conquest. The result would be to undermine the sublime and to ruin the [natural scene]" (p. 125). In McKinsey's view, then, the problem of environmental degradation is a result of an inauthentic relationship to nature, fostered by a form of distorted or incorrect discourse.

Given the conception of the sublime as a perceptual screen, however, some questions arise concerning this approach. When, might one ask, is the sublime ever "authentic" or "genuine," since it is always a conventional screen? As we have seen, even the early explorers inherited a European culture that incorporated the sublime and passed it on to North Americans. Historian William Cronon suggests that even the "artists of the first encounter" were influenced by inherited narrative conventions, employing either a "booster" narrative or a more subtle narrative of the "expansion of knowledge" (1992, pp. 44–45).[9] The loss of an authentic sublime, whether it is seen as inaccuracy, exaggeration, or a departure from tradition, has always been part and parcel of its aesthetic theory.

The notion of sentimentalization, as well, bears examination. Is the process of sentimentalizing a "destruction of the sublime," a "breaking down" of it, or a more thorough enculturation of it? Rather than being "eclipsed" after the Civil War, the sublime appeared in many popular and high art forms from 1860 to the turn of the century. It even survives today in the form of Sierra Club calendars and Ansel Adams portfolios that are used, appropriately enough, to support various preservationist causes. The tropic or conventional nature of the sublime might be the essence of its effectiveness as a basis for environmental advocacy.

Finally, by labeling popularization a "sentimentalization" or even a "destruction," McKinsey introduces a distinction between "low" and "high" art forms but does not acknowledge the problems accompanying that distinction. Certainly, the quality of many mass-mediated popular art forms may be called into question. But if the sublime is always a convention, then to label popular culture destructive is to valorize one variety of that convention over another, based solely upon the size and breadth of the audience. Furthermore, historians such as

Lawrence W. Levine recently have questioned whether the popular and high arts were as divergent as McKinsey suggests (Levine, 1988). Not only did nineteenth-century artists and writers often accomplish the transfer from popular to high art, but many of them earned their living by creating art directly for mass reproduction (Sandweiss, 1992, pp. 117–18). For example, Muir's essays in *Picturesque California* (1888) were also published in his book *The Mountains of California* (1894), without significant revision. One would be hard-pressed to determine whether one version was more "genuine" or "sentimental" than the other.

Therefore, the issue of the sublime as a more or less "genuine" or "authentic" representation of nature develops quickly into a consideration of conventionality, popularity, and the effects of mass production upon the aesthetic object. These issues are far too complex to solve here, but they make historical consideration of the sublime as a conventional rhetorical trope more relevant to the constraints of twentieth-century communication. Even then, a problem arises that might serve to introduce our second issue, that of the political tendency of sublimity. If what I am suggesting is plausible, that there is no referential sublime experience and all versions of the sublime are conventional, then how can we account for the kind that exploits nature as distinct from the kind that participates in saving it?

This second issue concerns the sublime as a political stimulus and directly relates to the exploitation and consumption of natural resources. One way to conceptualize the issue may be seen as follows: Does the sublime encourage a lack of political involvement in issues concerning the environment? Worse, does it encourage an attitude of possession and consumption? Or does it provide a common ground, a set of commonplaces, upon which to make arguments for the preservation and conservation of nature? We can turn to McKinsey for a statement of the first position. She argues, "While the experience of sublimity itself elevated [Americans] to a higher moral plane, paying homage to God's grandest creations also bought them the privilege to use his other lands" (1985, p. 117). In this view, the purchase and use of pictures and descriptions of the sublime imply that the scenes they depict are ownable and consumable as well.

This theory, that the sublime encourages the consumption of nature, or at least a consumptive attitude toward nature, is fairly widespread. Richard N. Masteller has argued that in the nineteenth century, the popular consumption of sublime scenery in the form of cheap photographic reproductions, stereoscopic slides, and elaborately illustrated publications reinforced the concept of manifest destiny and the attitude of domination toward nature. By "owning" definitive reproductions of a scenic spot, Americans commodified and therefore devalued the original (Masteller, 1981, pp. 55–71). Masteller's argument once again proposes that the mass communication of sublime conventions devalues what is "authentic" and "real."

Translating the issue into contemporary terms, one might consider whether an appreciation for natural scenery does or does not reinforce the ecological goals of habitat maintenance, species preservation, or wilderness management. Observing the wilderness is, after all, a consumptive "use." For example, a recognized area of conflict between scenic and preservationist values is to be found in the mandate for the national park system, which is to set aside pleasuring grounds and points of interest for recreational purposes as well as to retain the land for future generations. Alfred Runte makes the conflict clear: "Especially because most Americans still seek out spectacular scenery and natural phenomena, environmentalists caution that the public has little understanding of the restraints on visitation needed to protect the diversity of the parks as a whole" (1979, p. 1–2). And Roderick Nash notes that a rationale for wilderness preservation does not always accompany an ideology of scenic appreciation (1982, p. 121).

The sublime, in its emphasis upon the limitless and grand aspects of nature, might also reinforce the impression that the environment cannot be substantially damaged by human efforts. From the mid–nineteenth century on, evolutionary theory, combined with the aesthetics of the sublime, has suggested that the passage of eons of time would eliminate any evidence of human presence, including human damage, upon the Earth. Two great scientist-environmentalists of the nineteenth and twentieth centuries, John Muir and Rachel Carson, both held this attitude at the beginning of their careers. But sometime later, particular experiences of human depredation changed their minds. For Muir, it was sheep grazing in the Yosemite; for Carson, the killing of songbirds by pesticides. Carson even gave a well-distributed speech in which she stated, "I clearly remember that in the days before Hiroshima I used to wonder whether nature. . . actually needed protection from man. . . . Surely the sea was inviolate. . . . But I was wrong. Even these things, that seemed to belong to the eternal verities, are not only threatened but have already felt the destroying hand of man" (1962, p. 8).[10] Subsequent to their experiences, and in part because of them, both Carson and Muir became environmental activists.

Such accounts of the individual's subjective process of coming to environmental consciousness suggest that the sublime can indeed stimulate political action, or at least not forestall it. This view implies that rather than being entirely apathetic, audiences might be brought to recognize at least inconsistency between the aesthetic appreciation of nature and its impending destruction. Every icon of sublime nature, because of its implicit connection to what is natural and therefore "real," leads its viewer to assume that the absence of its referent would be not only undesirable but inconceivable. This prospect of absence and the permanent deferral of desire might produce sufficient incentive for fueling political initiatives.

From a rhetorical viewpoint, as well, activism requires at the minimum a shared basis of understanding and a practice that implements that understanding. Regardless of its other uses, the sublime did establish a set of beliefs about nature in the nineteenth and twentieth centuries that were then used as major premises by environmental activists to argue for the preservation of nature. These activists exploited the tension between the presence of the sublime wilderness and its imminent destruction. As O'Brien writes: "The existence of a regional aesthetic, its root causes in the nineteenth century and its consequences in this century, [helps us understand] the historical progression of land use in the United States from romanticized, idealized wildscape to . . . preservation" (1981, p. 9). Muir, for example, turned his natural history essays using sublime discourse into appeals for preserving Yosemite based upon that same sublimity. His readers were grati-fied to know that the nature Muir depicted still existed, even though they might never see it themselves. Moreover, by writing a letter or making a contribution, readers could ensure that the sublime scene would continue to exist, and thus they would obtain further gratification. The experience of the sublime, as always, re-mained a vicarious one. But it could, and did, produce political action.

To posit a political outcome for sublimity, then, addresses the issue of viewer consumerism and apathy. Yet the solution poses as many problems as it solves. How, specifically, does a set of codified signs depicting a scene lead to po-litical action? Is it necessary that the scene "exist" in some positive way? And if rhetorical efficacy depends upon shared beliefs, then how can we account for the emergence of alternative or even subversive views? Although we may not yet have answers to these questions, the study of the effects and consequences of the sub-lime promises to bring us to the core of what it means to be political in an environmental sense.

Language concerning our environment necessarily expresses an attitudinal orientation toward the natural world, while informing us of its material condition. Because of this dual nature of environmental discourse, the rhetorical convention of the sublime has enormous persuasive force. Its force has not diminished, but ex-tends up to the present day. From Rachel Carson's *Silent Spring* to Jonathan Schell's *Fate of the Earth* and Bill McKibben's *End of Nature,* the sublime is the founding narrative—the primary trope—in the rhetoric of environmentalism. On the one hand, technical and pragmatic discourse that ignores the sublime does so at its peril, because the sublime provides the perceptual foundation upon which we view the natural scene. On the other hand, sublime discourse that avoids environ-mental activism denies its own rhetorical power. Only when the two are combined does the sublime attain its full potential. We cannot have one without the other.

Notes

The writing of this chapter was made possible by a University of Utah Humanities Center Fellowship. I wish to thank Lowell Durham, director of the center, and Dean of Hu-manities Patricia Hanna for their support.

1. The use of the term *discourse* to denote all forms of communication, including nonverbal signs, can be attributed to the influence of post-structuralism. A characteristi-cally post-structuralist use of the term can be found in the writings of Michel Foucault, in particular *The Archaeology of Knowledge* (1972).

2. Compare, for example, the photograph by Charles Leaner Weed, *The Valley, from the Mariposa Trail* (1864), with the photograph by Carleton E. Watkins, *The Yosemite Valley from the "Best General View"* (1866); see Robertson, 1984, pl. 6, p. 15, and pl. 23, p. 71. These photographs show a tree that possibly has been denuded of its lower branches to afford a better view of Yosemite Valley (p. 16).

3. See the discussions of this painting by Cronon (1992, pp. 40-43) and by Sears (1989, p. 56). For a reproduction of the painting and a review of Cole's artistic career, see Powell, 1990, frontispiece, *The Oxbow (The Connecticut River near Northhampton)* (1836).

4. For images of sublimity used for the purposes of boosterism, see Goetzmann, 1988, particularly *The New Empire: Oregon, Washington, Idaho* (n.d.), p. 104. For images of industrialization as a form of the sublime, see "California's Vanishing Landscape," 1992, particularly Carleton E. Watkins, *Hydraulic Mining at the Malakoff Pit, North Bloomfield Mining Company* (ca. 1870), p. 184.

5. See, for example, Clarence King's first-person description of the Sierra crest: "Rising on the other side, cliff above cliff, precipice piled upon precipice, rock over rock, up against sky, toward the most gigantic mountain-wall in America, culminating in a noble pile of Gothic-finished granite and enamel-like snow. How grand and inviting looked its white form, its untrodden, unknown crest, so high and pure in the clear strong blue! I looked at it as one contemplating the purpose of his life; and for just one moment I would have rather liked to dodge that purpose or to have waited, or have found some excellent reason why I might not go; but all this quickly vanished, leaving a cheerful resolve to go ahead" (1872, pp. 54-55).

6. The story of the Mountain of the Holy Cross can be found in Kinsey, 1992, chaps. 8-9; and Runte, 1991, pp. 6, 7. For another discussion of how qualities of character and moral values attributable to the wilderness influenced the establishment of a national park, see Weaver, chap. 7 in this volume.

7. Monk calls this sequence "essentially the sublime experience from Addison to Kant" (1935, p. 58). See also Weiskel, 1976.

8. See, for example, Opie & Elliot, chap. 1 in this volume; they discuss the Puritans' characterization of wilderness as a spiritual wasteland and therefore something to be feared. Sears describes the site of an avalanche in the White Mountains that killed a family and later became a tourist attraction because of its gruesome appeal (1989, chap. 4, pp. 72-86). But these representations come early in the national experience. Later, as in King's description of the Sierra crest in note 5, the only negative emotion is a temporary lack of resolve to plunge ahead.

9. Even Meriwether Lewis, in describing his first impressions of the Great Falls of the upper Missouri River, "wished for the pencil of Salvator Rosa . . . that I might be enabled to give to the enlightened world some just idea of this truly magnificent and sublimely grand object" (quoted in Kinsey, 1992, p. 3).

10. See also a discussion of Muir's early beliefs in the ability of nature to regenerate despite human damage, in Oravec, 1979, pp. 92-94.

References

Anderson, N.K. (1992). "Curious historical artistic data": Art history and Western American art. In Prown et al., 1992, pp. 1-35.

Andrews, M. (1989). *The search for the picturesque: Landscape aesthetics and tourism in Britain, 1760–1800.* Stanford, Calif.: Stanford University Press.

Aristotle. (1941). *Rhetorica.* In R. McKeon (Ed.), *The basic works* (pp. 1325-1451). New York: Random House.

California's vanishing landscape: Nineteenth-century images from the collection of the California Historical Society. (1992, Summer). *California History, 71,* 2, pp. 178-91.

Carson, R. (1962, July). Of man and the stream of time. *Scripps College Bulletin,* pp. 1-11.

———. (1987). *Silent spring.* 25th anniversary ed. Boston: Houghton Mifflin.

Cronon, W. (1992). Telling tales on canvas: Landscapes of frontier change. In Prown et al., 1992, pp. 37-87.

Foucault, M. (1972). *The archaeology of knowledge.* A.M. Smith (Trans.). New York: Pantheon.

Goetzmann, W.H. (1966). *Exploration and empire: The explorer and the scientist in the winning of the American West.* New York: W.W. Norton.

———. (1988). *Looking at the land of promise: Pioneer images of the Pacific Northwest.* Pullman: Washington State University Press.

Kant, I. (1951). *Critique of judgment.* J.H. Bernard (Trans.). New York: Hafner.

King, C. (1872). *Mountaineering in the Sierra Nevada.* Boston: J.R. Osgood.

Kinsey, J.L. (1992). *Thomas Moran and the surveying of the American West.* Washington, D.C.: Smithsonian Institution Press.

Koerner, J.L. (1990). *Caspar David Friedrich and the subject of landscape.* New Haven, Conn.: Yale University Press.

Levine, L.W. (1988). *Highbrow, lowbrow: The emergence of cultural hierarchy in America.* Cambridge, Mass.: Harvard University Press.

Longinus. (1985). *On the sublime.* J.A. Arieti & J.M. Crossett (Trans.). New York: Edwin Mellen Press.

Masteller, R.N. (1981). Western views in eastern parlors: The contribution of the stereograph photographer to the conquest of the West. In J. Salzman (Ed.), *Prospects: The Annual of American Cultural Studies* (no. 6, pp. 55-71). New York: Burt Franklin.

McKinsey, E. (1985). *Niagara Falls: Icon of the American sublime.* Cambridge: Cambridge University Press.

Monk, S.H. (1935). *The sublime: A study of critical theories in eighteenth-century England.* New York: Modern Language Association.

Muir, J. (Ed.). ([1888] 1976). *West of the Rocky Mountains.* Rpt. of *Picturesque California.* Philadelphia: Running Press.

———. ([1894] 1977). *The mountains of California.* Berkeley, Calif.: Ten Speed Press.

Nash, R. (1982). *Wilderness and the American mind.* 3rd ed. New Haven, Conn.: Yale University Press.

O'Brien, R.J. (1981). *American sublime: Landscape and scenery of the Lower Hudson Valley.* New York: Columbia University Press.

Oravec, C.L. (1979). Studies in the rhetoric of the conservation movement in America, 1865-1913. Unpublished doctoral dissertation, University of Wisconsin–Madison.

Powell, E.A. (1990). *Thomas Cole.* New York: Abrams.

Prown, J.D. (1992). Introduction to *Discovered lands, invented pasts.* In Prown et al., 1992, pp. vii-xv.

Prown, J.D., Anderson, N.K., Cronon, W., Dippie, B.W., Sandweiss, M.A., Schoelwer, S.P., & Lamar, H.R. (1992). *Discovered lands, invented pasts: Transforming visions of the American West.* New Haven, Conn.: Yale University Press.

Robertson, D. (1984). *West of Eden: A history of the art and literature of Yosemite.* Yosemite National Park, Calif.: Yosemite Natural History Association.

Runte, A. (1979). *National parks: The American experience.* Lincoln: University of Nebraska Press.

———. (1991). *Public lands, public heritage: The national forest idea.* Niwot, Colo.: Roberts Rinehart, in cooperation with the Buffalo Bill Historical Center.

Sandweiss, M.A. (1992). The public life of Western art. In Prown et al., 1992, pp. 117-33.

Sears, J.F. (1989). *Sacred places: American tourist attractions in the nineteenth century.* New York: Oxford University Press.

Teale, E.W. (1954). *The wilderness world of John Muir.* Boston: Houghton Mifflin.

Weiskel, T. (1976). *The romantic sublime: Studies in the structure and psychology of transcendence.* Baltimore: Johns Hopkins University Press.

Wyatt, D. (1986). *The fall into Eden: Landscape and imagination in California.* Cambridge: Cambridge University Press.

4

Perceiving Environmental Discourse: The Cognitive Playground

James G. Cantrill

As the story is told, three baseball umpires were standing around home plate one day when the first official asserted, "I calls 'em as I *see* 'em." Not to be outdone, the second umpire exclaimed, "Well, I calls 'em as they *are!*" At this point, their colleague let out a little chuckle and declared, "Sorry, guys, they ain't *nothin'* till I calls 'em."

In some respects, the ways in which modern environmentalists treat the promotion of policy and public awareness are similar to the stances taken by the umpires in our mythical ballpark. Most environmental advocates try to give their target publics the "facts" about ecological health and decay and typically rely upon the ability of such audiences to accept the given interpretation of evidence as if it was both gospel truth and easily understood. Research conducted by Baglan, Lalumia, and Bayless (1986) suggests that these well-intentioned advocates generally turn to ostensibly "rational" appeals when constructing arguments. Unfortunately, and as Roszak observes, "environmentalists often work from poor and short-sighted ideas about human motivation; they overlook the unreason . . . at the core of the psyche" (1992, p. 38). Few of the environmental vanguard recognize that the "facts" about ozone depletion, toxic waste, and the overconsumption of limited resources do not stand alone in the minds of most people; rather, "facts" are interpreted in light of preexisting notions about the world and are modified by all manner of information-processing biases. Thus, the shortcomings of human cognition often preclude the worth of many logical arguments to protect our planet.

Since environmental discourse does not get generated in a psychological vacuum, it is important to consider the extent to which cognitive factors mediate the reception of communication and advocacy. Indeed, a variety of scholars have argued that people understand the nature and extent of environmental problems only insofar as the social construction of what *is* problematic comes to be accepted and appreciated by common individuals (e.g., Bird, 1987; Cantrill & Chimovitz, 1993; Sachsman, chap. 11 in this volume). That is, while grave threats may in fact exist in the environment, the perception of such danger, rather than the reality thereof, is what moves people to action (Rowan, 1991). If an environmental advocate is to enjoy any measure of success in a campaign to save the various corners of the Earth, then she or he will need to know what factors influence perception, where these cognitive biases originate, and how they affect the processing of communication. If, on the other hand, a public is taken to be more sophisticated and rational than it truly is, the advocate is likely to hear only the implicit call of "It ain't nothin' that really matters to us."

How the social and cultural milieu is represented in consciousness, and how that landscape functions in the processing of environmental messages, are the focus of this chapter. As researchers have begun to probe the substrata of human thoughts about the environment, an impressive array of findings has emerged that may provide a foundation for constructing more effective environmental discourse. In the following pages I hope to provide a summary of this psychology of environmental advocacy. In doing so, I will not focus on studies that concern the perception of physical environments (e.g., Craik, 1973) or analyses that adopt a relatively humanistic, psychoanalytic approach to environmental psychology (e.g., Roszak, 1992). Instead, I will briefly survey research exploring the various motivational and cognitive processes that influence the perception of environmental discourse, identify those biases that may result in the greatest distortion of environmental communication, and argue that these distorting perceptual mechanisms may ultimately account for the general ineffectiveness of environmental advocacy. In this manner, I hope to highlight the banter between language and cognition and the important role it plays in the most serious contest of modern environmentalism.

What the Game Looks Like from Home Plate

Although research into the psychological bases for perceiving environmental discourse represents a rather large field of play (Cantrill, 1993), previous studies can be positioned so as to represent one of three complementary families of empirical investigation. Each of these research foci reflect perspectives on the link between thinking about the environment, attending to environmental problems, or acting upon one's convictions regarding those beliefs and cognitive responses.

The Structure of Environmental Thought
It is somewhat ironic that the most fundamental basis for understanding how people perceive environmental advocacy—the manner in which individuals represent the universe of our environment—has received the least attention. Most early studies associated with thinking about the environment focused on identifying typical "environmental concerns." In this tradition, investigations have sought to establish which issues are considered legitimate foci of environmental advocacy in the minds of the public (Van Liere & Dunlap, 1981, 1983) and to determine the dimensions by which such concerns influence action, either social (e.g., Buttel & Johnson, 1977) or personal (e.g., Seligman et al., 1979). Taken together, these findings generally suggest that issues associated with pollution, natural resource management, population, and environmental activism anchor people's perceptions about environmental problems. That is, when asked to report what they consciously recognize as legitimate topics for environmental discourse, people turn to broad yet discrete categories of human impacts and behaviors.

Another approach, one that does not rely upon a person's awareness of particular environmental issues, was taken by Cantrill and Chimovitz (1993). In their study, subjects' descriptions of free associations for the word *environmental* were compared, revealing six general categories for perceiving the environment. When people mentally represent the environment, elements of that picture get parsed into the dimensions of physical spaces, tangible conditions, characteristics, ecological relations, psychological states, and human activities. A comparison of means for the average number of descriptors in each category of this "environmental schema" demonstrated that the most significant dimension of perception dealt with various ecological features (e.g., biosystemic connectedness). Supporting the worth of mass-mediated and institutionalized education, the structure of environmental thought seems weighted in favor of a naive appreciation for how pollution may affect elements in the world. Also, in addition to the physical aspects of the environment identified in this and other research (e.g., Tversky & Hemenway, 1983), a number of conceptual and social factors were associated with environmental schema. There was a noticeable relationship between the integration of environmental beliefs and the unique differentiation of a person's reaction to environmentally oriented terms. Those with a more "environmental" orientation seemed prone to an integrated worldview, resulting in slightly more complex perceptions of the environment.

In sum, the little we know about the cognitive representation of the environment suggests that people are relatively simple-minded when it comes to perceiving our world. This even seems to be the case for those who are more environmentally sophisticated. The reason for the oversimplification of an exceedingly complex environment has been established in a long line of cognitive studies (e.g., Freedle, 1981; Streufert & Streufert, 1978). William James's observations notwithstanding, the "blooming, buzzing confusion of the world" *is* too much for us to comprehend completely. This is especially true of our much maligned and increasingly alien biosphere, wherein we must form fairly tenuous schema based upon what we are told by the nightly news anchor or what we find in various interpersonal contexts (Shapiro & Lang, 1991; cf. Cantrill, 1993). Interestingly, Sears and Funk report that media tend to parcel out information in discrete chunks, resulting in "an information environment that itself is coded in terms of abstractions corresponding to the social constructs most common among attentive ordinary citizens" (1991, p. 14). Hence, the schema we use to demystify our threatened environment may be symbiotically linked to the social architects of those mental structures.

Making Sense of Environmental Problems

A second trend in research on environmental perception focuses on investigations of the various cognitive styles people use in reasoning about ecological problems. Stamm and Grunig's study (1977) showed that people's perceptions of distinct issues and situations account for observed differences in preferences either to reverse technological trends or to substitute more benign technology when confronted by environmental crises. Subsequent research (Grunig & Stamm, 1979) demonstrated that people will "hedge" their preferences for environmental solutions (i.e., adopt both reversal and substitution options) if they accept partial responsibility for existing problems but will only support reversals in policy (i.e., "wedge") if they believe institutions are to blame for environmental ills (see also

Miller, 1982). Others have explored the relationship between environmental knowledge, issue importance, and personal responsibility for ameliorative action (e.g., Abramowitz, 1979; Hamilton, 1986; cf. Seligman et al., 1979). In general, studies of cognitive style indicate that self-interest and knowledge of environmental issues mediates individuals' reactions to environmental problems. Grunig and Stamm (1979) suggested that recognition of whether a target population hedges or wedges may dictate the optimal choice of influence tactics in the environmental arena. For instance, Milbrath (1984) asserted that the "rearguard" of environmentalism is more willing to take risks in light of economic incentives, while the "vanguard" of the movement is more cautious and reserved in the use of technology to solve environmental problems.

While it is generally assumed that opinions are at least partially based upon information gleaned from public advocacy, evidence suggests that the provision of information alone is least likely to promote sensitivity to environmental degradation. In fact, a number of studies indicate that as persons become more informed about environmental problems, they may also become much more passive in their concern for that environment (e.g., Allen & Weber, 1983; Bart, 1972; Honnold & Nelson, 1979; Lowe, Pinhey, & Grimes, 1980; Milstein, 1977; Novic & Sandman, 1974; Ramsey & Rickson, 1976). Instead, people tend to redefine their obligations to solve environmental problems as being simply the need to be informed about such issues.

An obvious conclusion to be drawn from previous studies of perceiving environmental problems is that "the self" is a potent mediator in the process. Basing his analysis on the work of cognitive psychologists (e.g., Rogers, 1982) and others specifically interested in environmental education (e.g., Gigliotti, 1990; Ham, 1983), Cantrill (1992) suggested that our mental representations of where we are positioned *in* the environment act as perceptual filters when recognizing ecological problems. Even to the extent that Cantrill and Chimovitz (1993) showed that most people implicitly understand elements in the web of life, "the environmental self" still skews perceptions of environmental decay and may hinder appreciation for the human role in the process (Cantrill, 1992, p. 39; cf. Roszak, 1992). This is generally because, given one interpretation, our self schema evolve via tacit narratives we create to make sense of life events in systematic, self-serving ways (Gergen & Gergen, 1988). And because these internal autobiographies are the product of social interchanges in and about the environment, since they are continually in the process of being re-created, the environmental self is chronically accessible and always implicated in perceiving environmental problems (cf. Bargh, 1989).

Perceptions of the self in relation to the environment can affect appraisals of problems in a number of ways. For example, we know that the self exerts a strong referential effect in focusing attention and stimulating memory (e.g., Kihlstrom et al., 1988). Tichenor's analysis (1988) suggests that when we encounter, say, a survey of the perceived gravity of environmental problems, we conclude that the respondents must have reasons for their opinions, which we envision and compare with our own beliefs. Such self-interest anchoring encourages classic assimilation and contrast effects based on implicit, if perhaps faulty, social comparison processes (Niedenthal, Cantor, & Kihlstrom, 1985). An even more pointed analysis of self-interest grounded in local praxis was provided by Portnoy:

> ... decisions to site [hazardous waste] facilities constitute clashes between "experts," who want to site a facility in a specific location based on objective analysis, and the "public," which does not want it there. People who live in the affected community . . . believe implicitly that the scientific analysis does not reflect an understanding of their community or their way of life. . . . The experts, on the other hand, often perceive that they are conducting the best possible analysis of the risks and costs associated with the specific site. In their minds, the public is totally irrational in its opposition to the facility. . . . The point is that one or both parties in such conflicts exhibit rather strict adherence to the idea that they are correct and the other is incorrect, and that such strict adherence is rooted in deep differences of opinion of what should be valued. [1992, p. 16]

Thus, in light of these and other similar analyses (e.g., Lange, 1993; Miller, 1982; Peterson & Peterson, chap. 9 in this volume; cf. Tybout & Yalch, 1980), the perception of solutions as well as problems seems hemmed in by the way we picture ourselves in relation to our environments.

The Impulse for Environmental Conduct

Relatively early in the study of environmental advocacy, researchers began to probe the link between attitudes and behavior. In this tradition, some research and theory has been devoted to modeling the manner in which society molds our behavior by affecting our attitudes. Sandman (1976) advocated an approach wherein environmental beliefs are hierarchically arrayed and conceptually linked to ecological appeals to the extent permitted by education, socioeconomic status, and local conditions. His was a relatively underdefined model that left many elements in the process (e.g., values, worldviews) obscure. A year later, Grunig (1977) presented a model of information-seeking behaviors resulting from environmental public relations campaigns. In turn, this work laid the foundation for more contemporary analysis (e.g., Grunig, 1989). As with other market segmentation theories, Grunig's perspective allows for multiple target publics, each of which is influenced by a unique constellation of demographic forces. His work demonstrates when people will consider environmental issues, the extent to which those thoughts are organized, and the lengths to which people will go to act upon their beliefs. Others have implicitly (e.g., Honnold & Nelson, 1979; Milstein, 1977) or explicitly (e.g., Cable, Knudson, & Theobald, 1986; Syme et al., 1987) presented formulations and data consistent with the tenets of Fishbein's Theory of Reasoned Action (Fishbein & Ajzen, 1975). All of these types of models place heavy emphasis on the social-normative component of environmental motivations and pay relatively scant attention to specific attitudes per se (cf. Schmidt & Gifford, 1989). Overall, this family of analyses supports deHaven-Smith's finding (1988) that specific behaviors toward the environment are grounded in perceptions that are more narrow, normatively based, and self-relevant in nature than those organized around more abstract constellations of environmental assumptions. That is, people will act upon their environmentalist attitudes to the extent they believe those actions meet their immediate, socially supported interests (cf. Sears & Funk, 1991). Thus, self-interest and social expectations may again override the sorts of pro-environmental stances optimistically reflected in any number of attitudinal surveys (e.g., Dunlap, Gallup, & Gallup, 1993).

By the mid-1980s it had become apparent that there was more to the environmental belief-attitude-behavior link than could be accounted for by reference to socio-motivational pressures. In a general indictment, Taylor (1981) argued that a person's capacity to make and act upon judgments is not very orderly and is prone to all manner of information-processing biases. While we may, indeed, be motivated to act in our self-interest, that impulse is often not what drives our behavior. As Fischoff, examining the reasons why people did not act to prevent global warming, put it: " . . . people respond to problems as they see them rather than to problems as they are. . . . Both the content and quality of our response hinge on the validity of our (cognitive) understandings of what is happening to us and our world" (1981, p. 166). He went on to argue that the ecology of the mind, in oversimplifying the ecology of the Earth, compels people to take mental shortcuts in reasoning about the environment. We use these mental strategies, what I call environmental default mechanisms (EDMs), as bunkers for inactivity as well as staging areas for behavior in and toward the environment. Traditionally, researchers have turned to the notion of heuristics to categorize a variety of cognitive shortcuts, stipulating that these biases allow us to reason without much mental effort. I will be using the EDM construct, however, because some of the biases associated with the perception of environmental discourse (e.g., the default tendency to oppose environmental policies even after they have been shown to have little effect on ecology or the economy) may result in fairly elaborate rationalizations requiring significant cognitive load, while others may indeed function as time-saving heuristics.

The assumption that cognition mediates and biases environmental action has been established in a variety of contexts, including risk analysis (e.g., Keeney & von Winterfeldt, 1986; Otaway & Thomas, 1982), littering (e.g., Cialdini, Kallgren, & Reno, 1991) and recycling (e.g., Kok & Siero, 1985). Thus, in addition to oversimplifying the complexity of the world and developing self schema that may separate us from that environment, we tend to be cognitively impaired in acting upon even the limited knowledge and misgivings we possess. In the next section I will examine the way in which social and cognitive forces serve to forge and manage three especially dysfunctional EDMs.

Rules of Thumb and Environmental Information Processing

One way to account for the general finding that the targets for advocacy are quite limited in the processing of environmentalist discourse is to draw upon the work of cognitive psychologists. After more than two decades of explorations into the cognitive underpinnings for acting in the social and physical world, an extensive body of literature has evolved that discusses the bases for various EDMs. Some of this literature focuses on the nature of human inference (e.g., Nisbett & Ross, 1980), the grounding for social cognition (e.g., Fiske & Taylor, 1991), and cognitive responses to various persuasive appeals (e.g., Petty & Cacioppo, 1986). To build upon such general approaches, and to cast the material in relation to environmental communication, I will focus on three types of EDMs that individuals seem to use habitually in apprehending the subject matter of environmentalism. Specifically, the rules of this discursive game get twisted by our reliance on schema-based, automatic inferences, our tenacious retention of private and public commitments, and our overreliance on the salient ideologies we embrace. As we shall see, these shortcomings may account for the curious ballpark errors of the human mind reviewed in the previous section.

Environmental Mindlessness

As suggested earlier, to the extent we confront media that constantly reference the health of the planet, people possess environmental schema that allow them to understand terms dealing with ecology (e.g., *toxic waste*), to process arguments relating to the environment, and to position themselves in relation to those advocating various actions or policies. A contemporary tradition of research and analysis reveals that our conception of the natural environment is framed by the categories of our experience (e.g., Fiske & Taylor, 1991; Mervis & Rosch, 1981; Tversky & Hemenway, 1983; Ward & Russell, 1981). Further, various studies indicate that the processes leading up to impressions of social and physical environments are tacit and well below the threshold of normal awareness (e.g., Langer, 1989; Nisbett & Wilson, 1977; cf. Morris, 1981). Unfortunately, when we attend to environmental discourse, we all too often mindlessly rely upon such environmental schema in deciding how to respond to advocates rather than taking the time and effort to evaluate consciously what is being communicated. In this day and age, the novelty of environmentalist discourse has worn thin, and the simple sound bites of competing voices in the fracas do not require us to pay much attention to draw inferences.

When people encounter environmental communications, they implicitly conjure up the schema they have for that subject and/or the source of the appeal. In drawing forth the schema, individuals are efficient insofar as they focus first on the defining elements of the subject or source and use these characteristics to generate rapid inferences regarding how they themselves are related to the communication (via self schema) and how they should act in light of it. Sometimes, education and experience allow a person to reason inductively without having to expend much cognitive effort in the processing of discourse (e.g., a smog alert means a threat to health, mercury in fish signals avoidance). At other times, however, people use their experience-bound schema to infer characteristics of more social elements in the environment (e.g., the trustworthiness of activists or industrialists, the effect of logging bans on economic conditions) without recognizing the relatively amorphous nature of human interactions in the environment. Nonetheless, the configured perception (rather than the subject or source) is treated as if it were real, and the individual, considering past experiences, acts accordingly. And precisely because environmental issues are so complex, people tend to simplify the perceptual process of attending to physical and social stimuli, capture the essence of conditions and communicators, and end up using simple "fuzzy" schema that entail equally simple directives for behavior (Rothbart & Taylor, 1992). Thus, targets for environmental appeals automatically either do or do not pay attention, heed, and act.

One of the more intractable problems associated with environmental issues is that they generally evoke strong affective responses when discussed in the public sphere. Typically, the issue and its representative spokespersons threaten the quality and health of lives, and therefore strong emotion gets generated in heated disputes. The emotion itself may be triggered by the schematic representation of the subject or advocate, which further intensifies the polarization of the interactants and their self-interests (Fiske & Pavelchak, 1986). As Rothbart and Taylor demonstrate, the schema-driven hostility between environmental publics need not be based upon either a history of dispute or a "real" difference in position: "What one needs is self-defined groups (however arbitrary) and competition over limited resources" (1992, p. 23). In short, as long as environmental schema suggest that self-interests are threatened vis-à-vis an issue or advocate, people are more likely to steal themselves away reflexively, in the dugouts of existing beliefs and behaviors, than to consider reflectively how the environment should be approached.

Environmental Dogmatism

Of the more enduring ideas in social psychology, the notion that people are motivated to remain consistent in thought and deed has certainly received a great deal of empirical attention. We often commit ourselves to people and principles, which means that, unless circumstances arise beyond our control, we will follow through on our promises to self and other. And a significant impetus in such commitment-consistency drives is rooted in the desire to appear to others to be true to our word, even when we might wish to behave otherwise or, in fact, do so. Cialdini notes that the repeated application of a "commitment principle" will provide "a valuable shortcut through the density of modern life. Once we have made up our minds about an issue or have decided how to act in a given situation, we no longer have to process all of the relevant information when subsequently presented with the same (or highly similar) issue or situation. All we need to do is recall the earlier decision and respond consistently with it. It allows us a convenient, relatively effortless method of dealing with our complex environments that make severe demands on our mental energies and capacities" (1987, p. 169).

The use of a commitment and consistency EDM is clearly a driving force behind a variety of responses to environmental advocacy. When we are presented with an appeal or analysis of an environmental issue, we use our own past behaviors and thoughts in that regard to guide current appraisals. If in the past we have, say, always believed that citizens of the United States are inclined to the overconsumption of natural resources and have tried to compensate for such gluttony by conserving energy, then we are likely to follow a miserly dictum in turning off lights and burning calories instead of petroleum. On the other hand, if we have been convinced that the world is for those who can use it and have acted as if there were no tomorrow, we are not likely to curb our hedonic impulses. And in either case, the fact that we *are* consistent in our attitudes and actions makes our commitments quite available for guiding conduct in all sorts of environmental arenas (Fiske & Taylor, 1991).

A significant and potent inducement involving the origin and maintenance of commitments is the role played by social norms. In an intriguing set of nine studies, Cialdini, Kallgren, and Reno (1991) demonstrated the extent to which a commitment EDM could be activated and affect littering behavior. Distinguishing between "descriptive" and "injunctive" norms (i.e., what seems typical of others in society versus moral precepts), Cialdini and his associates discovered that people model their environmental behaviors after the actions of others, especially when the salience of those behaviors becomes more pronounced; that injunctive norms against littering can be unobtrusively activated; and that these injunctive norms persist, even when people find themselves in situations in which others are behaving at odds to the norm. In other words, once people become aware of what they should do in light of how others behave, they will remain consistent to that normative commitment even when others do not.

Although littering may not be as significant a threat to biospheric integrity as ozone depletion, groundwater contamination, or overpopulation, the implication of the Cialdini, Kallgren, and Reno study is clear. Individuals will commit themselves to and remain consistent with a disposition toward the environment when they believe that others will act accordingly *and* that others think it is the right thing to do. Note that the use of this EDM is independent of whether the action is ecologically destructive or benign. Also, one must be aware of his or her commitments to remain consistent with those implicit promises. Thus, it may only require convincing people that they are committed to preferencing quality of life over quantity of life to compel them to focus on short-term gains instead of long-term survival.

Environmental Ideology

Assuming that the EDMs described above mediate citizens' responses to advocacy begs the question of why we do not possess the schema and commitments to behave consistently in the best interests of the planet. To answer this question, we must consider the sorts of resources that fundamentally orient our acquisition of experiences, culminating in environmental schema and attitudinal or behavioral allegiances—the ideological contracts we forge between ourselves and dominant worldviews. In the most vicious of circles, these tacit assumptions remain salient in the processing of environmental communication because they are continually reified in the sources of information we use cognitively to manage our experiences and commitments. Ideology, as an ever-present shortcut in environmental decision making, is essentially a very real analogue for the last umpire at home plate.

In the years since the first Earth Day, various scholars have focused on the ideological underpinnings for environmental conduct and thought. Black (1970) focused on the differences between a "dominant western world view" and an "ecological world view," Drengson (1980) contrasted "technocratic" and "personplanetary" perspectives, Catton and Dunlap (1980) compared the "human exemptionalist paradigm" with the "new ecological paradigm," and others have proposed that society is shifting from a belief in growth, limitless resources, private property rights, and technological salvation to a "new environmental paradigm" associated with growth restrictions, resource conservation, biocentric order, and ecosystem integrity (e.g., Deluca, 1992; Dunlap & Van Liere, 1978; Geller & Lasley, 1985). More recently, Colby (1991) reframed such dichotomies as ideological conflicts between pro-development interests and those seeking restraint. Regardless of the label being attached, research suggests that these internal beliefs regarding the relationship between humans and the natural world exert a strong influence on how people position themselves vis-à-vis environmental discourse and that, indeed, most individuals continue to hold to an ideology of economic and scientific prowess. Cotgrove and Duff (1981) contend that people distinguish between idealized and practical environmental values and rationalize their choices as being, in fact, environmentally sound regardless of the actual consequences. Thus, setting aside the rhetoric of ardent environmentalism, we can safely assume that our culture still supports a traditional paradigm of growth, progress, and anthropocentrism.

Far from being abstract entities divorced from the lives of ordinary women and men, ideological perspectives are incorporated into and supported by all manner of human activities. A number of scholars have argued that ideologies such as those associated with environmental perception are organized so as to orient publics to act in concert with the prevailing norms and expectations of the day (e.g., McGee, 1978; Williams, 1980). Though various competing ideological orientations may coexist, a central social thesis will always govern the thoughts and deeds of a given people (cf. Fraser, 1990). We do not generally reflect upon the ideological foundation for our society, and our reliance thereupon thus becomes submerged in consciousness and obscured by the immediate pursuit of particular goals and/or objectives.

Because ideologies are continually being relied upon to process information about the systems and structures of society, they can come to be treated as EDMs that allow people a "quick and dirty" way of evaluating action and discourse. For example, Williams (1980) observed that, in the United States, the dominant ideology reifies an underlying economic dogma. Decisions about interpreting or actualizing policy are generally made by focusing on the influence of those policies on capitalist gain. Various environmental arenas are governed by an almost automatic reliance on this ideology: Forest Service planning consistently turns to economic utilitarianism to drive decision making (Allin, 1987); both the producers of energy and their consumers reason and act to minimize economic losses instead of maximizing financial gain, even while average consumers have a difficult time understanding how conservation serves to decrease the cost of using electricity and natural gas (Yates & Aronson, 1983); and those concerned with the preservation of species or general biodiversity too often blindly follow doctrines of financial evaluation (Peterson & Peterson, chap. 9 in this volume).

If there is one factor that promotes the use of ideology as an EDM, it is the mass media. As Altheide and Snow put it, "Culture is not simply mediated through mass media; rather, culture—in both form and content—is embodied in mass media" (1988, p. 196). Olien, Tichenor, and Donohue (1989) established that media coverage of environmental issues generally reflects the dominant ideology of a particular media market. Though the media may pair opposing views to provide some semblance of balanced coverage (Atwater, Salwen, & Anderson, 1985; Dunwoody & Rossow, 1989), Allen and Weber (1983) noted that the media relies upon the efforts of industry (which we can safely assume are economically motivated) more than upon the public relations efforts of environmental groups to cast interpretive frames for the public. Typically, this ideological spin reinforces the dominant economic, progress-oriented, and technologically situated expectations of mainstream society. Even reasonably "liberal" reporting generally reflects a conservative, institutional tilt (Schlechtweg, chap. 12 in this volume). Thus, the accepted symbolic universe of the dominant worldview gets perpetuated in society, and by default, much of our reasoning about environmental options gets skewed by ideological forces of which we are barely aware.

When all is said and done, our ideologies of progress and technology and consumption support the other EDMs that plague us, and these worldviews give rise to the most bitter of animosities. Fishman echoes this position in discussing how ideologies constrain the ability of people with different perspectives to communicate effectively and unbiasedly: "Divisiveness is an ideological position and it can magnify minor differences; indeed, it can manufacture differences in language as in other matters almost as easily as it can capitalize on more obvious differences. Similarly, unification is also an ideologized position and it can minimize seemingly major differences or ignore them entirely, whether these be in the realm of languages, religion, culture, race, or any other basis of differentiation" (1968, p. 45). It would not be stretching the imagination too far to find in Fishman's comments the image of a confrontation between any number of environmentalists, industrialists, and citizens among us.

Perceptual Shortcuts and Action in the Environment

The EDMs of environmental schema, commitments, and ideologies can account for the various research findings reviewed earlier. For example, we know that the general public understands something about the nature of our environment and the problems we face and that this awareness, mirrored in the shape of media reporting, can be represented by fairly narrow schema. Nonetheless, the structure of such environmental thought is fragmented, and most individuals do not see these issues as being very relevant to their self-interests. Consequently, environmental schema are not integrated with one another, and when one approaches environmental issues, the perception of such is not prone to a rich, longer-lasting self-referential encoding of the information. And the narrow focus of perception means that we tend automatically to filter out those appeals that are not seen as being immediately pertinent to the self. Everything else becomes part of "the other"—that is, someone else's concern—and not linked to any previously in-grained commitments (cf. Kihlstrom et al., 1988). Finally, the voice of tacit ideology continually reinforces the idea that, if there is a self-relevant problem, then technology will allow us to continue pursuing the consumption of resources in the here and now.

In a similar vein, when we pause to consider environmental problems, their complexity forces us to simplify and to dichotomize the world into a series of either/or propositions. By default, we thus believe that these oversimplistic choices are paired with equally naive solutions. Unfortunately, some of these "simple things you can do to save the Earth" (see Killingsworth & Palmer, chap. 10 in this volume), such as carpooling or reusing resources, can be seen as violating immediate self-interests or running counter to how most citizens position them-selves in the environment. This is a fundamental barrier that many environmental advocates have failed to appreciate, as is the fact that, while people may be con-ceptually committed to environmentalism, they do not have an ingrained habit of consistent action to support their aspirations. Even worse, the worldview of "progress" mandates that any solutions to environmental degradation must not result in a retreat from technology. Rather, we are driven to find end-of-pipe solu-tions to what often, in reality, are problems of overproduction for an overcon-sumptive people.

Finally, insofar as most individuals' anemic environmental schema are not well developed, most people do not know how to act upon their good intentions or how to assess the effectiveness of those actions. Instead, many turn to the orienta-tion provided by the media, which constantly asserts both our lack of unity with nature and the fact that our species is generally in opposition to it. In short, though the environmental self may not sense the wisdom of Pogo—"We have met the enemy and he is us"—it certainly allows people to register and fall in line with nor-mative influences. And what is "normal" is that the planet's ecology is out of whack, that there is not much that women and men of goodwill can do about it so long as the economy and lifestyles depend on consumption and pollution, and that the best one can do is to keep abreast of the maelstrom in order to maximize personal gain.

It may seem as if the biasing effects of EDMs, in concert and conspiracy with mainstream sources of information about the environment, lead to the conclusion that we are doomed to pass on to the future a corrupted ecosystem. On the surface, an understanding of how the game of environmentalism plays itself out on the field of cognition reinforces the futility of doing anything more than watching the environment spin out of control. However, as in the game of baseball, the turn of the century may be similar to a period known as the seventh-inning stretch, when strategies can be reassessed in light of unfolding events. Perhaps it is time for us to pause and reconsider why our EDMs tend to be so dysfunctional and to try to adapt to the dynamics of the game before it is over.

Implications for the Practice of Environmental Discourse

Although there exists a pregiven, physical world of polluted and pristine land-scapes, our understanding of the environment is not prespoken. That is, in giving voice to our perceptions, we draw upon and largely reinforce the meanings society has for the environment. These social meanings, filtered through various EDMs, mediate our reactions to discourse and advocacy, which in turn can trigger a continued reliance on shortsighted strategies for coping with a complex environment. It seems reasonable, then, to conclude that language serves a pivotal role in creating and activating the dysfunctional biases we employ in the processing of environmental communications. In particular, I am going to argue that specific symbols, constituted in mass-mediated portrayals of environmental problems and publics, currently foster a less than optimal basis for perceiving environmental discourse.

There can be little doubt that the content of our thoughts is the product of communication. Because we are symbol-using animals, we categorize and label what we experience, and these labels exert a powerful effect on the processing of information. This is especially true for environmental perception, where configured social and physical data get mapped onto existing schema grounded in culture and upbringing. For Mugerauer, the environment we experience is constructed and reified in the use of language: "We always find ourselves in the midst of an already interpreted environment; from this placement we both deepen our understanding and make mistakes. Further, both the primary interpretive experience and the secondary scientific abstraction are themselves possible only because *the environment and people always and already are given together in language*" (1989, p. 51; cf. Carbaugh, chap. 2 in this volume). In short, the perception of environmental discourse depends on the language used socially to construct the social and physical world around us.

Any language community will, over time, develop a shared set of linguistic norms for describing and acting in the environment. Such conventions become inextricably linked with the EDMs mentioned earlier. Equally important, however, is the fact that one can find embedded in language persistent themes that lay out the fundamental cultural commitments of a society (Agar, 1979; Steward, 1955). Reflexively, these commitments grow out of how language is used in reference to the environment. The effect of such discursive practices has been neatly summarized by Scott: "What cannot be done away with in a community is commitment to the norms of that community. Commitment and rhetoric stand in a reciprocal relationship: commitment generates rhetoric, and rhetoric generates commitment" (1976, p. 263). Thus, one viable method for circumventing dysfunctional EDMs may be to identify and mitigate those symbolic constructions that continually reinforce counterproductive environmental perceptions.

To use language effectively to manipulate environmental cognitions and commitments might require a reconceptualization of how language functions in the structure of thought. Tajfel and Forgas (1981) indicted the view that cognitive categories simply reflect reality; rather, our schema are the result of active constructions in sense making, given the social backdrop of experience. Moscovici (1981) contended that these social representations function as focal agents in producing and perceiving language; they also dynamically evolve, as social and normative contexts develop around us. As applied to environmental discourse, some scholars have couched the world-as-represented in the form of "condensational symbols" (Kraft & Wuertz, chap. 5 in this volume; Mitchell, 1984). These are tidy perceptual collections of various thoughts and feelings about the environment that come to be represented by powerful terms or phrases. When people use condensational symbols in discourse, they may reflect back to their audience the oversimplistic environmental schema it uses to structure the world. As a consequence, when a public attends to advocacy, certain potent themes—such as "progress," "toxic waste," or "endangered species"—trigger symbolic associations, forged by life experiences and/or cultural values, supporting or opposing environmental policies.

Recognition of the power of language in promoting and manipulating environmental perception entails a number of implications. Freedle argued that "by virtue of being in a language community for many years, a large repertoire of complex linguistic . . . schemata can be learned and come to function in an almost automatic way" (1981, p. 38). Furthermore, some research establishes that similar "symbolic predispositions" can override the bias of self-interest unless one's personal stake in the issue is large, clear, and immediately certain, which rarely occurs in most depictions of environmental harm (e.g., Sears & Funk, 1991). Hence, people may be unknowingly reacting to the fit between discourse and how they represent the environment in consciousness rather than the wisdom or worth of specific arguments.

In the arena of "symbolic politics" (Saxer, 1993), the power of language-bound predispositions has not been lost on hacks and practitioners. For example, Newt Gingrich, as U.S. House of Representatives minority whip, effectively used focus groups to discover which words evoked the most positive responses for Republican candidates (e.g., *reform* or *prosperity*) and which terms cast Democratic candidates in the worst light (e.g., *cynical* or *taxes*). According to Sternberg (1993), Gingrich trained a number of political operatives to use these symbolic twists to mobilize forces and increase campaign contributions around the country. Agents found they could use language features to prompt somewhat mindless actions, much like activating an EDM.

It seems reasonable to expect that, if politicians can effectively play upon symbolic predispositions, then environmental advocates should likewise be able to use language to *avoid* the debilitating effects of EDMs. It may not be all that simple, however. First, since the media control the access to and the framing of most environmental communication, advocates may not be able to manipulate the public's "rules for the recognition, retention, organization, and presentation of information and experience" (Altheide & Snow, 1988, p. 212). Second, since reporters generally turn to industrial authorities and scientists to understand complex issues, environmental activists and groups typically are not the "primary definers" of what even should be considered ecological concerns (Hansen, 1991, p. 449). For example, the concept of biodiversity, while being central to the preservation of habitat and species via federal and state policy, is routinely presented by resource consumers (e.g., those interested in game extraction or logging) as a ploy to lock up land for purely aesthetic values. And third, not only does the stridency of environmentalist rhetoric prompt well-financed countermeasures from corporate and free market sponsors of the "Dominant Social Paradigm" (Pirages, 1977), but, in addition, their "habitual reliance on gloom, apocalyptic panic, and the psychology of blame takes a heavy toll on public confidence" (Roszak, 1992, p. 35). Thus, until environmentalists gain some measure of media control and learn not to violate the bedrock values of the general public, as well as discover which symbols can activate loyalty to human survival on an endangered Earth, the game will forever be rigged against them.

In the end, when we survey all we know about the psychology of environmentalism, we still have much to learn about how the game should be played given the shortcomings of human cognition. Just as baseball is a game that has been emulated around the world, however, it should be possible for environmental activists in the United States to adopt a novel approach for discussing environmental issues, to promote more benign behavior by understanding what it means to live in such a complex environment, and to influence the less enlightened at home and around the globe. Such would be in keeping with the promise of the "American century" and wholly consistent with a national identity carved out of wilderness and progressive reform. If we are to be saviors, though, it is time we began appreciating the thoughts of those we are trying to save and reinvesting in studies designed to discover and exploit their weaknesses. They say that "the game's not over till it's over," and in this case, let us hope for perpetual extra innings.

References

Abramowitz, S.I. (1973). Internal-external control and social-political activism: A test of the dimensionality of Rotter's internal-external scale. *Journal of Consulting and Clinical Psychology, 40,* pp. 196-201.

Agar, M. (1979). Themes revisited: Some problems in cognitive psychology. *Discourse Processes, 2,* 11-31.

Allen, C.T., & Weber, J.D. (1983). How presidential media use affects individuals' beliefs about conservation. *Journalism Quarterly, 60,* pp. 98-104.

Allin, C.W. (1987). Park Service v. Forest Service: Exploring the differences in wilderness management. *Policy Studies Review, 7,* pp. 385-94.

Altheide, D.L., & Snow, R.P. (1988). Toward a theory of mediation. In J.A. Andersen (Ed.), *Communication Yearbook* (no. 11, pp. 194-223). Newbury Park, N.J.: Sage.

Atwater, T., Salwen, M.B., & Anderson, R.B. (1985). Media agenda-setting with environmental issues. *Journalism Quarterly, 62,* pp. 393-97.

Baglan, T., Lalumia, J., & Bayless, O.L. (1986). Utilization of compliance-gaining strategies: A research note. *Communication Monographs, 53,* pp. 289-93.

Bargh, J.A. (1989). Conditional automaticity: Varieties of automatic influence in social perception and cognition. In J.S. Uleman & J.A. Bargh (Eds.), *Unintended thought* (pp. 3-51). New York: Guilford.

Bart, W.M. (1972). A hierarchy among attitudes toward the environment. *Journal of Environmental Education, 4,* pp. 10-14.

Bird, E.A. (1987). The social construction of nature: Theoretical approaches to the history of environmental problems. *Environmental Review, 11,* pp. 255-66.

Black, J. (1970). *The dominion of man: The search for ecological responsibility.* Edinburgh: Edinburgh University Press.

Buttel, F.H., & Johnson, D. (1977). Dimensions of environmental concern: Factor structure correlates and implications for research. *Journal of Environmental Education, 9,* pp. 49-64.

Cable, T.T., Knudson, D.M., & Theobald, W.F. (1986). The application of the theory of reasoned action to the evaluation of interpretation. *Journal of Interpretation, 11,* pp. 11-25.

Cantrill, J.G. (1992). Understanding environmental advocacy: Interdisciplinary research and the role of cognition. *Journal of Environmental Education, 24,* pp. 35-42.

———. (1993). Communication and our environment: Categorizing research in environmental advocacy. *Journal of Applied Communication Research, 21,* pp. 66-95.

Cantrill, J.G., & Chimovitz, D.S. (1993). Culture, communication, and schema for environmental issues: An initial exploration. *Communication Research Reports, 10,* pp. 47-58.

Catton, W.R., & Dunlap, R.E. (1980). Environmental sociology: A new paradigm. *American Sociologist, 13,* pp. 41-49.

Cialdini, R.B. (1987). Compliance principles of compliance professionals: Psychologists of necessity. In M.P. Zanna, J.M. Olson, & C.P. Herman (Eds.), *Social influence: The Ontario symposium* (Vol. 5, pp. 165-84). Hillsdale, N.J.: Lawrence Erlbaum.

Cialdini, R.B., Kallgren, C.A., & Reno, R.R. (1991). A focus theory of normative conduct: A theoretical refinement and reevaluation of the role of norms in human behavior. In M.P. Zanna (Ed.), *Advances in experimental social psychology* (Vol. 24, pp. 201-34). San Diego: Academic Press.

Colby, M. (1991). *Environmental management in development: The evolution of paradigms.* Washington, D.C.: World Bank.

Cotgrove, S., & Duff, A. (1981). Environmentalism, values, and social change. *British Journal of Sociology, 32,* pp. 92-110.

Craik, K.H. (1973). Environmental psychology. *Annual Review of Psychology, 24,* pp. 403-22.

deHaven-Smith, L. (1988). Environmental belief systems: Public opinion on land use regulation in Florida. *Environment and Behavior, 20,* pp. 176-99.

Deluca, J. (1992). Articulation, progress, and the environment. In C.L. Oravec & J.G. Cantrill (Eds.), *The conference on the discourse of environmental advocacy* (pp. 10-18). Salt Lake City: University of Utah Humanities Center.

Drengson, A.R. (1980). Shifting paradigms: From the technocratic to the person-planetary. *Environmental Ethics, 3,* pp. 221-40.

Dunlap, R.E., Gallup, G.H., & Gallup, A.M. (1993). *Health of the planet.* Princeton, N.J.: Gallup International Institute.

Dunlap, R.E., & Van Liere, K.D. (1978). The "new environmental paradigm": A proposed measuring instrument and preliminary results. *Journal of Environmental Education, 9,* pp. 10-19.

Dunwoody, S., & Rossow, M. (1989). Community pluralism and newspaper coverage of a high-level nuclear waste siting issue. In L. Grunig (Ed.), *Environmental activism revisited: The changing nature of communication through organizational public relations, special interest groups, and the mass media* (pp. 5-21). Troy, Ohio: North American Association for Environmental Education.

Fischoff, B. (1981). Hot air: The psychology of CO_2 induced climatic change. In J.H. Harvey (Ed.), *Cognition, social behavior, and the environment* (pp. 163-84). Hillsdale, N.J.: Lawrence Erlbaum.

Fishbein, M., & Ajzen, I. (1975). *Belief, attitude, intention, and behavior.* Reading, Mass.: Addison-Wesley.

Fishman, J.A. (1968). Nationality-nationalism and nation-nationism. In J.A. Fishman, C.A. Ferguson, & J.D. Gupta (Eds.), *Language problems in developing countries* (pp. 54-86). New York: Wiley.

Fiske, S.T., & Pavelchak, M.A. (1986). Category-based versus piecemeal-based affective responses: Developments in schema-triggered affect. In R.M. Sorrentino & E.T. Higgins (Eds.), *Motivation and cognition* (pp. 167-203). New York: Guilford.

Fiske, S.T., & Taylor, S.E. (1991). *Social cognition.* 2nd ed. Reading, Mass.: Addison-Wesley.

Fraser, N. (1990). Rethinking the public sphere: A contribution to the critique of actually existing democracy. *Social Text, 25-26,* pp. 56-80.

Freedle, R. (1981). Interaction of language use with ethnography and cognition. In J.H. Harvey (Ed.), *Cognition, social behavior, and the environment* (pp. 35-60). Hillsdale, N.J.: Lawrence Erlbaum.

Geller, J.M., & Lasley, P. (1985). The new environmental paradigm scale: A reexamination. *Journal of Environmental Education, 17,* pp. 9-12.

Gergen, K., & Gergen, J. (1988). Narrative and the self. In L. Berkowitz (Ed.), *Advances in experimental social psychology* (Vol. 21, pp. 17-56). San Diego: Academic Press.

Gigliotti, L.M. (1990). Environmental education: What went wrong? *Journal of Environmental Education, 38,* pp. 9-12.

Grunig, J.E. (1977). Review of research on environmental public relations. *Public Relations Review, 3,* pp. 36-58.

———. (1989). A situational theory of environmental issues, publics, and activists. In L. Grunig (Ed.), *Environmental activism revisited: The changing nature of communication through organizational public relations, special interest groups, and the mass media* (pp. 50-82). Troy, Ohio: North American Association for Environmental Education.

Grunig, J.E., & Stamm, K.R. (1979). Cognitive strategies and the resolution of environmental issues: A second study. *Journalism Quarterly, 56,* pp. 715-26.

Ham, S.H. (1983). Cognitive psychology and interpretation: Synthesis and application. *Journal of Interpretation, 8,* pp. 11-27.

Hamilton, J.P. (1986). Environmental locus of control as a function of the perceived importance of environmental problems and environmental knowledge. *Journal of Interpretation, 11,* pp. 15-31.

Hansen, A. (1991). The media and the social construction of the environment. *Media, Culture, and Society, 13,* pp. 443-58.

Honnold, J.A., & Nelson, L.D. (1979). Support for resource conservation: A predictive model. *Social Problems, 27,* pp. 220-34.

Keeney, R.L., & von Winterfeldt, D. (1986). Improving risk communication. *Risk Analysis, 6,* pp. 275-81.

Kihlstrom, J.F., Cantor, N., Albright, J.S., Chew, B.R., Klein, S.B., & Neidenthal, P.M. (1988). Information processing and the study of the self. In L. Berkowitz (Ed.), *Advances in experimental social psychology* (Vol. 21, pp. 145-80). San Diego: Academic Press.

Kok, G., & Siero, S. (1985). Tin recycling: Awareness, comprehension, attitude, intention, and behavior. *Journal of Economic Psychology, 6,* pp. 157-73.

Lange, J.I. (1993). The logic of competing information campaigns: Conflict over old growth and the spotted owl. *Communication Monographs, 60,* pp. 239-57.

Langer, E.J. (1989). Minding matters: The consequences of mindlessness-mindfulness. In L. Berkowitz (Ed.), *Advances in experimental social psychology* (Vol. 22, pp. 137-73). San Diego: Academic Press.

Lowe, G.D., Pinhey, T.K., & Grimes, M.D. (1980). Public support of environmental protection: New evidence from national surveys. *Pacific Sociological Review, 23,* pp. 423-45.

McGee, M.C. (1978). "Not men, but measures": The origins and import of an ideological principle. *Quarterly Journal of Speech, 64,* pp. 141-54.

Mervis, C.B., & Rosch, E. (1981). Categorization of natural objects. *Annual Review of Psychology, 32,* pp. 89-115.

Milbrath, L. (1984). *Environmentalists: Vanguard for a new society.* Albany: State University of New York Press.

Miller, A. (1982). Tunnel vision in environmental management. *Environmentalist, 2,* pp. 223-31.

Milstein, J.S. (1977). Attitudes, knowledge, and behavior of American consumers regarding energy conservation with some implications for governmental action. In W.D. Perreault (Ed.), *Advances in consumer research* (Vol. 4, pp. 315-21). Atlanta: Association for Consumer Research.

Mitchell, R.C. (1984). Public opinion and environmental politics in the 1970s and 1980s. In N.J. Vig & M.E. Kraft (Eds.), *Environmental policy in the 1980s: Reagan's new agenda* (pp. 51-74). Washington, D.C.: Congressional Quarterly Press.

Morris, P. (1981). The cognitive psychology of self-reports. In C. Antaki (Ed.), *The psychology of ordinary explanations of social behaviour* (pp. 185-203). New York: Academic Press.

Moscovici, S. (1981). On social representation. In J.P. Forgas (Ed.), *Social cognition* (pp. 181-210). New York: Academic Press.

Mugerauer, R. (1989). Language and the emergence of environment. In D. Seamon & R. Mugerauer (Eds.), *Dwelling, place, and environment* (pp. 51-70). New York: Columbia University Press.

Niedenthal, P.M., Cantor, N., & Kihlstrom, J.F. (1985). Prototype matching: A strategy for social decision making. *Journal of Personality and Social Psychology, 48,* pp. 575-84.

Nisbett, R., & Ross, L. (1980). *Human inference: Strategies and shortcomings of social judgment.* Englewood Cliffs, N.J.: Prentice-Hall.

Nisbett, R., & Wilson, N. (1977). Knowing more than you can tell. *Psychological Review, 84,* pp. 231-59.

Novic, K., & Sandman, P.M. (1974). How use of mass media affects views on solutions to environmental problems. *Journalism Quarterly, 50,* pp. 448-52.

Olien, C.N., Tichenor, P.J., & Donohue, G.A. (1989). Media and protest. In L. Grunig (Ed.), *Environmental activism revisited: The changing nature of communication through organizational public relations, special interest groups, and the mass media* (pp. 22-39). Troy, Ohio: North American Association for Environmental Education.

Otaway, H., & Thomas, K. (1982). Reflections on risk perception and policy. *Risk Analysis, 2,* pp. 69-82.

Petty, R.E., & Cacioppo, J.T. (1986). *Communication and persuasion: Central and peripheral routes to attitude change.* New York: Springer-Verlag.

Pirages, D.C. (Ed.). (1977). *The sustainable society: Implications for limited growth.* New York: Praeger.

Portnoy, K.E. (1992). *Controversial issues in environmental policy.* Newbury Park, N.J.: Sage.

Ramsey, C.E., & Rickson, R.E. (1976). Environmental knowledge and attitudes. *Journal of Environmental Education, 8,* pp. 10-18.

Rogers, T.B. (1982). A model of the self as an aspect of the human information processing system. In J. Suls (Ed.), *Psychological perspectives on the self* (Vol. 1, pp. 193-214). London: Lawrence Erlbaum.

Roszak, T. (1992). *The voice of the Earth: An exploration of ecopsychology.* New York: Simon & Schuster.

Rowan, K.E. (1991). Goals, obstacles, and strategies of risk communication: A problem-solving approach to improving communication about risks. *Journal of Applied Communication Research, 19,* pp. 300-329.

Rothbart, M., & Taylor, M. ((1992)). Category labels and social reality: Do we view social categories as natural kinds? In G.R. Semin & K. Fiedler (Eds.), *Language, interaction, and social cognition* (pp. 11-36). London: Sage.

Sandman, P.M. (1976). Persuasion and communication theory. *Northern Rockies Action Group Papers, 1,* pp. 1-18.

Saxer, U. (1993). Public relations and symbolic politics. *Journal of Public Relations Research, 5,* pp. 127-51.

Schmidt, F.N., & Gifford, R. (1989). A dispositional approach to hazard perception: Preliminary development of the Environmental Appraisal Inventory. *Journal of Environmental Psychology, 9,* pp. 57-67.

Scott, R.L. (1976). On viewing rhetoric as epistemic: Ten years later. *Central States Speech Journal, 27,* pp. 258-66.

Sears, D., & Funk, C. (1991). The role of self-interest in social and political attitudes. In M.P. Zanna (Ed.), *Advances in experimental social psychology* (Vol. 24, pp. 1-92). San Diego: Academic Press.

Seligman, C., Kriss, M., Darley, J.M., Fazio, R.H., Becker, L.H., & Pryor, J.B. (1979). Predicting summer energy consumption from homeowners' attitudes. *Journal of Applied Social Psychology, 9,* pp. 70-90.

Shapiro, M.A., & Lang, A. (1991). Making television reality: Unconscious processes in the construction of social reality. *Communication Research, 18,* pp. 6685-705.

Stamm, K.R., & Grunig, J.E. (1977). Communication situations and cognitive strategies in resolving environmental issues. *Journalism Quarterly, 54,* pp. 713-20.

Sternberg, W. (1993). Housebreaker. *Atlantic Monthly, 276,* 1, pp. 26-28, 37-42.

Steward, J.H. (1955). *Theory of culture change: The methodology of multilinear change.* Urbana: University of Illinois Press.

Streufert, S., & Streufert, S.C. (1978). *Behavior in the complex environment.* Washington, D.C.: Winston & Sons.

Syme, G.J., Seligman, C., Kantola, S.J., & MacPherson, D.K. (1987). Evaluating a television campaign to promote petrol conservation. *Environment and Behavior, 19,* pp. 444-61.

Tajfel, H., & Forgas, J.P. (1981). Social categorization: Cognitions, values, and groups. In J.P. Forgas (Ed.), *Social cognition* (pp. 113-40). New York: Academic Press.

Taylor, S.E. (1981). The interface of cognitive and social psychology. In J.H. Harvey (Ed.), *Cognition, social behavior, and the environment* (pp. 189-211). Hillsdale, N.J.: Lawrence Erlbaum.

Tichenor, P.J. (1988). Public opinion and the construction of social reality. In J.A. Andersen (Ed.), *Communication Yearbook* (no. 11, pp. 547-54). Newbury Park, N.J.: Sage.

Tversky, B., & Hemenway, K. (1983). Categories of environmental scenes. *Cognitive Psychology, 15,* pp. 121-49.

Tybout, A.M., & Yalch, R.F. (1980). The effect of experience: A matter of salience? *Journal of Consumer Research, 6,* pp. 406-13.

Van Liere, K.D., & Dunlap, R.E. (1981). Environmental concern: Does it make a difference how it's measured? *Environment and Behavior, 13,* pp. 651-76.

———. (1983). Cognitive integration of social and environmental beliefs. *Sociological Inquiry, 53,* pp. 333-41.

Ward, L.M., & Russell, J.A. (1981). The psychological representation of molar physical environments. *Journal of Experimental Psychology, General, 110,* pp. 121-52.

Williams, R. (1980). *Problems of materialism and culture.* London: Verso.

Yates, S.M., & Aronson, E. (1983). A social psychological perspective on energy conservation in residential buildings. *American Psychologist, 38,* pp. 435-44.

5

Environmental Advocacy in the Corridors of Government

Michael E. Kraft & Diana Wuertz

Words are magical in the way they affect the minds of those who use them. "A mere matter of words," we say contemptuously, forgetting that words have power to mold men's thinking, to canalize their feeling, to direct their willing and acting.

Aldous Huxley, "Words and Their Meanings"

As Aldous Huxley noted so well, language can be exceptionally powerful in shaping our views of the world and our responses to it. Words can influence beliefs, attitudes, and perceptions, which in turn affect the way we characterize social problems and proposed solutions. This is as true in the political system as it is in other institutional settings.

To communicate effectively, individuals and groups must use language that can persuade, and mobilize, their intended audience. Doing so is a demanding task in public policy and a challenge that environmentalists and their adversaries know well. After twenty-five years of often contentious activism—and significant progress in achieving goals for environmental quality—further expansion of environmental policy will be difficult, costly, and controversial. It will almost certainly be stoutly resisted by interest groups that bear the direct short-term costs. Moreover, ideological opposition to the role of government in environmental regulation will continue to be pronounced in an era in which conservatism flourishes.

The bitter debates in the U.S. Congress in 1995 over provisions of the Republican Contract with America that aimed at weakening environmental protection laws made those trends apparent. National environmental groups and their local affiliates initially were caught off guard, but they soon realized that they needed to revamp their communication strategies if they were to have any hope of battling what they termed a war on the environment within the House of Representatives. The leading groups quickly designed a counterstrategy to mobilize their members in a massive grassroots lobbying campaign. Among other actions, they assembled "Breach of Faith," a report that documented the devastating impact of the Contract on major environmental laws. They distributed one version of it via electronic mail, to reach thousands of environmentalists nationwide within days of its completion (Natural Resources Defense Council, 1995). They also made use of all the usual techniques of direct lobbying in the corridors of government, including an effort to persuade President Clinton to use his veto power to block the Republican assault if acceptable legislative compromises could not be fashioned.

As the intense disputes over environmental policies in the 104th Congress illustrate, effective policy making is not based solely on credible scientific research and policy analysis. It turns at least as much on the capacity of diverse interest groups to frame the terms of policy debate, using whatever evidence or arguments are at hand, and to build a supportive public opinion. To this end, environmental groups and others formulate explicit advocacy strategies. Both sides in this contest attempt to affect the policy process with the construction and communication of ideas and images. Through such communication, organized groups hope to persuade the general public and especially the critical policy community in Washington, D.C., and elsewhere to adopt their view of environmental problems and issues and to accept their policy recommendations.

Environmental policy is greatly affected by this process of interest representation and agenda setting. The communication activities of environmental and other groups are part of a continuous democratic dialogue concerning the policy goals and means that constitute the environmental agenda. This dialogue may not be fair, balanced, or even rooted in scientific fact. Indeed, it may be intensely political, with highly prized values riding on the outcome. As the broad-based attack on environmental policy in Congress in 1995 demonstrated vividly, even anecdotal evidence in well-told stories of alleged bureaucratic excesses can win the day despite solid scientific evidence supporting a contrary perspective (Kenworthy, 1995).

The importance of these activities is evident. Yet we have little systematic knowledge about how the dialogue takes place, how group activities affect political agendas and policy decisions, and how well-placed policy entrepreneurs shape the process (Berry, 1989; Ingram & Mann, 1989; Kraft, 1995; Petracca, 1992; Schlozman & Tierney, 1986). In this chapter we explore how environmental organizations attempt to influence policy making through the use of political communication and other agenda-setting strategies and also suggest some avenues for further inquiry to help fill gaps in our knowledge.

First, we suggest that the study of environmental communication strategies can be anchored in a framework derived from studies of political agenda setting and policy change. Second, we briefly examine the ways in which environmentalists have developed advocacy strategies during the past two decades in an effort to shape environmental policy change, and we address the conflicts among environmental groups. We then turn to an assessment of environmental advocacy in government, particularly in the U.S. Congress. What do environmental groups do to promote public discourse on issues of environmental protection and natural resources? How effective are they in raising the typically low salience of these issues for the public and policy makers? And to what extent are they able to convert broad, yet shallow, public concern for the environment into a potent political force? In addressing these questions, we give special attention to the symbolic aspects of political communication both in government and at the grass roots.

Agenda Setting, Political Communication, and Policy Change

The communication strategies of environmental organizations are tied intimately to the larger processes of agenda setting and policy change. Agenda setting is a stage in the policy process that refers to the perception and definition of problems, the development of public opinion about them, and the organization of public concerns and new policy ideas to demand action by government. Agenda setting comprises all those activities that bring environmental problems to the attention of the public and political leaders and that shape the ideas and policy alternatives considered in government (Kingdon, 1984). Successful agenda setting is widely viewed as a prerequisite to policy adoption, and the process can be greatly affected by the activities of policy entrepreneurs. Included among those entrepreneurs are leading environmental activists and their allies in government.

John Kingdon defines the agenda as "the list of subjects or problems to which government officials, and people outside of government closely associated with those officials, are paying some serious attention at any given time" (1984, p. 3). The governmental agenda may be influenced by the larger societal agenda, the problems that most concern people, as is to be expected in a democracy when the public is mobilized around salient issues. It may also be shaped by the diffusion of ideas among policy communities (or policy elites) or by a change in the political climate. Such a change occurred with the election of Ronald Reagan in 1980 and Bill Clinton in 1992, and even more strikingly with the success by Republicans in taking over both houses of Congress following the 1994 midterm elections.

In Kingdon's model, the agenda-setting process depends on three relatively independent "streams" of activities that flow through the political system, with the assistance of policy entrepreneurs and the media. These three streams affect the perception and definition of problems (the problem stream), the proposed courses of action that receive attention and are continuously tested by the policy community (the policy stream), and the political climate or public mood at any given time (the politics stream). All three may be affected by the activities of organized groups and the communication strategies they employ. Often the choice of such strategies represents an explicit effort to shape the way problems are understood and defined (e.g., the extent of health or environmental risks posed) and the virtues of policy alternatives (e.g., regulation, market-based incentives, public education, or privatization). Whether directed at the mass public, policy elites, or both, such communication strategies also may be intended to alter the larger political climate by stimulating and mobilizing public outrage over environmental problems or, from the conservatives' perspective, over governmental mismanagement of programs and the imposition of unreasonable costs on the public.

Studies of cases in agenda setting and policy change suggest the centrality of these political processes. For example, Christopher Bosso (1987), who studied changes in federal pesticide policy from the 1940s to the 1980s, attributed the shift to stronger regulation largely to challenges by environmental and consumer groups to the once dominant agricultural chemical industry and its allies in Congress and the Agriculture Department. The older "politics of clientelism" and its reigning paradigm based on the benefits of pesticide use, Bosso argued, collided head on in the 1960s and 1970s with new concerns for safety and environmental quality advanced by environmentalists and others. In short, the definition of the problem was altered, as were the implications for public policy. The change occurred because new public attitudes about chemical risks were emerging, and environmental advocacy groups were able to advance the case for regulation in a Congress that by the 1970s increasingly reflected such new public concerns. The result has been a gradual shift away from the old pesticide policy, with frequent bouts of policy gridlock as the two coalitions of interests do battle in Congress and in the agencies.

Similarly, Frank Baumgartner and Bryan Jones (1993) have highlighted the impressive and rapid decline of the formerly influential nuclear power subgovernment from its prominence in the 1950s. By the late 1960s the rise of antinuclear environmental groups had effectively altered the salience of nuclear power issues. By the 1970s those groups posed an insurmountable challenge to the long-dominant alliance of the nuclear power industry, the old Atomic Energy Commission (AEC), and their allies in Congress (primarily on the Joint Committee on Atomic Energy). Much of this change was attributable to a shift from a largely positive to a negative image of nuclear issues conveyed in the media. That in turn reflected the efforts by environmental groups to stress the adverse health and environmental effects of reliance on nuclear power. The image of nuclear power changed "from one of progress and efficiency to one of danger and waste" (Baumgartner & Jones, 1993, p. 69). Political conflict expanded outward from the long-dominant policy subsystem as AEC technical staff began to question the agency's record of safety. Eventually the change in the definition of the issues provided an incentive for policy makers to respond to the growing antinuclear sentiment among the public.

Few political scientists have focused explicitly on the role of communication strategies in such cases of agenda setting and policy change. One of the leading models of policy change, however, incorporates as a key element "policy-oriented learning" within policy networks such as the environmental community (and its adversaries). This learning occurs through changes in perceptions, beliefs, and values of advocacy coalitions, that is, through the interactions of interest groups organized on one side or the other of policy disputes. New knowledge is integrated with the basic values and causal assumptions that comprise the core beliefs of the advocacy coalitions, contributing to policy change over an extended period of time (Sabatier, 1993).

Each of these models and studies holds important implications for the analysis of environmental advocacy in government. Environmental communication is a part of the broader effort to define problems and acceptable policy alternatives, and hence to build public and political support for desired policy action. Doing so has important implications for both direct and indirect (grassroots) lobbying strategies that environmental and other groups employ. Assessment of the present-day activities of these advocacy groups, and their effectiveness, first requires a brief look backward to understand the development of the groups since the late 1960s.

Origins and Development of Environmental Advocacy

Environmental groups have been politically active for more than a century: the Sierra Club began in 1892, and the National Audubon Society in 1905. Their policy goals and political strategies have evolved significantly over time, most notably between the late 1960s and the mid-1990s. During this period the groups that constitute the modern environmental movement greatly increased the size of their memberships, expanded their financial resources, and gained impressively in access to the centers of political power at all levels of government (Bosso, 1994; Dunlap & Mertig, 1992). Notwithstanding a recent decline in membership, financial strength, and especially political influence, these changes are remarkable. They also help to explain the building of U.S. environmental policy during the past twenty-five years. By the late 1980s most of the leading national groups—for example, the so-called Group of Ten—had adopted a more professional and politically pragmatic orientation to environmental policy issues and had turned increasingly to the same lobbying techniques used by other Washington-based interest groups (Mitchell, 1990).[1] Nevertheless, the political styles and communication strategies that environmental groups rely on continue to reflect their distinctive origins and the unique qualities of environmental issues, for better or worse.

Both scholars and activists have long observed that environmental groups, much like groups at the center of other social movements, are a diverse lot. Some, such as the Wilderness Society and the Sierra Club, trace their origins to battles for natural resource conservation in the late nineteenth century. Others were born only in the late 1960s or 1970s, when accumulating scientific information and public alarm spurred the development of a new generation of groups that focused on concerns such as public health risks, energy consumption, population growth, food production, biological diversity, and climate change. As the environmental movement matured over its first two decades, new organizational tensions and challenges arose that have yet to be fully resolved.

Despite their many shared values and political goals, environmentalists began to splinter into factions with conflicting styles and political strategies. The differences became so great that some analysts have asked whether they could be subsumed under the same label at all. For example, writing in 1991, Robert Mitchell noted that the unity of environmental groups was "tempered by a diversity of heritage, organizational structure, issue agendas, constituency, and tactics" and that the groups competed with each other "for the staples of their existence—publicity and funding" (p. 83). Some of these divisions were noticeable even in the early 1970s.

For these reasons, students of the environmental movement such as Mitchell and Bosso have suggested that the activities of environmental groups can best be understood by dividing the groups into three categories: mainstream, green (or radical), and grassroots (see Bosso, 1991, 1994). Yet one can also distinguish the large, multi-issue national groups like the Sierra Club from the smaller and more focused ones like Friends of the Earth. These in turn can be differentiated from groups that are largely preoccupied with education, policy analysis, and scientific research—such as Resources for the Future, the Worldwatch Institute, and the World Resources Institute—or that are more concerned with private land conservation than with policy advocacy (e.g., the Nature Conservancy).

Whatever typology is used, the conclusion about group differences is much the same. Diversity in organizational beliefs and purpose is striking. And it is great enough to override the common ground that exists in the support of a new environmental paradigm or a long-term environmental agenda that most environmentalists share. Analysts sometimes focus on the similarities and sometimes on the differences. Thus political theorists like Robert Paehlke (1989) see much commonality among environmentalists in their core values and political goals, while others wonder whether we can speak usefully about the environmental movement "as a single, identifiable entity" (Bosso, 1994). As these observations suggest, generalities about strategies for environmental advocacy must be tempered with recognition of the considerable variability among environmental organizations, past and present.

Aside from differences in philosophy and political agendas, environmental groups vary significantly in size and organizational resources. Here too changes over the past two decades are notable. In the 1960s and 1970s membership in environmental groups grew at an extraordinary pace, representing what Jeffrey Berry (1989) has called the "advocacy explosion" in Washington-based interest groups and the law and public relations firms on which they rely for lobbying. The Sierra Club saw its membership grow from 15,000 in 1960 to 136,000 in 1972; the membership in the National Audubon Society increased from 32,000 in 1960 to 232,000 in 1972; and the Wilderness Society grew from 10,000 in 1960 to 51,000 in 1972 (Mitchell, Mertig, & Dunlap, 1992). New groups, such as Environmental Action, the Natural Resources Defense Council, and the Environmental Defense Fund, also grew appreciably during the 1970s.

Growth and Challenge in the 1980s

The growth in membership translated into greatly improved budgets, which in turn enhanced organizational resources that could be used to lobby for environmental policies. Before 1970, environmentalists were at a competitive disadvantage relative to business groups in access and lobbying prowess. They became far more politically adept and successful in the 1970s and 1980s, though still less well financed and less capable than their corporate peers. Nevertheless, they learned how to compensate politically. They staked out the moral high ground, used the media to good effect, and mobilized their troops when needed. As a result, much of the "environmental gridlock" in Congress in the Reagan era can be traced to the ability of environmentalists and business groups to block one another on issues ranging from clean air to reform of pesticide laws (Kraft, 1994).

As is well known, the Reagan years severely tested the foundations of the environmental movement. President Reagan believed that he had been elected to bring regulatory relief to the business community and to spur economic growth. Thus environmental regulation became a frequent target during his tenure. Environmentalists regarded the Reagan Administration as the most environmentally hostile in half a century, and they spent much of their energy and resources in the 1980s defending the legislative and administrative achievements of the Environmental Decade from critics in the business community and the executive branch (Vig & Kraft, 1984). The ferocious battles with a Republican Congress in 1995 indicate that the old Reagan agenda was not as thoroughly vanquished in the 1980s as some environmentalists believed. It has returned with a vengeance as a new generation of conservatives seek to accomplish what they failed to achieve during the Reagan presidency.

Ironically, environmental groups prospered under Reagan. They effectively used the president's agenda and the activities of his Interior secretary, James Watt, and his initial Environmental Protection Agency (EPA) administrator, Anne Burford, to appeal to a public that continued to display remarkable devotion to environmental values. Public concern about the environment grew strongly in the 1980s despite the persistent efforts of the Reagan Administration to deflect it. As Riley Dunlap has noted, the environment became a consensual issue once again, as it had been in the early 1970s: "A majority of the public supports increased environmental protection efforts and a *huge* majority appears opposed to weakening current efforts" (1987, p. 11).

One message here is clear. Even the most carefully crafted antienvironmental communication efforts of a popular president could not convince a public committed to environmental values. Although the salience of environmental issues was low in the early to mid-1980s, concern for environmental protection continued to grow. A variety of survey data leaves little doubt about public attitudes, even if the policy implications are less clear (Dunlap, 1991). Consistent with rising public concern for the environment, group memberships continued on the growth curve of the 1970s, thanks in part to the effective use of direct mail recruitment techniques. The groups' budgets and staffs grew in parallel with membership rolls (Baumgartner & Jones, 1993, p. 189). For example, as shown in table 5.1, the Sierra Club's membership increased from 181,000 in 1980 to 600,000 in 1990, and the Wilderness Society saw an increase in membership from 35,000 in 1980 to 350,000 in 1990. Even the environmental law groups, which are not, strictly speaking, membership organizations, saw their rolls more than triple between 1980 and 1990. Perhaps the best demonstration of the wide appeal of environmental values is provided by Greenpeace. It began its U.S. direct mail membership campaign in 1979, and by 1990 it could claim 2.3 million members (Mitchell, Mertig, & Dunlap, 1992), reflecting a growth rate of well over 100 percent per year. By one count, in 1990 alone Greenpeace raised about sixty million dollars from direct mail contributions (Bosso, 1994).

Such developments in advocacy strategies paid handsome dividends for environmental groups. The mainstream groups in particular learned to use the natural advantage that environmentalists have in their campaigns for condensational symbols, that is, concepts with emotionally charged and powerful images. These symbols include *clean air, clean water, toxic substances, wilderness, endangered species, environmental protection,* and *public health.* As Mitchell (1984) has

Table 5.1 Memberships and Budgets of Selected National Environmental Advocacy Groups, 1960-1990

Organization	Year Founded	Membership						1990 Budget ($ million)
		1960	1970	1975	1980	1985	1990	
Sierra Club	1892	15,000	150,000	170,000	181,000	364,000	600,000	28
National Audubon Society	1905	32,000	105,000	275,000	400,000	550,000	575,000	40
National Parks and Conservation Association	1919	15,000	50,000	50,000	31,000	45,000	100,000	4
Wilderness Society	1935	10,000	44,000	50,000	35,000	150,000	350,000	14
National Wildlife Federation	1936	NA	3.1 mil.	3.7 mil.	4 mil.	4.5 mil.	5.8 mil.[a]	79
Environmental Defense Fund	1967	—	b	30,000	46,000	50,000	150,000	15
Natural Resources Defense Council	1970	—	b	35,000	42,000	50,000	125,000	13.5

Sources: Bosso, 1994, p. 36. The 1990 estimates of organizational budgets are drawn from Kriz, 1990, p. 1828.

Note: Membership figures are notoriously hard to pin down. All figures reported here should be considered estimates and used only to illustrate growth over time.

[a]NWF membership figures include the large number of affiliated members. Without counting affiliates, NWF membership was approximately 975,000 in 1992.

[b]In 1970, neither the Environmental Defense Fund nor the Natural Resources Defense Council was a membership organization.

argued, such symbols replaced older terms such as *conservation* and *wise use*. The new symbols arouse strong protective feelings on the part of a public that survey research consistently indicates is broadly sympathetic to environmental values and policy goals (Bosso, 1994; Dunlap, 1991; Mitchell, 1990).

In contrast, the symbols invoked by the opposition, such as *growth, development, use,* and *progress,* had lost much of their previous appeal by the 1970s. Indeed, as public concern for issues pertaining to quality of life rose, industry groups that emphasize development and growth found themselves battling an image of being dirty and dangerous (Harrison, 1983). Yet these older symbols remain powerful during times of economic recession or when loss of jobs or other adverse economic consequences in a community or region become important.

The invocation of such symbols reached a new height in early 1995, when the 104th Congress debated reform of environmental regulation. Environmental policies were frequently and sharply criticized for their alleged adverse economic effects. Critics focused particularly on the burden and costs imposed on industry and thus indirectly on consumers for what was described as ineffective, inefficient, or needless regulation. Environmental laws, especially the Endangered Species Act, were also faulted for what their detractors asserted was an illegal "taking" of private property without compensation. Yet there is little evidence that the electorate in November 1994 favored policy actions embodied in the Republican Contract, and there is virtually no indication that the public wanted to weaken environmental policies (Greenberg, 1995; Shute, 1995).

Thus, despite the many advantages that environmentalists enjoy in public support and the appeal of their values and symbols, by the early 1990s they often found themselves on the defensive. Internal divisions weakened them politically, and membership and revenues leveled off or declined somewhat. Their opponents, such as the Wise Use and property rights movements, grew far more aggressive and found new success by imitating the grassroots organizing and communication strategies that had worked so well for environmentalists for twenty years (Bosso, 1994; Carney, 1992; Poole, 1992). One effect of these changes was declining corporate donations to the major environmental groups, a consequence of both the recession and a shifting political climate that was less positive for environmentalists. This development may have led some groups to temper their overt criticism of the business community for fear of losing funds (Schneider, 1991). The suspicion, difficult to prove, is that groups cannot afford to lose corporate sponsorship of television broadcasts and other expensive activities. This issue was raised in 1991, for example, when General Electric withdrew support from *The World of Audubon* documentaries following consumer boycott threats from groups associated with the National Inholders Association, a lobbying group for loggers, ranchers, and property owners in or near national forests and parks. One of the programs had criticized cattle grazing for its environmental impact on the land (Bosso, 1994, pp. 36–37). Some leading environmental groups have also received generous donations from industries eager to improve their corporate images. For example, in the early 1990s, Waste Management Incorporated, one of the nation's largest hazardous waste disposal companies, contributed one million dollars annually to groups such as the World Wildlife Fund, the National Wildlife Federation, the National Audubon Society, the Sierra Club, the Izaak Walton League of America, and the World Resources Institute (Bosso, 1994, p. 36). In a time of declining organizational budgets, some groups may be reluctant to risk the loss of such funds.

Another important change in the late 1980s and early 1990s was in the salience of environmental issues. Even though strong public support continued, environmental issues only rarely proved to be a determining factor in congressional or presidential elections. After rising to new highs in 1989 following the *Valdez* oil spill, the salience of the environment for the public returned to lower levels by the early 1990s. In January 1992, for example, a *New York Times*–CBS News poll found that only 4 percent of a national sample mentioned the environment in response to a question about the issues they would like to see the presidential candidates emphasize in their campaigns, "besides issues like war and peace, the national economy, and jobs." Environmental issues tied for fifth place, along with the elderly, abortion, and drugs (Bosso, 1994, p. 30).

Assessing Environmental Advocacy in Government

The core challenge to environmental groups in the late 1990s remains what it has been throughout much of the previous two decades: how to convert diffuse public support into effective political clout. It is for that reason that we need to ask about advocacy strategies. What do environmental groups do to further their political agendas, and how effective are they?

Interest Group Strategies

Interest group activity is a vital if often unappreciated component of democratic politics. It is largely through the actions of organized groups that the public's opinions shape policy making and thus help assure that policies are responsive to popular preferences. These group activities occur throughout the policy process, from setting both the societal and governmental agendas to influencing policy formulation and implementation. The general pattern also applies to environmental groups. They contribute to the critical process of policy legitimation by creating and sustaining a dialogue that promotes effective problem solving and ensures that policy will reflect the public's views. Ultimately, the success of environmental policy turns on the quality of this dialogue and on the interaction and accommodation of pertinent interest groups.

It is more important to ask how environmental groups affect this dialogue through their involvement in the various stages of the policy process than it is to try to determine their precise influence on decision making. The latter is a task beset with conceptual and methodological difficulties. Decision making is a highly complex process, and the outcome is affected by multiple variables, such as personal political beliefs, legislative and executive leadership, public opinion, and the role of professional staff. Most of these variables are hard to isolate and even harder to measure. Thus, causal relations are exceptionally difficult to establish with much confidence (Furlong, 1992; Kingdon, 1984, 1989). Students of environmental advocacy would be better advised to focus on the big picture. We need to learn more about how the dialogue on environmental policy takes place and what its consequences are. Who are the serious participants? What values and expectations do they bring to the dialogue? How does the process affect the ideas and policy proposals that are proffered?

During the past twenty-five years, mainstream environmental groups have relied on all of the usual advocacy strategies and techniques found in U.S. politics: direct lobbying (both formal and informal), indirect or grassroots lobbying, public education, electioneering, and litigation. These approaches to political influence, particularly for legislative strategies, have been described at length by Edward Schneier and Bertram Gross (1993), among others. A focus on environmental policies does not appreciably alter the nature of such campaigns to influence public policy. It is also evident that over the past two decades environmental groups have become as technologically sophisticated and politically astute as other interest groups. Environmentalists use television extensively, advertise frequently in the nation's print media, operate large telephone banks for fundraising, and regularly use direct mail as well as newer techniques such as computer bulletin boards and electronic mail to spread the word on emerging policy battles.

When environmental organizations attempt to influence governmental decisions, the specific lobbying strategies are sensitive to context. They vary with the organizational and political situation. Most advocacy strategies can be classified, however, as either direct or indirect lobbying, with the latter playing an important role in shaping direct appeals to policy makers.

Direct Lobbying

Individuals and groups who use direct lobbying attempt to influence members of Congress or officials in agencies of the executive branch through personal contact. In a legislature in particular, the degree of consensus for policy action among the public makes a difference to elected officials. The greater the public consensus, the easier it is for Congress to reach agreement on environmental policy and to resist pressure from opposing interest groups. Even though the public strongly supports environmental protection, its influence on the details of public policy is often very weak. The reason is that its understanding of the issues (and of Congress) is limited, its views are sometimes inconsistent, and its communication with Congress is infrequent and often ineffective. Environmental groups in effect act as surrogate representatives of the public.

Among the key actions of these groups that fall into the category of direct lobbying are personal visits with members and staff in a variety of settings and the provision of technical, constituency (i.e., public opinion), and political or strategic information. For example, environmental groups routinely testify at hearings on proposed legislation to document the scope of environmental threats and to urge specific policy action. It would be difficult to find many hearings on Capitol Hill concerning the renewal of the major environmental statutes (e.g., the Clean Air Act, the Clean Water Act, Superfund, and the Endangered Species Act) that do not include such testimony by leading environmental organizations.

In 1995, however, environmentalists identified some cases that were notable for being unusual. Industry lobbyists were given abundant opportunities to work closely with the Republican chair of the House Committee on Transportation and Infrastructure, Bud Shuster (R-Pa.), as the committee revised the Clean Water Act. They even sat at the head table in the committee's hearing room and worked side by side with the staff director (Engelberg, 1995). Yet environmentalists complained bitterly that they were largely ignored in the process. Carol Browner, administrator of the EPA, raised similar concerns, arguing that she and other EPA officials had "essentially . . . not been allowed to see [the bill] or to be part of the process." She reflected the indignation voiced by environmentalists in saying that it was "absolutely outrageous that these kinds of bills are being produced with no input from the agency that will ultimately implement them" (Cushman, 1995, p. 1). In another case, in the Senate's rewriting of the Endangered Species Act (ESA), Senator Slade Gorton (R-Wash.) relied heavily on a group of Washington, D.C., lawyers representing timber, mining, ranching, and utility interests, who have been among the most severe critics of the act. Press accounts indicated that the draft bill was largely written by those lawyers. At the same time, Gorton openly acknowledged that he did not consult with environmentalists at this early stage of policy formulation: "I already know what their views are, and they don't want any changes" in the law, he said (Egan, 1995). He did promise, however, that environmentalists' views would be heard when hearings were scheduled on the bill.

Interest groups also are often key participants in administrative processes in regulatory agencies. Indeed, as shown in table 5.2, Kay Schlozman and John Tierney (1986) reported that their comprehensive 1981–82 survey of Washington-based interest groups found that 99 percent of advocacy organizations use testifying at either legislative or administrative hearings as a method to influence policy decisions. They also found that 98 percent of interest groups contact government officials directly to present their arguments. There are virtually no systematic studies of lobbying by environmental groups and its effect on either the legislative or the administrative process, although much anecdotal evidence is available in case studies and insider accounts that confirm these general patterns (e.g., Bosso, 1987; Bryner, 1996).

Several changes within Congress in the 1970s and 1980s increased the access and influence of environmental groups. One was the decision to decentralize power from committees to subcommittees and to give the subcommittees professional staff and jurisdictional autonomy. Decentralized decision making meant more authority for younger and more environmentally oriented members. It also increased the opportunities for environmentalists to work with policy entrepreneurs in Congress who shared their policy orientations, such as Henry Waxman (D-Calif.), who led the fight in the House for a strong Clean Air Act. A second change was the reform of rules and practices, which created a more open and participatory Congress that was more responsive to rank-and-file members and to emerging national issues such as environmental protection (S.S. Smith, 1985).

These changes underscore an important and well-understood characteristic of lobbying. Group contact with members of Congress and other public officials

Table 5.2   Lobbying Techniques Used by Washington-Based Advocacy
Organizations

Technique	Percentage of Organizations Using
Testifying at hearings	99%
Contacting government officials directly to present point of view	98
Engaging in informal contacts with officials, at conventions, over lunch, and in similar settings	95
Presenting research results or technical information	92
Sending letters to members of the organization to inform them about organization's activities	92
Entering into coalitions with other organizations	90
Attempting to shape policy implementation	89
Talking with people from the press and media	86
Consulting with government officials to plan legislative strategy	85
Helping to draft legislation	85
Inspiring letter-writing or telegram campaigns	84
Shaping the government's agenda by raising new issues and calling attention to previously ignored problems	84
Mounting grassroots lobbying efforts	80
Having influential constituents contact the offices of their congressional representatives	80
Helping to draft regulations, rules, or guidelines	78
Filing suit or engaging in litigation	72
Making financial contributions to electoral campaigns	58
Publicizing candidates' voting records	44
Engaging in direct-mail fund-raising	44
Running advertisements in the media on issue positions	31

Source: Schlozman and Tierney, 1986, p. 150.

involves two-way communication. Lobbying is instigated in part by policy makers themselves. They need environmental and other groups for the information they provide, for their political support, and for their assistance in furthering or blocking legislation or parallel action in the executive agencies. For these reasons, policy makers keep in constant communication with outside groups and frequently consult them before embarking on major policy initiatives, much as Senator Gorton did on renewal of the ESA in 1995. Indeed, political scientists have come to refer to the concepts of policy communities and issue networks to capture these relationships and their role in the policy process (Kingdon, 1984). It is not much of an exaggeration to say that the growth of a significant presence of environmental groups in Washington during the past two decades has placed representatives of the leading organizations at the center of federal policy making, even if the extent of their access and influence depends on the vagaries of the electoral process, as the 1994 congressional elections demonstrated.

Whether the lobbying effort is direct or indirect, the goal is to influence the outcome of the policy-making process. At the national level, this effort is focused on Congress. Once legislation is enacted, however, the emphasis shifts to bureaucratic agencies responsible for the drafting of regulations, implementation, and enforcement. Michael McCloskey, the Sierra Club's former executive director, noted the implications for environmentalists: "Faith began to wane that major lobbying efforts had much payoff. . . . Surely, the movement could still get statutes enacted, and it could win lawsuits over governmental refusal to implement them. But could we get them implemented properly in the final analysis? . . . What was the point of great lobbying campaigns in Congress if so little came of the enactments in the end?" (McCloskey, 1992, p. 83).

Thus the agencies, particularly the EPA and other administrative offices within the Interior and Agriculture Departments—such as the National Park Service, the Fish and Wildlife Service, the Bureau of Land Management, the Bureau of Reclamation, and the Forest Service—have become targets of advocacy groups (Culhane, 1981). Those who lose battles in the Congress often try again in the bureaucracy or at the White House, as business groups did in the Bush Administration when they objected to draft regulations prepared by the EPA to implement the Clean Air Act Amendments of 1990. They took their case to the Council on Competitiveness, housed in the office of Vice President Dan Quayle, and with the president's support, they won significant concessions from the EPA (Bryner, 1996).

The same strategic choices present in influencing Congress are found when groups lobby executive agencies. Both direct and indirect lobbying campaigns can be used to affect policy implementation. In fact, as shown in table 5.2, Schlozman and Tierney (1986) found that 89 percent of advocacy groups attempt to influence these bureaucratic processes, with fully 78 percent assisting in drafting regulations, rules, and guidelines. Since policy implementation continues to involve the Congress—for example, in appropriation decisions to fund environmental programs and in oversight and evaluation of program operations—lobbying of Congress continues at this stage as well.

No single lobbying campaign can convey the diversity of strategies and tactics used by environmental groups to influence policy making, in part because so many different kinds of issues are addressed, from use of public lands and protection of biological diversity to clean air and national energy policy. Groups also may act independently or collectively. In many of the most successful efforts of recent years, the environmental community has formed ad hoc coalitions to pool the groups' limited resources. For example, the National Clean Air Coalition, organized in the early 1980s under its chairman, Richard Ayres of the Natural Resources Defense Council, included the Sierra Club, the National Wildlife Federation, the National Audubon Society, and the Environmental Defense Fund, along with other church, labor, public health, and citizens' groups. The coalition fought persistently for nearly ten years to renew and strengthen clean air legislation. It coordinated congressional lobbying as well as grassroots letter-writing and telephone campaigns toward these ends. Environmentalists lost some key clean air battles over particular provisions in both the House and the Senate; they also won some. For most of a decade, the Clean Air Act's stature as the nation's premier environmental policy, the powerful symbolism of the term *clean air,* and abundant statistics on the serious health effects of air pollution proved insufficient to win the day (Bryner, 1996). Yet by 1989 the tide had turned. After thirteen years of gridlock, environmentalists could finally claim a measure of victory in the 1990 Clean Air Act.

Similarly, in the late 1980s and early 1990s the Campaign for Pesticide Reform, representing forty-one environmental, health, consumer, and labor groups, sought an array of changes in policies regulating the use of agricultural chemicals. While no significant policy reforms were enacted by the end of the 103rd Congress, the campaign was effective in standing up to the opposition, organized as the National Agricultural Chemicals Association (which in 1994 changed its name to the American Crop Protection Association). Even without gaining new legislation, the environmental coalition succeeded in raising issues concerning the safety of pesticides and the environmental effects of their use, which are likely to be thoroughly evaluated in future policy deliberations.

One case that illustrates as well as any the continued effectiveness—and the many limitations—of environmental advocacy in Congress is the effort in the early 1990s to modify President Bush's proposed National Energy Strategy (NES). Submitted to the Congress in February 1991, the president's NES relied heavily on increased energy production rather than conservation. It was sharply criticized by environmentalists on this ground. One proposal, especially irksome to the environmental community, recommended opening the pristine Arctic National Wildlife Refuge (ANWR) to oil and gas exploration. Environmentalists mounted a major media and lobbying campaign against opening the ANWR to such development, but they were unsuccessful in persuading the Senate Energy and Natural Resources Committee, led by J. Bennett Johnston (D-La.), to drop the provision. The committee's members were more conservative on environmental issues than the full Senate, and they received sizable campaign contributions from the oil, natural gas, and nuclear energy industries. Environmentalists then focused on the Senate itself.

As the committee's bill was readied for consideration on the Senate floor, three junior senators, Richard Bryan (D-Nev.), Paul Wellstone (D-Minn.), and Timothy Wirth (D-Colo.), representing a coalition of some eighty groups (chiefly environmentalists, consumers, and supporters of alternative energy sources), mounted a filibuster. In what the press called a stunning victory for environmental groups, Johnston lost the November 1, 1991, cloture vote to end the filibuster by a ten-vote margin. Afterward, he spoke openly of his admiration for the "political skill" that environmentalists exhibited, saying that they "wrote the text-book on how to defeat a bill such as this" (Kraft, 1994; U.S. Congress, 1991).

Environmentalists enjoy repeating the story, but they do not mention their failure to follow through on this battle. After the vote, Johnston, Wirth, and others agreed to rewrite the bill, without either the ANWR or the equally controversial Bryan proposal to increase sharply Corporate Average Fuel Economy (CAFE) standards for automobiles, and to add energy efficiency measures. What might have been a successful effort to accommodate environmental concerns fell apart late in 1991. The Gulf War receded in memory, energy issues no longer seemed so urgent, and environmentalists lost the momentum they had had in early November. They and their allies in the Senate failed to hold the anticipated meetings and redraft the bill. As adept as they were at derailing a proposal they found wanting, environmentalists could not put together a positive coalition on energy policy. Ultimately, Johnston rebuilt his own coalition and regained control of the energy bill. The ANWR proposal was absent from the Energy Policy Act of 1992, as were new CAFE standards. Yet the final measure was closer to Johnston's original bill than to environmentalist preferences.

One lesson in such cases is that communication strategies are, of course, only one element in lobbying. However appealing the language and images environmentalists create in these campaigns, the reality of powerful political opposition remains. With so much at stake, there are likely to be equally impassioned battles in the future. As one sign of that scenario, President Clinton's moderate energy tax proposal of early 1993, based on the British thermal unit (Btu) or heat content of fuels, came under fierce lobbying pressure from groups representing industries that produce energy-intensive products such as petrochemicals, steel, and cement. The National Association of Manufacturers played a key role; other active organizations included the American Petroleum Institute, the Chemical Manufacturers Association, the U.S. Chamber of Commerce, and the American Farm Bureau Federation. The leading groups vowed to kill the tax, and others were intent on altering it to remove some or all of the burden that they would otherwise face (Greenhouse, 1993; Wines, 1993). One industry coalition, operating under the banner of the American Energy Alliance, spent more than one million dollars on simulated grassroots lobbying, to encourage telephone calls and letters from the public and community leaders in some twenty key states in opposition to the tax.[2] The campaign was successful, in part because the White House did little to defend its proposal. Warned by leading Democratic senators that he faced defeat on the measure, the president by early June had backed off from his Btu tax in favor of a smaller fuel tax.

As might be expected, the messages conveyed in letters and advertisements in such legislative lobbying often exaggerate or misrepresent the impact of policy proposals. For example, print advertisements by the American Energy Alliance alleged that Clinton's Btu tax would cost the United States some four hundred thousand jobs and would seriously hinder the nation's economic recovery

(American Energy Alliance, 1993). In contrast, environmentalists maintained that reducing the nation's dependence on fossil fuels and encouraging energy conservation would create jobs and help the economy in the long run (Kriz, 1993). Comparable claims and counterclaims characterized debate over the probable environmental effects of adopting the North American Free Trade Agreement in 1993. It is unclear what impact such "facts" have on policy makers, although many would discount such messages in light of the obvious bias of the source. Policy makers also are likely to hear from both sides, particularly in the most important or most salient policy battles, and thus can compare conflicting arguments and data and reach their own conclusions. When genuine competition of this kind exists, the effect that any single group will have on policy decisions will be moderated. Unfortunately, such competition is not always present.

As the cases noted here suggest, environmental groups lack the capacity to lobby as effectively in the legislature, bureaucracy, and courts as their counterparts in business. With limited resources, they must choose carefully which of the hundreds of issues each year merit priority attention. In this sense they are still no match for industry and resource development interests, which are far better funded and prepared to continue their fights over many years and in many different arenas. The imbalance in political influence was evident in early 1995, as the House of Representatives moved with uncharacteristic speed in approving nearly all the provisions of the Contract with America within the first hundred days of the 104th Congress. Those provisions included far-reaching regulatory reform measures that would significantly weaken implementation of environmental laws (Benenson, 1995). Echoing the sentiment among environmentalists, EPA administrator Carol Browner complained that the "special interests and the polluters won—the big corporate lobbyists" ("After 100 Days," 1995).

Despite the striking growth of environmental groups—and other public interest groups—during the past three decades, then, major obstacles remain. Political scientists continue to find a strong advantage for business groups in lobbying government (Berry, 1989; Furlong, 1992; Plotke, 1992). In one of the most thorough assessments of group resources, Schlozman and Tierney concluded that E.E. Schattschneider's characterization of the bias of the interest group system in 1960 remained true in the 1980s. As they noted, "Taken as a whole, the pressure community is heavily weighted in favor of business organizations: 70 percent of all organizations having a Washington presence and 52 percent of those having their own offices represent business" (Schlozman & Tierney, 1986, p. 68). In contrast, the figures for citizens' groups were 4.1 percent and 8.7 percent, respectively. More recent research indicates that these figures remain much the same in the 1990s (Furlong, 1992). This unequal division of financial and other organizational resources has important implications for the advocacy strategies used by environmental groups. Recognition of these limitations leads many environmental groups to turn increasingly to grassroots campaigns, consumer education, and lobbying at the state and local level, where their success is more easily assured (John, 1994).

Indirect Lobbying and Media Use

In contrast to conventional group representation in policy making, indirect lobbying attempts to influence members of Congress (or the executive branch) by mobilizing the public and crucial constituencies. Environmental organizations—and increasingly their opponents in agriculture, mining, forestry, and other business interests—have found such grassroots campaigns to be an effective way to increase the political salience of their issues. They regularly remind elected officials that critical votes are likely to appear in the year's "scorecard" on environmental voting; a prominent index, for example, is compiled annually by the League of Conservation Voters. As noted in table 5.2, 92 percent of advocacy organizations send letters to their members to inform them about legislative and other activities, 84 percent of these groups encourage and organize letter-writing or telegram campaigns, and 80 percent mount grassroots lobbying efforts. Electoral activity, from publicizing a candidate's voting behavior to making financial contributions, is much less common, in part because of Internal Revenue Service rules.

By all accounts, reliance on grassroots lobbying by both environmental and business organizations increased substantially in the early 1990s, as these groups discovered direct approaches to be less effective in the contemporary political climate. The new grassroots campaigns are more sophisticated and seemingly more effective in shaping environmental policy than early forms of indirect lobbying. For example, extensive use is now made of computerized fax machines, electronic mail via personal computers, and satellite network connections for rapid communication and public mobilization. Telemarketing companies are able to transfer calls to their toll-free 800 numbers directly to congressional offices, allowing them to screen citizen opinions and channel callers to their elected representatives (Brinkley, 1993; Browning, 1994). One consequence of these recent developments is that congressional offices are increasingly flooded with communication from constituents on environmental and other public policy concerns, complicating an already difficult task of determining "real" opinion on what are often low-salience issues.

Virtually all environmental groups other than those concerned chiefly with scientific and policy research also make regular use of the mass media, whether print or electronic. They have found journalists to be fairly receptive to their efforts, and there is some evidence of a continued "greening of the press," which suggests even more favorable coverage in the future (Stocking & Leonard, 1990). Yet media coverage of environmental issues has often been criticized for a narrow focus on particular controversies and "hot" topics, and for a lack of adequate scientific and political analysis (LaMay & Dennis, 1991). These critiques may resemble the usual complaints that journalists offer shallow or biased coverage of difficult and controversial topics. There are, however, important differences with environmental policy coverage. One is a greater potential for distortion of the issues because of their technical complexity. Journalists may provide inaccurate or insufficient technical information, or they may present it in a manner that hinders public comprehension and thus the making of informed decisions by citizens (Mazur, 1988). Much of the debate in recent years over the inadequacy of risk communication is a case in point. The public often becomes unnecessarily alarmed about minor environmental and health risks while ignoring more important ones (U.S. Environmental Protection Agency, 1990). Another difference is that environmental policy is critically dependent on public participation. Yet limited analysis of political conflicts and policy choices may discourage citizens from active involvement, particularly at the local level, where their participation can significantly shape the policy process.

Despite these weaknesses, groups outside the mainstream—the greens or more radical organizations—have found media-based strategies especially appropriate for their needs. There is no coherent green movement in the United States, and generally the greens are less politically active than mainstream groups. Many have little faith in the political system, and they studiously avoid conventional tactics such as lobbying, electoral activity, and litigation. Groups such as Greenpeace instead emphasize protest activities designed to attract media coverage and thus to reach a mass public with their message. Such actions help to keep environmental issues on the agenda even when there are no impending public policy decisions (Anderson, 1991).

There is a political risk, however, in emphasizing direct action strategies and the assorted "stunts" that groups like Greenpeace use. Although the action may appeal to journalists who value coverage of confrontation and dramatic issues, the groups' credibility may suffer. Alison Anderson (1991) reports, for example, that nine out of twelve journalists and broadcasters questioned did not trust information provided by Greenpeace. The effect may also extend to unaffiliated environmental groups that are not distinguished from the greens. If that occurs, then lobbying effectiveness suffers. Political scientists have found that credibility is critically important in legislative lobbying, where groups need to develop a reputation for providing reliable and accurate information to policy makers (Schlozman & Tierney, 1986, p. 298). Even mainstream environmental groups such as the Sierra Club expressed concern in 1995 that the term *environmental activist* had taken on a new image of "extremist," allowing opponents in Congress and other venues a plausible reason to reject their arguments, despite the findings of survey research that showed that the American public continued to favor strong environmental protection measures (Wilson, 1995). Although the reasons for such a shift in perceptions among policy makers cannot easily be identified, the actions of the greens (and some other grassroots groups) would seem to be part of the changing image of environmentalism in the mid-1990s.

The mainstream, and some grassroots, environmental groups approach media coverage in a different way than do the greens. Most have become intimately familiar with the art of the press release, and many have regular contacts with journalists whom they use to reach their publics. Some have developed exceptionally calculated information campaigns through television media, often to the dismay of both industry and public officials. One of the most dramatic was the effort by the Natural Resources Defense Council (NRDC) in 1989 to push the EPA to ban the chemical growth regulator Alar (daminozide). The NRDC used the CBS news program *Sixty Minutes* and the talents of actress Meryl Streep to present its assessment of the potential health risks of using Alar on apples. They focused on the risks to children and won a sympathetic coverage in the press. Even though both industry and the EPA sharply criticized the validity of the NRDC study, the resulting public furor and demand for Alar-free products forced growers to stop using the chemical. Eventually its manufacturer withdrew it from the market.

Environmental groups often use the media to reinforce an image of corporate greed and carelessness. They have found such a strategy especially effective in a period in which the public exhibits little confidence in business organizations in comparison with earlier decades (Lipset & Schneider, 1987). This strategy is aided by the media's preference for stories with identifiable victims and perpetrators. One of the best recent examples is the *Valdez* oil spill, which raised the usually low salience of environmental issues to new heights (Mitchell, 1990). In a case of local press coverage that was duplicated nationwide, the *Anchorage Daily News* ran photographs of oil-soaked birds five times in several weeks. In this and many other media venues, such photographs helped to evoke sympathy for these helpless victims of the spill and anger at Exxon (Daley & O'Neill, 1991). Exxon's bungling of the cleanup and its inept defense of its actions in the media did little to help its case. Although the causal association between media coverage and the activity of environmental groups is hard to prove, a good many groups seem to have used the coverage of the oil spill well, both to raise funds and to increase membership rolls. Moreover, within a year of the spill, the U.S. Congress enacted oil spill prevention legislation that had languished for years in committee without sufficient public pressure to force action (Kuntz, 1990).

Lobbying as Communication

To succeed in the kinds of advocacy activities discussed above, environmental groups and others often develop explicit communication strategies. Although the effects of communication, both content and form, on policy decisions cannot easily be determined, the relationships merit examination nonetheless.

Raymond Bauer, Ithiel Pool, and Lewis Dexter argued in their classic work *American Business and Public Policy* (1961) that the essence of lobbying is communication, communication of policy ideas and information. Any well-developed lobbying strategy will require decisions about what kinds of information and images to convey to the public and policy makers, and how best to do so. A great deal of research in fields as diverse as public opinion, communication psychology, and legislative behavior has long suggested, however, that the impact on policy makers will be indirect and constrained by policy makers' previously established positions on issues (Dexter, 1969, pp. 131–50; Key, 1963). Indeed, there is no shortage of scholarly studies that find that lobbying campaigns chiefly reinforce and mobilize members who already support an organization's positions rather than convert individuals to new positions (e.g., Bauer, Pool, & Dexter, 1972; Kingdon, 1989; Whiteman, 1985). In short, advocacy groups normally preach to the converted, but this is nevertheless an important political act.

Yet as Richard Smith has noted in regard to the U.S. Congress, "The influence of lobbyists on establishing and altering (rather than just reinforcing) the positions of members of Congress has been underestimated. By recognizing the significance of interpretations, a whole new process of influence becomes clear" (1984, p. 59). Similarly, Mark Petracca underscored the importance of language in shaping such interpretations of political reality: "Language is the medium through which we come to experience reality, and hence, politics. . . . The process by which agendas are formed involves both the construction of reality by individuals or groups and the translation of that construction into specific demands and issues with particular definition" (1992, p. 1). While the usual explanations for the success of interest groups in Congress emphasize membership size, financial resources, staff skills and motivation, technical expertise, and political skills and reputation, it is also apparent that the form and content of communications are significant variables.

Unfortunately, the impact of such communication on policy makers is difficult to study. Too many variables affect the process, and researchers cannot control for all of them in an empirical study. There is little question, however, that the relationship between advocacy groups and policy makers is important in environmental policy making, as well as other kinds. We referred to this relationship earlier as a form of democratic discourse. If this discourse were to be studied systematically, several questions could be addressed: What ideas and information are communicated? How does this communication occur? What kind of dialogue occurs between advocacy groups and policy makers? Does this dialogue make a difference, especially for policy decisions? Studies of agenda setting concentrate on the communication process, and they often emphasize the importance of policy ideas—the varied ways in which the causes and consequences of public problems are conceptualized and the available policy alternatives are evaluated. Such ideas circulate in the pertinent policy communities (in which environmental groups are now key actors), and over time they can profoundly affect the perceived legitimacy and feasibility of policy alternatives (Baumgartner & Jones, 1993; Kingdon, 1984). Students of environmental advocacy should find a rich body of theory and empirical evidence in this literature that could be applied to studies of environmental communication and its political impact.

More attention seems to have been paid to the use of communication strategies at the grassroots level than in Washington. Grassroots communication often involves the use of the condensational symbols whose power derives from the public's strong commitment to environmental values. Survey data also confirm the public's belief that environmental conditions pose significant public health risks and that environmental quality will worsen in the future (Dunlap, 1991; Mitchell, 1990). Such beliefs create opportunities for environmental groups to mount grassroots campaigns that are increasingly vital to their survival, both for membership and operating revenues. Membership attrition can be extremely high, often 20 to 30 percent per year, thus requiring a continuing effort to bring in new members, especially if the group's core issues are declining in salience. As might be expected, members recruited through direct mail tend to be only marginally committed to the environmental organizations they "join" by returning a check (Godwin, 1988, p. 65). Competition among the groups for potential members and funds should also inspire creative use of language and graphic symbols. That is, the groups should be motivated by these "market" conditions to consider carefully how they can best appeal to the public.

We are not aware of any systematic study of the design and effectiveness of such appeals, whether for fund-raising, member recruitment, or grassroots lobbying directed at public policy makers. Such a study would require a sample of the solicitations used by diverse groups, some evidence of comparative response rates (percentage of successful appeals), and some measures of effectiveness of these communication efforts for achieving organizational goals, especially changes in public policy. It would seem, however, that many of the most experienced environmental groups (and the public relations firms they employ for these purposes) develop a fairly good sense of what works, after years of experience with different letter (or television) formats and issues.

A review of a decidedly limited and unrepresentative sample of direct mail solicitations suggests that many environmental groups give careful thought to the use of symbols and images being conveyed to the public even if they cannot be sure what kind of effect these communications have on different segments of the public. For example, a Union of Concerned Scientists letter from 1992 invites readers to become "Earth citizens," helping to save the planet from the threats of global warming, environmental degradation, overconsumption, and population growth. The letter plays on emotions related to community well-being and individual responsibility. Some forty environmental groups have tried to institutionalize this kind of appeal in their Earth Share campaign, modeled on United Way fund-raising through payroll deductions. Many similar examples could be cited. A 1992 Environmental Defense Fund letter is typical. It describes the disposal of plastics in the oceans using condensational symbols: "As quickly as the oceans fill with plastics, they will empty of marine life." The issue of ocean disposal is linked emotionally to an image of victimized marine life.

Business organizations use similar strategies. The nuclear energy industry, anxious to rebuild its lost public support, combines pictures of trees with the sentence "Nuclear plants don't pollute the air." In the 1990s the industry discovered the value of appealing to an environmentally conscious audience worried about both air pollution and global warming that may result from the burning of fossil fuels.

Grassroots lobbying campaigns by both environmental and industry groups may be carefully planned, particularly when the stakes are high, though such effort is no guarantee of success. For example, in 1991 the nuclear industry initiated an elaborate and costly campaign in Nevada to persuade citizens to accept the nation's first high-level nuclear waste repository at Yucca Mountain. This was likely to be a difficult task in light of intense public fears about nuclear power and nuclear waste and overt hostility to the siting of a repository in the state. Organized by the American Nuclear Energy Council (ANEC), the campaign was predicated on the belief that a "media and public relations effort would convince a majority of the state's citizens to support construction of the site in their state." Plans called for extensive television, radio, and newspaper ads as well as the training of teams of government scientists ("truth squads") to respond to unfavorable media reports (Rosa, Dunlap, & Kraft, 1993, pp. 312–15). The impacts of the ANEC public relations effort were what one would expect. Although surveys indicated that 72 percent of Nevada residents reported having seen or heard the advertisements, the campaign failed in its primary goal of altering public attitudes. The extent of repository opposition among state residents remained unchanged (Flynn, Slovic, & Mertz, 1993).

Parallel efforts by environmental groups are common. For example, in May 1993 a coalition of environmental groups, including the Rainforest Action Network, the Ecoforestry Institute, and Save America's Forests, published a full-page advertisement in the *New York Times* in an effort to affect President Clinton's forestry plan, which was to be announced within weeks. The ad criticized what the groups called "industrial forestry" (tree farms and clear-cutting practices) as inconsistent with protection of biological diversity. The concept of ecoforestry was described and defended at length and offered as a "practical alternative." The ad attempted to affect the policy agenda through provision of new perspectives (Ecoforestry Institute, 1993).

The fundamental premise of many media campaigns of this kind, particularly those organized by industry, that provision of technical information or new perspectives will cause a shift in public attitudes, flies in the face of decades of social scientific research that demonstrates the difficulty of changing intensely held opinions in this manner. Not surprisingly, there is little evidence to suggest that pro-nuclear campaigns of the kind mounted in Nevada have been successful (Flynn, Slovic, & Mertz, 1993). One could speculate that much the same may be true for some of the environmentalists' efforts to win public support, although here, at least, the sponsoring groups usually begin with an audience sympathetic to them.

Whether initiated by industry or environmental groups, such communications are most likely to affect those already supportive of the organizations' goals. Under these conditions, the most significant short-term impact may well be the reinforcing of beliefs. Longer-term effects on public beliefs and attitudes cannot be discounted, however, and this is surely an objective of organizations sponsoring educational campaigns such as that by the Ecoforestry Institute. Even a modest short-term effect, such as stimulating a portion of the audience to take action or to donate money, may be sufficient to justify the cost of the campaign.

Environmental Advocacy and Policy in the Late 1990s

The rest of the 1990s will present environmentalists with an array of new challenges and dilemmas that flow from their success over the past two and a half decades. The mainstream groups are now a political fixture in Washington and in many state capitals, but they and other environmentalists will face difficult choices in the years ahead over organizational priorities and approaches. They are confronted by newly energized opponents, not only in the Wise Use and property rights movements but also in industry. Traditional industry critics have long attacked environmental policies as unnecessarily costly and burdensome. Republicans in both the 103rd and 104th Congresses have been highly responsive to these critical voices, as some of the legislative decisions in 1995 demonstrated. Mainstream environmentalists also have been losing ideological fervor to the greens and the emerging grassroots environmental groups. Internal disagreements over goals, strategies, and political styles—and competition for members and funds—continue to divide the environmental community. These conditions suggest a clear need to rethink how these diverse groups can contribute most effectively to the nation's political discourse on the environment.

Although the mainstream groups have suffered diminished political effectiveness over the past several years, environmentalism still needs them. Grassroots environmentalism has a certain appeal in its emphasis on citizen empowerment and democratic participation, particularly in light of new attention to issues of environmental equity or justice that affect poor and minority communities. But grassroots groups have limited capabilities for analysis and action on state, national, and international issues such as population growth, climate change, and loss of biological diversity, where the opposition to public policy proposals remains powerful. Sometimes they are not even well equipped to deal with technical issues at the local level, at least not without considerable assistance. Many grassroots organizations also tend to emphasize a negative agenda. They may organize in opposition to local activities to which they object, but they have difficulty building broad coalitions and fashioning acceptable public policies (Mazmanian & Morell, 1994).

Despite the considerable liabilities they face, some local and regional environmental groups have been at the forefront of successful efforts to promote the goal of sustainable communities. They have encouraged a broad vision of environmental sustainability and have contributed in important ways to the search for new approaches to environmental protection, such as incentive-based initiatives, public education, and public-private partnerships. The results can be seen in sustainable development efforts in cities as distinctive as Seattle, San Francisco, Los Angeles, and Chattanooga, and in promising plans for ecosystem management in the Great Lakes and the Pacific Northwest, as well as other areas of the nation.[3]

For all these reasons, environmental discourse in the late 1990s and in the early twenty-first century will necessarily involve local and regional groups. Yet it will also continue to depend on the national—and international—organizations with the capabilities and resources to carry on the varied public dialogues so essential to defining the environmental agenda and formulating policy responses. The grassroots groups and the greens are vital in another respect. They contribute significantly to maintaining and enhancing the popular base of the environmental movement, which is responsible for much of the political influence the movement has enjoyed during the past twenty-five years.

These tasks of environmental advocacy and mobilization are likely to be more difficult in the next two decades, as environmental problems become even more complex and formidable. Solutions will likely become more costly and contentious, public acceptance will be less readily achieved, and opponents will be at least as well prepared to do battle as they have been over the past few years. The sharp challenge to environmental policies in the 104th Congress have made that scenario clear.

In the process of making their case, however, environmentalists must be careful not to undercut their natural base of public support. In recent years they have faced what some analysts have termed a "compassion fatigue" that afflicts the public when it is bombarded with emotional and apocalyptic messages of repeated crises that demand emergency response and public sacrifice. Messages of this kind in grassroots lobbying campaigns may well have contributed to the new image of environmental groups as hostile to technology and industry, and even to improved standards of living (Lewis, 1992; Roszak, 1992; Schneider, 1994). It is an image that is easily exploited by opponents of environmentalist policy.

Environmental groups at all levels, then, will have to give fresh thought to their political strategies and communication efforts. They need to ask how well they will serve both their organizational goals and the larger need for effective dialogue on policy choices and environmental change in the late 1990s.

Notes

We would like to thank Scott Furlong and the anonymous referees for their willingness to read a draft of this chapter and for their helpful comments on it.

1. In the mid- to late 1980s the Group of Ten comprised the Environmental Defense Fund, Friends of the Earth, the Izaak Walton League of America, the National Audubon Society, the National Parks and Conservation Association, the National Wildlife Federation, the Natural Resources Defense Council, the Sierra Club, the Wilderness Society, and the Environmental Policy Institute.

2. This approach is often called astroturf campaigning because it is said to create the illusion of public opposition, when in fact most of the letters and telephone calls are arranged by public relations firms employed by industry for this purpose.

3. A book now being edited by Daniel Mazmanian and Michael Kraft, *Toward Sustainable Communities: Protecting the Environment through Local and Regional Strategies,* will provide extensive studies of these encouraging developments and the role of grassroots environmental groups in advancing them.

References

After 100 days, a "legacy of unfairness" or a "bolder direction"? (1995, April 9). *New York Times*, natl. ed., p. 12.

American Energy Alliance. (1993, June 26). 2,118 reasons why the Btu energy tax (or any son-of-a-Btu) is *still* a bad idea. *Congressional Quarterly Weekly Report*, pp. 1702-3. Advertisement.

Anderson, A. (1991). Source strategies and the communication of environmental affairs. *Media, Culture, and Society, 13,* pp. 459-76.

Bauer, R.A., Pool, I., & Dexter, L.A. (1972). *American business and public policy: The politics of foreign trade.* 2nd ed. Chicago: Aldine-Atherton.

Baumgartner, F.R., & Jones, B.D. (1993). *Agendas and instability in American politics.* Chicago: University of Chicago Press.

Benenson, B. (1995, June 17). GOP sets the 104th Congress on new regulatory course. *Congressional Quarterly Weekly Report,* pp. 1693-97.

Berry, J.M. (1989). *The interest group society.* 2nd ed. Glenview, Ill.: Scott, Foresman.

Bosso, C.J. (1987). *Pesticides and politics: The life cycle of a public issue.* Pittsburgh: University of Pittsburgh Press.

———. (1991). Adaptation and change in the environmental movement. In A.J. Cigler & B.A. Loomis (Eds.), *Interest group politics,* 3rd ed. (pp. 151-76). Washington, D.C.: Congressional Quarterly Press.

———. (1994). After the movement: Environmental activism in the 1990s. In Vig & Kraft, 1994, pp. 31-50.

Brinkley, J. (1993, November 1). Cultivating the grass roots to reap legislative benefits. *New York Times,* natl. ed., pp. A1, 14.

Browning, G. (1994, October 24). Zapping the Capitol. *National Journal,* pp. 2446-50.

Bryner, G.C. (1996). *Blue skies, green politics: The Clean Air Act of 1990.* 2nd ed. Washington, D.C.: Congressional Quarterly Press.

Carney, E.N. (1992, February 1). Industry plays the grass-roots card. *National Journal,* pp. 281-82.

Culhane, P.J. (1981). *Public lands politics: Interest group influence on the Forest Service and the Bureau of Land Management.* Baltimore: Johns Hopkins University Press.

Cushman, J.H., Jr. (1995, March 22). Lobbyists helped the G.O.P. in revising Clean Water Act. *New York Times,* natl. ed., p. 1.

Daley, P., & O'Neill, D. (1991). "Sad is too mild a word": Press coverage of the *Exxon Valdez* oil spill. *Journal of Communication, 41,* pp. 42-57.

Dexter, L.A. (1969). *The sociology and politics of Congress.* Chicago: Rand McNally.

Dunlap, R.E. (1987, July-August). Public opinion on the environment in the Reagan era. *Environment, 29,* pp. 6-11, 32-37.

———. (1991, October). Public opinion in the 1980s: Clear consensus, ambiguous commitment. *Environment, 33,* pp. 9-37.

Dunlap, R.E., & Mertig, A.G. (Eds.). (1992). *American environmentalism: The U.S. environmental movement, 1970-1990.* Philadelphia: Taylor & Francis.

Ecoforestry Institute. (1993, May 24). The ecoforestry alternative. *New York Times,* natl. ed., p. A7. Advertisement.

Egan, T. (1995, April 13). Industry reshapes Endangered Species Act. *New York Times,* natl. ed., p. A9.

Engelberg, S. (1995, March 31). Business leaves the lobby and sits at Congress's table. *New York Times,* natl. ed., pp. A1, 11.

Flynn, J., Slovic, P., & Mertz, C.K. (1993). The Nevada Initiative: A risk communication fiasco. *Risk Analysis, 13,* pp. 497-502.

Furlong, S.R. (1992, September). Interest group influence on regulatory policy. Paper presented at the 1992 annual meeting of the American Political Science Association, Chicago.

Godwin, R.K. (1988). *One billion dollars of influence: The direct marketing of politics.* Chatham, N.J.: Chatham House.

Greenberg, S.B. (1995, March 9). Mistaking a moment for a mandate. *New York Times,* natl. ed., p. A15.

Greenhouse, S. (1993, May 6). Manufacturers and farmers join in opposing a tax on energy. *New York Times,* natl. ed., p. A14.

Harrison, B.E. (1983). Green lobby communication: Why is it so effective? *Public Relations Journal, 39*, pp. 8-9.

Ingram, H.M., & Mann, D.E. (1989). Interest groups and environmental policy. In J.P. Lester (Ed.), *Environmental politics and policy: Theories and evidence* (pp. 135-57). Durham, N.C.: Duke University Press.

John, D. (1994). *Civic environmentalism: Alternatives to regulation in states and communities.* Washington, D.C.: Congressional Quarterly Press.

Kenworthy, T. (1995, March 27–April 2). Letting the truth fall where it may. *Washington Post National Weekly Edition,* p. 31.

Key, V.O., Jr. (1963). *Public opinion and American democracy.* New York: Alfred A. Knopf.

Kingdon, J.W. (1984). *Agendas, alternatives, and public policies.* Boston: Little, Brown.

———. (1989). *Congressmen's voting decisions.* 3rd ed. Ann Arbor: University of Michigan Press.

Kraft, M.E. (1994). Environmental gridlock: Searching for consensus in Congress. In Vig & Kraft, 1994, pp. 97-119.

———. (1995). Congress and environmental policy. In J.P. Lester (Ed.), *Environmental politics and policy: Theories and evidence,* 2nd ed. (pp. 168-205). Durham, N.C.: Duke University Press.

Kraft, M.E., & Vig, N.J. (1994). Environmental policy from the 1970s to the 1990s. In Vig & Kraft, 1994, pp. 3-29.

Kriz, M. (1990, July 28). Shades of green. *National Journal,* p. 1828.

———. (1993, April 17). A green tax? *National Journal,* pp. 917-20.

Kuntz, P. (1990, July 28). Long-delayed oil-spill measure closer to reaching president. *Congressional Quarterly Weekly Report,* pp. 2401-3.

LaMay, C.L., & Dennis, E.E. (1991). *Media and the environment.* Washington, D.C.: Island Press.

Lewis, M.W. (1992). *Green delusions: An environmentalist critique of radical environmentalism.* Durham, N.C.: Duke University Press.

Lipset, S.M., & Schneider, W. (1987). *The confidence gap: Business, labor, and government in the public mind.* Rev. ed. Baltimore: Johns Hopkins University Press.

Mazmanian, D.A., & Morell, D. (1994). The "NIMBY" syndrome: Facility siting and the failure of democratic discourse. In Vig & Kraft, 1994, pp. 233-49.

Mazur, A. (1988). Controversial technologies in the mass media. In M.E. Kraft & N.J. Vig (Eds.), *Technology and politics* (pp. 140-58). Durham, N.C.: Duke University Press.

McCloskey, M. (1992). Twenty years of change in the environmental movement: An insider's view. In Dunlap & Mertig, 1992, pp. 77-88.

Mitchell, R.C. (1984). Public opinion and environmental politics in the 1970s and 1980s. In Vig & Kraft, 1984, pp. 51-74.

———. (1990). Public opinion and the green lobby: Poised for the 1990s? In N.J. Vig & M.E. Kraft (Eds.), *Environmental policy in the 1990s* (pp. 81-99). Washington, D.C.: Congressional Quarterly Press.

———. (1991). From conservation to environmental movement: The development of the modern environmental lobbies. In M.J. Lacey (Ed.), *Government and environmental politics: Essays on historical developments since World War Two* (pp. 81-113). Baltimore: Johns Hopkins University Press.

Mitchell, R.C., Mertig, A.G., & Dunlap, R.E. (1992). Twenty years of environmental mobilization: Trends among national environmental organizations. In Dunlap & Mertig, 1992, pp. 11-26.

Natural Resources Defense Council. (1995, February). Breach of faith: How the contract's fine print undermines America's environmental success. New York: Natural Resources Defense Council.

Paehlke, R.C. (1989). *Environmentalism and the future of progressive politics.* New Haven, Conn.: Yale University Press.

Petracca, M.P. (1992). Issue definitions, agenda-building, and policymaking. *Policy Currents, 2,* pp. 1, 4.

Plotke, D. (1992). The political mobilization of business. In M.P. Petracca (Ed.), *The politics of interests: Interest groups transformed* (pp. 175-98). Boulder, Colo.: Westview Press.

Poole, W. (1992, November-December). Neither wise nor well. *Sierra,* pp. 59-61, 88-93.

Rosa, E.A., Dunlap, R.E., & Kraft, M.E. (1993). Prospects for public acceptance of a high-level nuclear waste repository in the United States: Summary and implications. In R.E. Dunlap, M.E. Kraft, & E.A. Rosa (Eds.), *Public reactions to nuclear waste: Citizens' views of repository siting* (pp. 291-324). Durham, N.C.: Duke University Press.

Roszak, T. (1992). *The voice of the Earth: An exploration of ecopsychology.* New York: Simon & Schuster.

Sabatier, P.A. (1993). Policy change over a decade or more. In P.A. Sabatier & H.C. Jenkins-Smith (Eds.), *Policy change and learning: An advocacy coalition approach* (pp. 13-39). Boulder, Colo.: Westview Press.

Schlozman, K.L., & Tierney, J.T. (1986). *Organized interests and American democracy.* New York: Harper & Row.

Schneider, K. (1991, December 23). Falling company gifts test environmentalists. *New York Times,* natl. ed., p. A8.

———. (1994, November 6). For the environment, compassion fatigue. *New York Times,* natl. ed., p. E3.

Schneier, E.V., & Gross, B. (1993). *Legislative strategy: Shaping public policy.* New York: St. Martin's Press.

Shute, N. (1995, Winter). Capitol shakeup. *Amicus Journal, 16,* pp. 18-25.

Smith, R.A. (1984). Advocacy, interpretation, and influence in the U.S. Congress. *American Political Science Review, 78,* pp. 44-63.

Smith, S.S. (1985). New patterns in decisionmaking in Congress. In E. Chubb & P.E. Peterson (Eds.), *The new direction in American politics* (pp. 203-33). Washington, D.C.: Brookings Institution.

Stocking, H., & Leonard, J.P. (1990, November-December). The greening of the press. *Columbia Journalism Review,* pp. 37-44.

U.S. Congress. (1991, November 1). *Congressional Record,* p. S15755.

U.S. Environmental Protection Agency. (1990). *Reducing risk: Setting priorities and strategies for environmental protection.* Washington, D.C.: Environmental Protection Agency, Science Advisory Board.

Vig, N.J., & Kraft, M.E. (Eds.). (1984). *Environmental policy in the 1980s: Reagan's new agenda.* Washington, D.C.: Congressional Quarterly Press.

——— (Eds.). (1994). *Environmental policy in the 1990s.* 2nd ed. Washington, D.C.: Congressional Quarterly Press.

Whiteman, D. (1985). The fate of policy analysis in congressional decision making: Three types of use in committees. *Western Political Quarterly, 38,* pp. 294-311.

Wilson, A. (1995, April). How to organize, chapter by chapter. *Planet: The Sierra Club Activist Resource,* p. 3.

Wines, M. (1993, June 14). Tax's demise illustrates first rule of lobbying: Work, work, work. *New York Times,* natl. ed., pp. A1, 11.

6

Retalking Environmental Discourses from a Feminist Perspective: The Radical Potential of Ecofeminism

Connie Bullis

Ecofeminism has, for the past two decades, evolved as an alternative radical environmental discourse. My purpose in this chapter is to illustrate the potential value of ecofeminism to environmental discourse. Because ecofeminism has evolved largely through its relationship with feminism, and feminism has emphasized the destabilization of patriarchal modes of discourse, ecofeminism is historically situated in a way that is not grounded in the modernist, patriarchal paradigm. In this essay I briefly describe ecofeminism, deep ecology, and social ecology, and then explore the radical potential of ecofeminist discourse to a continued conversation among them. Ecofeminism is used as a (plural) voice or standpoint from which to consider problems that limit the potentials of deep and social ecology to serve as radical, transformative discourses. From an ecofeminist perspective, deep ecology and social ecology are deeply rooted in the system they purport radically to critique and transform. In some ways, then, they reproduce, rather than critique and transform, the dominant paradigm.

This analysis focuses mainly on radical environmental voices as they purport to change the ways in which people of the industrialized Western world talk and live. While ecological crises are global in scope, it is commonly agreed that the West (or "first world") and its industrial-scientific development is more responsible for current crises than is the rest of the world (c.f. Capra, 1982; Merchant, 1992; Rifkin, 1992; Waring, 1988). Therefore, in examining the potentials of radical discourses for transformation, I focus on Western discourse in the late twentieth century because of its connection with the environmental crises that risk bringing about the death of life itself. It is the historically situated need to avert this end that animates the generation of and conversations among radical environmental discourses. Radical environmental discourses share the claim that reform environmentalism is inadequate to salvaging life. Treating the environment as an issue within a dominant discourse is inadequate because the dominant discourse inherently perpetuates the environmental destruction responsible for the current crisis. Instead, alternative discourses not grounded in the current dominant discourse are essential for adequate transformation.

This critique is best considered as a polemical comment in an ongoing conversation among radical environmental discourses. Such an ongoing conversation should enable a radical shift from the status quo while avoiding a transformation that results in an equally univocal future state. Instead, diversity and conversation need to continue to be incorporated into radical environmental discourse as means of continuing ongoing reflexivity and self-critique. A healthy critical conversation among various voices strengthens radical environmental discourses. Each serves to temper potential excesses of the others. In this chapter I specifically focus on the contributions of ecofeminism as it helps to identify such potential excesses of deep ecology and social ecology. Given the propensity of deep and social ecology either to subsume or to ignore ecofeminism (c.f. Biehl, 1991; Bookchin, 1994; Devall, 1988; LaChapelle, 1988), it is important for ecofeminist voices to continue to perpetuate critical conversation. Drawing on the ecological focus on diversity as it perpetuates the dynamic stability of systems, we should encourage ongoing diversity within radical environmental discourse as it perpetuates a sustained and diverse critical discourse. Therefore, in this chapter I emphasize difference over commonality.

Ecofeminism

Just as feminism is more accurately called feminisms, ecofeminism is best described as ecofeminisms, in that there are a variety of ecofeminist voices. The description I offer here is based on a synthesis of my own reading, personal experiences, and hopes and is not necessarily identical to descriptions of ecofeminism offered by others. D'Eaubonne initially coined the term *ecofeminism* in 1974 when she claimed that the male system itself (neither socialist nor capitalist) has "seized control of the soil, thus of fertility (later, industry), and of woman's womb (thus fecundity)," resulting in "this double peril, menacing and parallel: overpopulation (a glut of births) and destruction of the environment (a glut of products)" (1994, p. 177). More recently, three key ecofeminist anthologies, coupled with more academic writings in the philosophy journal *Environmental Ethics,* may be considered to constitute a "center" of ecofeminist writings. The first anthology, *Reclaim the Earth,* was edited by Caldecott and Leland in 1983. The second, *Healing the Wounds,* edited by Judith Plant, appeared in 1989, and the third, *Reweaving the World,* edited by Irene Diamond and Gloria Orenstein, was published in 1990. Since then, a number of books have emerged to continue the evolution of ecofeminism (Adams, 1993, 1994; Diamond, 1994; Gaard, 1993; Mies & Shiva, 1993; Norwood, 1993; Warren, 1994).

Ecofeminists assume that the oppressions of women, races, classes, and nonhuman nature are interconnected parts of the same dynamic. All are involved in what Karen Warren (1990) describes as a "conceptual framework" of oppression. Such frameworks explain, justify, and maintain relationships of domination and subordination. Since the same logic of domination pervades the oppressions, all are best conceived as integrally related. The logic common to the oppressions is that differences exist, differences warrant hierarchical ordering of differences, and hierarchical ordering of differences warrants subordination of those placed lower on hierarchies by those placed higher.

Historically, the womanizing of nature and the naturizing of woman have evolved as ways to oppress both. Carolyn Merchant (1980), for example, traces the language used to authorize the change from an organic paradigm to a mechanistic, scientific paradigm. Through emphasizing feminized images of nature such as "harlot" and "wild" as opposed to alternative images such as "mother," the advent of a dominating worldview enabled "subduing" and "taming" nature rather than "honoring" nature. Susan Griffin (1978) traces the associated material oppression of women. For example, she juxtaposes events associated with the historical evolution of science with the simultaneous burning of witches. Similarly, Vandana Shiva (1988) and Maria Mies (Mies & Shiva, 1993) trace an understanding of ecofeminism by illustrating the material difficulties of subsistence for women in the South (sometimes termed the "third world"). These writers illustrate the literal, material consequences of the current dominant discourse in the lives of people in both the North and the South who live closest to the environmental destruction that makes their daily subsistence a struggle. Shiva and Mies point out that poverty rates in the United States are particularly high for women and children. Moreover, working women in the West are effected by industrial hazards in disproportionate numbers as their ability to reproduce healthy children is compromised. Joni Seager (1993) summarizes: Companies typically enact policies to exclude women rather than exclude the toxins causing reproductive problems. In the South, women struggle to attain the necessities of life for their families in the face of increasing destruction of the natural systems upon which they directly depend (Shiva, 1988; Waring, 1988).

In other words, the connection of women and nature articulated in philosophical discourse by Karen Warren is intricately connected to the realities of women's lives traced by a number of ecofeminist analyses. It is important to differentiate this focus on revealing previously ignored historical trends from current cautions against essentializing connections between women and nature. Although I believe it is naive to assume that biological differences should be completely ignored, the focus here is on discourse. To reveal how language has been used to connect and oppress women and nature and to examine material examples do not constitute a claim that such a connection is a "natural" or "essential" connection. To identify alternative transformative discourses (including myths and metaphors) is neither to assume an essential connection between women and nature nor to engage in a reversal of domination that would place women at the top of a modernist hierarchy. Rather, it functions as a material, historical grounding to encourage a better understanding of how current discourse evolved and how it oppresses. Warren concludes that ending naturism should be considered as "an integral part of any feminist solidarity movement to end sexist oppression and the logic of domination which conceptually grounds it" (1990, p. 132).

Similarly, ecofeminists claim that ending sexism, racism, and classism should be an integral part of any solidarity movement to end naturism. An important and sometimes overlooked term in ecofeminism is *solidarity* (c.f. Haraway, 1989). The local and specific dynamics of these dominations are varied, as are the historical and material circumstances of oppression. Therefore, solidarity, rather than unity, is a key concern. Solidarity is consistent with perhaps the most common ecofeminist image: the world is viewed as a "web of life" (Norwood, 1993), as an alternative to hierarchy. Feminists draw on this understanding not dualistically, to advocate egalitarianism as opposed to hierarchy, but rather to advocate heterarchy, in which no single individual or principle is in command. Temporary authority structures form and dissipate as appropriate (Marshall, 1989). Hierarchy, as integrated into the logic of domination described by Warren, enables the silencing of subordinate voices—the voices of minorities and women, and all of nature's voices that are nonhuman. This logic of domination is common to capitalism (Merchant, 1980, 1989) and socialism (Bookchin, 1982; Mies, 1993).

Related to the web imagery, ecofeminism articulates a view of relationship that is not dualistic. The web of life includes all forms of life. This web does not imply a lack of differences or diversity but rather implies that differences are integrally connected. For example, ecofeminism particularly emphasizes the physical and the spiritual as legitimate diverse voices that are connected. At the same time, ecofeminism embraces rationality and emotionality. Ecofeminism is particularly concerned with two damaging dualisms: culture/nature and male/female. The web of life includes valued but different and interdependent equals rather than dominants and subordinates. Ecofeminism, then, does not argue for replacing the current elevation of (male) science and reason with the oppositional (female) alternatives such as spirit, personal experience, or body. Rather, ecofeminism advocates a partnership society (c.f. Eisler, 1987) that eschews domination. Although academic writing is one form of ecofeminist practice, ecofeminism is not restricted to written essays and books. Ecofeminism is political practice. Ecofeminists take direct action to protect and enhance webs of life. Political practice includes such nonviolent actions as those undertaken at the U.S. Pentagon to oppose nuclear buildups (King, 1989) and the more widely known Greenham Common women's peace camp in England (Cook & Kirk, 1983). The Chipko movement in India and the greenbelt work in Kenya are also examples of ecofeminist practice. Practices, such as witchcraft, that are beneficial in creating and maintaining empowering connections are also ecofeminist practice (c.f. Starhawk, 1987). Women in Los Angeles rallied to prevent a nuclear incinerator from being built near their homes (Hamilton, 1990). A common thread of much ecofeminist work is in revealing, resisting, and re-visioning ways in which bodies, work, homes, and lives are sites where the intersecting domination of "others" and nature occurs.

These and many other writings and practices reveal immense and important commonalities among the three radical environmental languages. I see no quarrel between ecofeminism, deep ecology, and social ecology regarding most of these points. All three seek to transform a language that constructs nature as a set of passive, inert "resources" for human benefit. All three focus on the problem of hierarchical ordering. Moreover, as Merchant (1994) notes, all three are positioned to go beyond critical theory's analysis of the human domination of nature toward generating alternatives. It is clear to me that all three discourses appropriate language from one another as they respond to critiques. Yet again, continued debate among them should complement their awareness of commonalities.

Social ecology and ecofeminism are particularly closely related. In many cases, critiques of deep ecology from ecofeminist and social ecologist perspectives are indistinguishable. One key distinguishing factor, though, is that social ecofeminism, unlike social ecology, is complemented by a multitude of different voices, such as those of cultural ecofeminism, liberal ecofeminism, and socialist ecofeminism (Merchant, 1992), while social ecology claims a coherence (Bookchin, 1982) that ecofeminism eschews as a guiding criterion. A second distinction is ecofeminism's grounding in material lives and the surrounding circumstances producing those particular lives. Social ecology relies on a more distanced, analytic grounding. Finally, ecofeminism's consideration of gender as figure rather than ground provides a basis for a radical critique lacking in the others. Although there are a variety of ongoing debates and critiques among the three discourses, I will not summarize all of them here (c.f. Biehl, 1991; Chase, 1991; Doubiago, 1989; Eckersley, 1992; Fox, 1989; List, 1993; Pepper, 1993; Plant, 1989; Zimmerman, 1990, 1994). Rather, I will turn to differences in emphasis that are most relevant to this critique.

Some ecofeminist discourse provides an alternative discourse that holds radical potential, not because it is more correct than social ecology or deep ecology but because it is historically timely and positioned to draw on contemporary feminist discourses in useful ways that deep ecology and social ecology ignore. Merchant (1980), in her historically grounded discussion of the machine and organic metaphors, argued that paradigms are historically situated and are useful to particular purposes within their historical moments. Merchant's analysis shows, however, that both the earlier organic metaphor that preceded the mechanistic view and the current mechanistic view were historically timely metaphors. They enabled specific human societies. For example, she describes how the mechanistic metaphor interacted with historical conditions in New England to enable the interactive evolution that has occurred (Merchant, 1989). There is widespread agreement that the modernist, mechanistic paradigm that has dominated the past several centuries is not a useful one for the near future, given both the need to live in more sustainable ways in order to avoid a predicted ecological catastrophe, and the decolonized historical moment. Change is needed. Ecofeminists have begun to draw on wider feminist and postmodern discourses that emphasize difference, relationship, and the creative potential of language needed to bring about serious change. Deep ecology and social ecology rely too heavily on patriarchal, modernist discourse that is historically linked to Western dominance.

Environmental discourses rooted in this dominating discourse are unable to provide the radical critique needed for transformation. The discourse of deep ecology can illustrate the differences with ecofeminism, because deep ecology holds greater current popularity and is more distinctly problematic from an ecofeminist view than social ecology. After examining deep ecology, I will turn to a brief exploration of social ecology.

Deep Ecology

While deep ecology is articulated slightly differently by different deep ecologists, Bill Devall and George Sessions have written an integrative book, *Deep Ecology* (1985), that quotes a variety of writings on the subject.

I have not seen challenges to the book in more recent works (c.f. Sessions, 1995). Rather, it appears to be more commonly accepted as a centerpiece of deep ecology than does Tobias's work published the same year (Tobias, 1985; c.f. LaChapelle, 1988). I am treating it here as a classic statement of deep ecology, although deep ecology frequently traces its roots to Arne Naess's explication (1973, 1989). The organization Earth First! is here viewed as a direct action branch of deep ecology. Deep ecology, like ecofeminism, criticizes mechanism and draws on an organic image in providing a critique of culture and society. As in ecofeminism, the web metaphor is central to deep ecology, as is the assumption of egalitarianism. While deep ecology mainly focuses on the need for a changed consciousness, the direct actions of Earth First! offer clear support to the claim that the emphasis on changing consciousness is integrally related to political practice. Deep ecology, then, is similar to ecofeminism in that proponents have moved beyond writing critiques and espousing scientifically grounded paradigm shifts to engaging actively in political practice (c.f. Capra, 1982).

Deep ecology is most often summarized as a philosophy comprising two "ultimate norms": "self-realization and biocentric equality." Through a deep questioning of human life, society, and nature, "as in the Western philosophical tradition of Socrates," people may come to understand and accept these ultimate norms (Devall & Sessions, 1985, p. 65). In other words, philosophical and religious questioning rather than scientific and technology-based questioning is needed to address the human role in the larger cosmos. Naess, a prominent deep ecologist, bases his ecosophy on a reconstructed Self (1989). The Self perceives and experiences self as intricately and necessarily identified with the whole of life (i.e., self-realization). Naess's philosophy encourages this deeper understanding of the Self and the interrelatedness of life. Such a deeper understanding leads to a paradigm within which altruism and morality are superseded by the deep understanding of self-interests as identified with the deep interests of all life.

Devall and Sessions similarly espouse "biocentric equality" as a broader, more holistic, more egalitarian view of life than the more common anthropocentrism, in which the natural world is objectified, with humans arrogantly assuming a dominant position. Roderick Nash's historical description (1989) embeds deep ecology within the evolution of liberalism, democracy, and religion, and deep ecology encourages a continued extending of the status of subject (which includes moral and legal rights) to all life, as rights have been increasingly extended to lower classes, to women, to ethnic groups, and to other previously objectified groups. Nash uses the abolition of slavery in the United States through the Civil War as a metaphor for environmental change. Nash's critique, according to deep ecologists, calls for the radical transformation of human society based on an ecocentric (or biocentric) set of assumptions rather than anthropocentric, humanistic assumptions. Using direct action, Earth First! seeks to facilitate the radical transformation described by deep ecology.

Social Ecology

Like ecofeminism and deep ecology, social ecology critiques the mechanistic worldview. While ecofeminism foregrounds the objectification and concomitant oppression of women, people of color, some classes of people, and nonhuman nature, and deep ecology emphasizes the domination of nature by humans, social ecology focuses first on mechanism's treatment of laborers as passive, atomized recipients of external control by elites. Like ecofeminism, it extends its concern to include various classes of people. Emphasizing the class advantage inherent in a mechanistic worldview, social ecology is motivated toward equality primarily for laboring classes. Compared with the relatively unified position of deep ecology and the relatively diverse ecofeminisms, social ecology lies between the two. There are two definable schools of social ecology, both grounded in Marxism. Social ecology may best be summarized as "an ecologically sensitive form of Marxism" (Merchant, 1992, p. 134). Social ecologists are concerned with social justice first, but their view of social justice is sensitive to ecological sustainability. Marxism provides a critique of capitalism as the root of ecological problems. Peddling the Earth for profit, according to Friedrich Engels, always creates side effects that undermine the ability to continue the profit-making activities. For example, when Italians cut the forests in the Alps to graze dairy cattle, they destroyed the watersheds on which their dairy industry depended (Merchant, 1992). Socialist ecology further emphasizes the inherent contradiction between the natural conditions on which capitalist production depends and increased production for profit. As natural resources are impoverished and workers' health is affected by pollution and toxic materials, production necessarily decreases. The resulting crises are predicted to result in increasing reliance on socialist responses.

The second school strays further from Marxism, refusing to retain the central Marxist position critiquing capitalism. Instead, anarchist social ecology focuses on hierarchy as it transcends capitalism and socialism. Murray Bookchin, the most frequently cited social ecologist, has authored several key books on the subject. Here I rely most strongly on his *Ecology of Freedom* (1982) and his *Remaking Society* (1989), as well as his recent comments in a dialogue with Earth Firster Dave Foreman (Chase 1991). Social ecology is based on an evolutionary view of nature that assumes that humans evolved through natural history into a species uniquely capable of rational choice and therefore of manipulating the natural environment. The social systems that humans have developed as a result of their unique abilities are the source of current ecological crises. In devising social systems, civilization took a wrong turn by veering away from increasing diversity and freedom within whole communities. Instead, humans have created systems of decreased diversity, freedom, and stability and increased hierarchy and domination. The transformation of human social systems, then, is intricately related to solutions to the ecological crisis. Hierarchy was historically established within human communities and expanded to human conceptions of human-nature relationships. An end to hierarchy within human social systems should lead to nonhierarchical relationships between humans and nature.

Social ecology insists that acknowledging the historical differences between humans and nature is not tantamount to assuming human superiority, but rather realistically accepting natural evolutionary facts. The natural evolutionary direction has been toward freedom. The solution to ecological crises lies in creating human communities that are consistent with this natural direction in enabling freedom. Anarchism is an integral basis of Bookchin's recommendation that decentralized communities that are ecologically sustainable and promote human freedom should be used as a model for transforming human societies. People should live in small communities where face-to-face interchanges are the locus of political decision making. Small communities may be related through a loose federation.

An Ecofeminist Critique

The radical potential of ecofeminism lies in its ability to critique several narratives within these environmental discourses. Ecofeminism, by drawing on feminism in a way that deep and social ecology have not, has the ability to provide an alternative voice. In this essay I summarize and extend this critique by examining particular instances of problematic themes and offering potential alternatives based on ecofeminism. This critique illustrates that deep ecology and, to a lesser extent, social ecology fail to escape the Western system they purport to critique. Rather, they unreflectively adopt the deepest assumptions of the system they attempt to critique and transform. In doing so, deep ecology in particular and social ecology in part unwittingly perpetuate the target of their criticisms. Ironically, this failure stems from the failure to question their own assumptions deeply. Several themes illustrate the problem: unity through transcendence and identification, biocentric egalitarianism, war as a metaphor for environmental activism, the Judeo-Christian fall from grace and the accompanying totalizing vision of returning to Eden, reliance on objectivity as a truth criterion, and dualistic thinking.

The Vision of Unity

Perhaps deep ecology's insistence on a totalizing vision of unity that is accomplished through transcendence and identification with the whole is its most troubling aspect from a feminist perspective. This vision is inherent in the espoused Self of deep ecology.

Deep ecology's description of self-realization through the transition from self to the Self is a totalizing vision of unity. Devall and Sessions invite their readers to "explore the vision that deep ecology offers" (1985, p. 7). One section is titled "Ecotopia: The Vision Defined." There the authors depict a vision of Eden. The vision is one of "the dance of unity of humans, plants, animals, the Earth" (p. ix). The vision is defined for its value as a yardstick to measure one's progress. This totalizing vision is transcendent and is gained through wider identifications with the whole web of life.

The reliance on identification with the whole has long been one of the "ultimate norms" purported by deep ecologists. The whole of the Earth, perhaps the whole of the universe, is the focus of identification. This identification transcends the individual self and becomes the Self that is all-encompassing. This identification erases boundaries (Devall & Sessions, 1985, p. 66). Unique diverse parts are valued because they are parts of the whole, contributing to the whole. The Earth and all of the parts become part of one's expanded Self. This expanded Self becomes the basis for protecting wild places and specific parts of the Earth. In protecting such places, one is protecting one's Self. Devall and Sessions cite the "Cathedral Forest Wilderness Declaration" in explaining this. The rationale for environmental advocacy is that "whatever we do to the earth we do to ourselves. If we destroy our remaining wild places, we will ultimately destroy our identity with the earth" (p. 196). In other words, "in a profound mature sense, one sees that such preservation is in one's self/Self interest" (p. 82). The true merging of individual and whole identities is evident in Devall and Sessions's "paraphrasing" of Aldo Leopold: "I dreamed I was thinking like a mountain but when I awoke I did not know if I was a man thinking like a mountain or a mountain thinking like a man" (p. 112). Similarly, Devall and Sessions quote John Rodman: "'Man' is . . . a microcosm of the cosmos who takes very personally the wounds inflicted on his/her androgynous body" (p. 194).

Devall and Sessions point out that this identification with the whole is necessary for maturity and growth. For example, in describing Gandhi's experience, they quote Gandhi as saying he served no one but himself, a Self that embraced the whole village. This, to Devall and Sessions, defines *maturity*. Deep ecology, then, relies on a standard for what is more and less mature. This creates a continuum of more and less developed beings by which to determine one's progress. Later, we will see that progress is assumed to be relative to life's business of achieving salvation. It assumes a unitary yardstick by which all people can be measured. Deep ecologists have the truth by which the yardstick is defined and others presumably can be judged. There is no room here for alternative measures. Rather, according to Paul Shepard, this model of development is both "universal" and "natural" as well as "psycho-genetic" (quoted in Devall & Sessions, 1985, p. 183). This definition of maturity through identification transcends the constraints of given historical and material situations; it is instead comprehensive and pertains to questions asked in all cultures and all ages (p. 65). Devall and Sessions point out that in order truly to identify with the whole, "we must see beyond our narrow contemporary cultural assumptions and values and the conventional wisdom of our time and place" (p. 67). They include a writing by Gary Snyder that appropriates Buddhism as a basis of "a planetary culture" (p. 251). In other words, deep ecology is not bound by culture, time, and place. Like the modernist discourse in which deep ecology is situated, it is an escape from the limits of historical time and space.

There are three main problems with this narrative. First, deep ecology's insistence on a totalizing vision that is achievable through transcendent identification is the point that social ecologists and ecofeminists have consistently challenged. For example, Bookchin points out that identifying with the whole erases differences that matter (Chase, 1991). By using this category of the whole, even as it applies only to humans, differences between the president of Exxon and a child in Harlem are erased. Similarly, by assuming similarity with all of nature, uniquely human attributes (especially reason) are ignored. Ecofeminist criticism focuses on a different problem. Kheel (1990), drawing on object relations theory (Chodorow, 1978; Dinnerstein, 1967), points out that the Self of deep ecology is a Self that, in expanding to identify with the whole of nature, simultaneously objectifies, symbolizes, and destroys particular instances of nature. Critiquing Leopold's work, she points out that when the Self becomes the whole, the whole takes precedence over the particular. Specifically, in Leopold's description of hunting, the merging of the hunter's Self with nature through the hunted animal occurs as the essence of the hunt, while the particular animal's death is an incidental by-product of the process of connecting with the whole.

Similarly, Cheney (1989) insists that the expanded Self is an extended egoism. For example, Gandhi, as described above, operated on an expanded self-interest. Plumwood (1991) similarly points out that deep ecologist Warwick Fox (1990) views the larger identifications as a way of avoiding the corrupting personal and particular (i.e., the feminine). Like the hunted animal in Leopold's scheme, particular individual bodies become objectified props for the key drama of men enacting their self-expansion.

From an ecofeminist view, deep ecologists' transcendent, expanded Self ignores the fundamental importance of relationship, difference, and materiality. Although deep ecologists claim concern with the web of life, they enact this web by assimilating organic difference into an expanded Self. This denies and destroys difference, a central point on which feminists in general, and ecofeminists in particular, have insisted. By relying on an expanded Self and on "clear principles," "laws," or "ultimate norms," deep ecologists have sacrificed the material, concrete relations through which webs are formed, maintained, and transformed. Zimmerman (1994) points out the fascistic consequences of this approach. The ecofeminist web of life insists on maintaining difference. Evolving and changing solidarity among differences, rather than unity through sameness, is the basis of this connection (Warren, 1990). Moreover, deep ecology's notion of the transcendent Self depends on an abstraction that denies rootedness in a particular historical context. Again, after paying tribute to (historically embedded) traditions from which the transcendent Self derives, deep ecologists leap to universalizing this Self, as if blatantly to deny the limitations of the historical situations they themselves acknowledge. Ecofeminists insist on a contextualized, relational, and changing understanding of self. For example, ecofeminists in the "first world" may adopt a "self" that embraces vegetarianism in honor of animals, because of their awareness of the availability of a multitude of choices of food in the first world and their desire to reduce their impact on the Earth. Yet an ecofeminist "self" in Africa works for better management of cattle, facilitating meat-eating as an aspect of regional development.

Second, while ecofeminists have not, to my knowledge, critiqued the developmental model declared by deep ecology, this critique follows from ecofeminism. Following Carol Gilligan's classic critique (1982) of Kohlberg's model of moral development, the ecofeminist position must critique the universal model of self-development adopted by deep ecologists. Like Kohlberg, deep ecologists offer a model that defines a desired goal of development, describing the self-realized or mature person as the person who adopts the view of reality espoused as a universal truth by deep ecologists and asking all people to measure themselves on that yardstick. This is modernity reproduced. An ecofeminist response must deny this universalist model of human development. Rather, selves are constructed, reconstructed, and transformed through and in particular but shifting connections. Self-development through time is necessarily accomplished in and through such particular instances. "Selves," then, are momentary constructions deeply embedded in tangible, material, historical circumstances that are diverse. A universally imposed set of assumptions of what counts as development and maturity must be criticized as patriarchal and modernist. Maturity, from a feminist view, is context-dependent and relationally defined. The "self" is fundamentally a self-in-relation. This is a self that is defined through caring, responsible relationships. The deep ecologist view of development and maturity is one Gilligan defines as masculine; it is based on achievement of universally defined stages.

Social ecology's developmental account of nature is problematic at a different analytic level. It relies on a Darwinian evolutionary view to describe and explain the historical emergence of the hierarchical, dominating systems that typify the present. Although this account clearly differentiates elites from diverse groups of "others," little acknowledgment is made of current alternative systems. Rather, the account is presented more as a universalizing narrative of the whole. By grounding this account in evolution, which claims to have the "correct" story, social ecology denies its historical situatedness in a limited paradigm. Specifically, social ecology is rooted in Darwin's account of evolution. Nature, in this view, dictates a universal progressive trend toward human autonomy, freedom, and self-realization (Zimmerman, 1994). Humans, when fully developed, are cooperative by nature. This analysis relies on the same modernist, developmental model as Kohlberg and deep ecology.

Third, deep ecologists rely on a totalizing vision as a central point in their discourse. Creating such a universal vision perpetuates the use of the gaze as a means of enforcement. Moreover, the vision depicted is essentially a return to a simplistic notion of the past. Social ecology suffers from the same problem, although its vision is one of decentralization. By acknowledging nature's ways and the wrong turn taken, humans can reconnect with nature's plan and create decentralized communities destined to enable human freedom and cooperation.

Ecofeminists, instead, insist that there is "no centralized rule in the universe, no simplicity that will save us all" (Keller, 1990, p. 263). Rather, we are historically and materially situated in a complex and diverse world in which simplistic, totalizing discourses are not adequate. Ecofeminism, instead, insists on complex and varied voices that are located in and stem from local and different circumstances. A "tapestry" of voices rather than "one picture based on a unity of voices" derives from ecofeminism (Warren, 1990). A title of a recent set of ecofeminist readings further clarifies this difference. Rather than "Re-vision the World," the title is *Reweaving the World* (Diamond & Orenstein, 1990). Weaving is a concrete, material activity rather than an abstract creation. Another example of ecofeminist critique is Garb's deconstruction of the monolithic image of the Earth from space that "has come to represent our age" (Garb, 1990, p. 265). Like the totalizing vision offered by deep ecologists, this vision of the Earth, Garb points out, is a univocal model, depicting a hegemonic version of one true story that leaves out diverse experiences and senses.

As Haraway (1991) observes, feminists need to reconstruct vision in a way that is not totalizing. For example, an ecofeminist vision might look more like a collage of multiple, particular, contesting, overlapping images (Warren, 1990). Ecofeminists use vision as a personal practice. For example, ecofeminists help each other to create individualized visions as a healing and empowering practice (c.f. Jacobs, 1990). In contrast to modernism, vision, from an ecofeminist view, is embodied. An embodied vision implies a particular locatedness. All visions, then, are positioned, perspectival, and grounded in particular times, places, and bodies.

Biocentric Egalitarianism

Western liberalism, with its emphasis on rationalism, democracy, and individual rights, undergirds the core principle of biocentric equality. Unity is attained by allowing each individual the right to develop and evolve as itself. For example, Fox asserts that "egalitarian attitude . . . allows all entities (including humans) the freedom to unfold in their own way unhindered by the various forms of human domination" (1989, p. 6). Similarly, Devall and Sessions explain: "The intuition of biocentric equality is that all things in the biosphere have an equal right to live and blossom and to reach their own individual forms of unfolding and self-realization within the larger Self-realization" (1985, p. 67). This ultimate norm enables Nash's integration of Western traditions to encourage expanding individual rights to the natural world (1989). It also enables an understanding of the natural world based on a projection of Western democratic ideals and individual rights. Social ecology places more emphasis on class differences in its analysis of problems. Like deep ecology, however, social ecology assumes that autonomy and freedom for each individual constitute the goal toward which we should strive. Again the liberal assumption of autonomous, rights-bearing individuals undergirds this goal.

Ecofeminists point out that this reliance on rights is, again, a construction of Western patriarchy. The language of individual rights ignores relationships as more fundamental than constituent parts. It projects a Western institution, democracy, and its accompanying assumption of autonomous individuals with inherent rights onto the natural world. As Cheney (1987) notes, this framework is inadequate since it derives from a system that assumes that individuals have rights, that rights inherently conflict, and that such conflict must be externally adjudicated based on principles of fairness. Some deep ecologists assume the adjudication principle to be what is best for the organic whole. Others assume such conflict is inevitable. Cheney describes a feminist sensibility involving relationships that are historicized and particular within moral communities as an alternative to a universal principle of individual rights. Similarly, Plumwood (1991) explains that a Western rationalist tradition undergirds the logic of rights.

According to Zimmerman (1994), Bookchin's emphasis on the individual is so focused that it undermines his ability to unify anarchism with ecology. In other words, the individual takes precedence over interconnectedness. In "ultimate norms," then, we find universalizing Western patriarchal assumptions. In a decolonized world, such discourses are not adequate. The feminist insistence on multiple voices and attention to the particular, partial, and historicized material relations within which all voices are embedded is more timely.

War as a Metaphor for Environmental Activism

War, in both deep ecology and social ecology, is constructed as a means of achieving peace and harmony with the planet. I need to qualify this point by emphasizing that I am drawing here mainly from material outside Devall and Sessions's work. Their work is not heavily peppered with war imagery. In the imagery of Earth First!, the active arm of deep ecology, however, activists are labeled eco-warriors. The well-known monkeywrenching book is entitled *Ecodefense* (Foreman & Haywood, 1987). Two other recent books are titled *Confessions of an Eco-warrior* (Foreman, 1991) and *Eco-warriors* (Scarce, 1990). And a recent book in which Murray Bookchin and Dave Foreman discuss their similarities and differences is titled *Defending the Earth* (Chase, 1991). In the dialogue between the two, Bookchin claims credibility by quoting Dasmann: "World War III has begun and it is waged by multinational corporations against the earth" (p. 40). He brags that he "was on the front lines . . . long ago" (p. 27). Later he suggests, "It is a shameful slander to even suggest that I do not support the struggles of Earth First! and its militants" (p. 127). Similarly, Devall and Sessions describe the global environmental crisis as "the ticking time bomb" (1985, p. 59). More consciously, Nash (1989) chooses the American Civil War as an analogy for the environmental movement. After considering several strategies, including patiently working within established institutions, appealing to the consciences of Americans, and separating from the exploiting state, he points out that military "coercion," in that comparative war, was the only successful strategy. On the other hand, Devall and Sessions, as well as Bookchin, also describe a peaceful relationship between humans and nature. Peace is offered as an alternative future that should derive from ecodefense.

Ecofeminist discourses offer alternatives. They question the means/ends relationship. A narrative and/or strategy that relies on war as a metaphor for creating a peaceful world is rooted in the same paradigm that has undergirded Western civilization for at least several centuries (c.f. Eisler, 1987). A war metaphor is related to the assumption of inherent individual rights. As such rights come into conflict, war is inevitable without an external adjudicator. The war metaphor, in addition to being violent, relies on several dualisms. For example, wars are fought by allies and enemies, winners and losers, those who are right and those who are wrong. As wars are waged, enemies typically are constructed as objects of hatred in order that they may be destroyed by whatever means are necessary. War-waging armies rely on a paradigmatic hierarchy as a way to organize. The popular cartoon *Captain Planet and the Planeteers,* for example, relies on this hierarchical way of organizing. War imagery, then, reproduces the dominant paradigm of dualistic, oppressive hierarchies in the extreme. This is the paradigm that environmental discourse needs to transform.

Marilyn Waring (1988) explains that the current world economic system has its historical roots in the need to maintain a war economy. Moreover, the economic system was developed in the West. The use of war-oriented discourse is not a way to transform such an economy. Instead, it perpetuates the status quo. Discourse that privileges peaceful subsistence offers hope. Some ecofeminist writers emphasize celebrating birth as an alternative to the war metaphors. Razak specifically asks, "Why don't we celebrate birth instead of war?" (1990, p. 172). Others (e.g., Cheney, 1987) emphasize decision making by consensus, embedded in a community of relationships of caring and friendship. Consensus is used to arrive at decisions among members of a community, whereas war is used to defend rights against aggressors. Other writers (e.g., Haraway, 1989) describe permanent struggles and contests that produce neither winners and losers nor peace.

The Fall from Grace

While deep ecologists and social ecologists have been careful to trace and acknowledge the Western intellectual traditions from which they draw, they stop short of acknowledging the limitations and boundaries that are also derived from their historical situations. Rather, they draw some disturbing conclusions that demonstrate their lack of awareness of such cultural and historical limitations. Rather than concluding that their understandings are necessarily partial and situated, they leap from a historical view to a totalizing view. Specifically, these narratives reflect their adoption of a version of the Judeo-Christian creation myth. They depict a Judeo-Christian fall from grace and a vision of a single way to return to the Garden.

Bookchin's version of the fall from grace is in his discussion of "second nature." He points out that humans have evolved from nature, and so we can be conceived as having both first and second natures. First nature is biological nature. Humans share this nature with the nonhuman community. Second nature is peculiarly human. Social ecologists cogently describe second nature: in their view, it is clearly not a basis for domination but rather an acknowledged product of evolution. They point out, however, that humans use second nature to create human social systems and that these social systems have gone awry, creating the current social ecological crisis. It is in this discussion of second nature that the fall from grace is evident. For example, Bookchin explains, "We can contribute to the diversity, fecundity, and richness of the natural world—what I call 'first nature'—more consciously, perhaps, than any other animal. Or our societies—'second nature'—

can exploit the whole web of life and tear down the planet in a rapacious, cancerous manner." He continues, explaining that this second nature "has taken a wrong turn. Society is poisoned" (Chase, 1991, p. 33). This is a Judeo-Christian narrative that assumes a single wrong moral choice was made long ago. Rather than Eve taking the apple from the serpent, society organized itself into dominating hierarchies. The result is the same, a destructive alienation from what is naturally right. We now must overcome the "wrong turn" in order to achieve salvation. Bookchin insists on small, face-to-face cooperative communities as the garden for which we must strive as we return to the correct path. Similarly, deep ecology relies on alienation and separation as the source of ecological crises. Alienation, in this view, estranges individuals from the whole web of nature. Wholeness, or reintegration with the whole through rediscovering the Self, is the route to salvation.

The descriptions offered by deep ecologists are strikingly similar to Judeo-Christian narratives. Consider the following examples presented by Devall and Sessions (1985). They call for "reenchantment" and then quote Theodore Roszak's claim that the "essential business of life" is "to work out one's salvation with diligence" (p. 10). The question is posed, "How can imperfect persons reach toward self-realization within the larger ultimate norms of self-realization and biocentric equality?" (p. 180). Much of Devall and Sessions's book is devoted to describing how to accomplish this salvation. Each life is viewed as a quest through which we discover our purpose (p. 205). Life's goal (salvation) is attained in the form of psychological maturity and identification with all life (p. 187). The book is peppered with narratives that clarify the point that maturity and identification with the whole are intricately related to a description of a return to Eden as the single means for accomplishing salvation. Nature becomes Eden. Untouched nature is constructed as the Garden. It is an escape from history. As Haraway notes, "If history is what hurts, nature is what heals." This Western narrative undergirds the uniquely Western concept of wilderness. Wilderness is constructed as timeless, separate, and pure, "untrammeled by man." This construction of nature hides the discursive fields in which it is constructed and contested through material practices. Rather, this is nature "resurrected as a dream of time before the fall" (Haraway, 1989, p. 194).

This concept is often discussed as Wilderness, a primary aim of deep ecology and the vision that animates the redemptive activity of preserving wilderness, a central focus of Earth First! For example, Devall and Sessions point out that people need to remember what was once known (1985, p. 80) or to "reawaken" an Earth wisdom (p. ix). This salvation involves "reinhabiting the land" (p. 84), learning to reenter the first world of nature (p. 152). They describe Paul Shepard's writings, showing all that there is to learn from primal cultures, which foster "the universal natural psycho-genetic development for humans" (p. 183). In contrast, industrial cultures encourage people to become "stuck" at a developmental stage of adolescence (p. 185). Separation, or the fall from grace, then, is a given condition of humans.

To reenter the Garden, humans must develop psychologically. This means that they must learn to identify with the whole of life. The way to accomplish this is through self-understanding. Since the Self is linked to the whole of life, self-understanding is a route to holistic identification. Earlier, purer cultures provide some examples because they were closer to nature than modern cultures. A variety of rituals to reconnect with the Earth are offered (c.f. LaChapelle, 1978, 1988). Through this connection, redemption is achieved. Naess thus sounds more like an evangelical preacher than a philosopher when he claims that people who search "will necessarily come to the conclusion that . . . " (quoted in Devall & Sessions, 1985, 11).

Edwin Pister's essay (1987) and lecture (1990) on type A and type B fisheries managers follow Leopold's earlier vision of salvation through realization. Leopold (1949) came to realize that he had done something wrong as he watched the eyes of a dying wolf. Pister describes his experience of coming to realize that fish are not merely a commodity to be grown for humans but are integral members of ecosystems as he observed Californians crowding fishing areas as if they were attending an entertaining event. Both Leopold and Pister reported an experience of discovering an absolute truth. Each then became evangelical in his need to convince others to share these truths. In other words, deep ecologists are saved, and others who listen to the saved and follow their ways may also achieve salvation. This is salvation through an absolute truth, which enables people to reenter the Garden.

Ecofeminists are concerned with this view in three ways. First, the story of the fall from grace is rooted in Western patriarchy. The authors of deep ecological texts are so thoroughly grounded in this myth that they are not able to acknowledge it, although they are so careful to acknowledge intellectual backgrounds to their points. Because of this unreflective rooting, deep ecologists repeatedly appropriate other cultural traditions as resources in the service of their totalizing discourse. For example, Eastern traditions such as Buddhism are misappropriated for use in this (Western) project. Not only does this appropriation falsely universalize the Western, patriarchal creation myth by obscuring difference, but it also masks the ethnocentricity of the deep ecologists, who believe they are treating Eastern traditions adequately. Asquith explains that a Buddhist view of unity is not an egalitarian, but a hierarchical, view of unity, which assumes domination and subordination (cited in Haraway, 1989, pp. 146–47). Similarly, primal cultures are appropriated by deep ecologists as resources for Western man's quest for salvation. I am concerned not so much that such appropriations occur but that they are not perceived as cultural appropriation. Rather, they are treated as truth.

Keller points out that, although there is value in drawing on Christian texts, "none of these ancient texts come free of their own sexism and nationalism" (1990, p. 261). It is this point that deep ecologists ignore. Sontheimer (1988) points out that women are commonly blamed for ecological problems. The narrative of the fall from grace clearly carries this misogynist implication.

Feminism, on the other hand, by focusing on material relationships and women's experiences, did not set up an abstract system in which those who disagree can be labeled immature along the road to salvation or shallow rather than deep. Instead, feminists launched a search for the experiences that would define women.

Although this quest turned out to be naive, because women's experiences differ as the specific conditions under which their experiences are shaped differ, when the search for "the defining female experience" was discovered to be flawed, feminism transformed itself into polyphonous feminisms. Feminists have subsequently adopted a "permanent refusal" of "master narratives" (Haraway, 1989, p. 286).

Also, the narrative of the fall from grace labels the entire human species as "guilty" by projecting a historically situated pattern onto a universal template. Ecofeminists, by drawing on feminism, refuse this universalizing move. Many marginalized people have not fallen from grace because their awareness and conceptions have not been so shaped by patriarchal categories. Historically marginalized groups, then, have been protected from some of the "crippling effects patriarchy has had on others" (Zimmerman, 1990, p. 143). This is particularly true of some women in the "third world." Ecofeminists refuse to inscribe all people with the guilt and separation of patriarchs. Most obviously, some third world victims are not guilty. For example, Waring (1988) describes Tendai, a girl who lives in Zimbabwe and works seventeen-hour days caring for the subsistence needs of her family by fetching water and wood, cooking, cleaning, and so on. Tendai is not guilty of alienation from the Earth, nor is she guilty of overconsumption, charges deep ecologists make about people in general. She has no time or energy to engage in the abstract quests for the Garden encouraged by deep ecologists. Similarly, patriarchal separation should not be projected onto the Chipko women in India who confront the international economic system in order to subsist (Shiva, 1988). Similarly, Dankelman and Davidson (1988) document many and varied instances of oppressions and resistances of women in the third world, while Vandana Shiva (1994) documents issues affecting many locations.

Similarly, ecofeminists are articulating the voices of marginalized Western people who are not alienated and do not need salvation. My depiction of Margaret, a wildlife activist, is one such description (Bullis, 1990). Nelson (1990) describes women's voices articulating women's health problems. Hamilton (1990) tells of Charlotte Bullok, a Los Angeles resident who worked to stop an incinerator that threatened her neighborhood. Ecofeminists claim that the alienation, or fall, which deep ecologists assume is universal, is a peculiarly Western male experience (Kheel, 1990). Ecofeminist writings treat multiple positions and particular experiences as figures in their discourses (e.g., Diamond & Orenstein, 1990; Plant, 1989).

Finally, Alison Jolly, in examining the cultural contexts of scientific studies of primates, has discovered that "the wild as a separate place, nature as a separate entity 'to go back to,' is the cumbersome Western concept" (quoted in Haraway, 1989, p. 273). Haraway emphasizes several alternative discourses. An Indian view has developed historically as an interactive paradigm. Feminists might consider a view in which natural status depends on the quality of relationship rather than noninterference. A Japanese discourse has historically treated nature as art. Some ecofeminist voices have treated nature and self as body (e.g., Griffin, 1978; Krall, 1994). This ecofeminist self acknowledges the self as a particular bodily instance of nature rather than as transcendent or primarily rational. At a historical moment when transcendence has been overvalued in Western evolution and bodies are in peril, it is appropriate to revalue the body. Environmental discourse needs to participate in continuously nonpermanent reconstructions of social relations and nature in a decolonized world.

The Centrality of Objectivity

Although deep ecologists and social ecologists claim to critique society fundamentally, objectivity is yet another category that they maintain as a true value. Bookchin (1982, 1989) offers his account of human development as an objectively "true" account. Biehl's recent critique of ecofeminism (1991) is based largely on ecofeminism's valuing of multiple, different voices and its lack of an objectively validated account of human history. In other words, this critical voice relies on its claim to a modernist objectivity as its claim to credibility.

Objectivity, to deep ecologists, is analogous to a "real" understanding. The goal is to be more objective than "subjective" economics, the "shallow science." John Muir is quoted for his quest "to find the law that governs the relations subsisting between humans and Nature" (Devall & Sessions, 1985, p. 111). This is an objectivity that is based on but better than traditional Western science. Such an objectivity allows people to discover and then act on "clear principles." Current land managers are encouraged to employ objective ecological criteria rather than subjective public opinion (p. 154).

It appears that deep ecologists' objectivity involves the same return to nature we have seen as the salvation story. For example, Devall and Sessions quote Robinson Jeffers: "'It is time for us to kiss the earth again.' He clearly rejected anthropocentric subjectivism, or what he called human solipsism, and strove instead to objective truth" (1985, p. 101). Devall and Sessions laud D.H. Lawrence, Robinson Jeffers, John Muir, and Gary Snyder for being "more objective" than some of the Romantic and Transcendentalist writers. These writers better suggest getting in touch with the land itself. They help us better to understand the early American primal peoples who "lived in harmony with the land" and provide a vision of how we might "reinhabit the land based upon the 'old ways'" (p. 84). Muir searched for "the law" by going to nature directly. Similarly, social ecology's vision of anarchism is grounded in the assumption that nature inherently (i.e., objectively) trends toward natural cooperation. This is offered as an "objective" understanding that humans simply need to acknowledge. Moreover, the analysis proceeds from the perspective of the rational, distanced observer who can more objectively understand.

From an ecofeminist perspective, such a reliance on timeless objectivity as a category is a problem for both social ecology and deep ecology. First, such claims to objectivity are denounced as naive in their blindness to their own positionality. For example, Warren asserts that "a feminist ethic makes no attempt to provide an 'objective' point of view, since it assumes that in contemporary culture, there really is no such point of view" (1990, p. 140). It is again a patriarchal concept, central to the system that social ecology and deep ecology claim to reject. By denying the partiality of knowledge, the claim to objectivity is used to privilege some voices over others.

Second, some feminists argue that objectivity should be reclaimed in new, transformed ways (Haraway, 1991; Harding, 1991). Aspiring to a more objective, more real, more transcendent objectivity through a return to nature in order to discover nature's laws, however, is not a transformation. A reclaimed objectivity is likely to include reflexivity and/or to decenter objectivity.

Dualisms

Deep ecology relies on a series of dualisms. First, anthropocentrism/biocentrism is a basis for deep ecology. Anthropocentrism is devalued as an assumption that nature exists solely for human use, while biocentrism is valued as a "correct" assumption because it values the natural world for itself. Ecofeminism does not make this distinction. The machine/organism worldviews form another dualism. The mechanistic worldview is wrong, while an organic worldview is right. Ecofeminism, too, calls on this distinction but considers an organic view for its appropriateness in the historical moment rather than for its inherent correctness. The third dualism is shallow/deep. Shallow is devalued, while deep is valued. Ecofeminism does not address this dualism. Present/past, West/East, and industrial culture/nature also operate as dualisms. As deep ecology is explained, these dualisms operate together, with biocentrism, organic, deep, past, East, and nature as one end and anthropocentrism, mechanistic, shallow, present, West, and industrial culture at the other. Deep ecology can be conceived as a critique based on these dualisms. It is a discourse that articulates dualisms, constructs one side as negative, and then elevates the positive. It is based, then, on reversals.

Social ecology has long eschewed a reliance on dualisms. Its theory of first and second nature, however, is vulnerable to being treated as a dualism. More clearly, social ecology is based on a reversal of common interpretations of Darwin's work. Specifically, it claims that competition has been emphasized instead of cooperation. Social ecology, then, elevates cooperation over competition so that competition becomes erased in the vision of anarchist community.

A critique based on reversals, according to Haraway (1989), is not a strong one. Rather, the destabilizing of narratives or categories offers a more convincing critique. Ecofeminism, by incorporating gender more centrally, is able to rely on intersections of multiple axes rather than on parallel dualisms. It potentially, then, should provide a stronger critique. Ecofeminist discourse has provided three metaphors that should offer alternatives to some key dualisms. Haraway (1989, 1991) illustrates that the nature/human boundaries have been blurred. Similarly, the organism/machine boundaries have also blurred. A more situationally, historically timely metaphor for the late twentieth century is the cyborg. Haraway (1991) introduces the cyborg as a combination of human, animal, machine, and cybernetic system. The cyborg indicates a contemporary overlap and interdependence among categories that have been treated as distinct. The cyborg denies nature, human, and machine as separate, distinct categories.

The cyborg should operate as a basis for alternative possibilities for environmental discourses. For example, Devall and Sessions offer advice that includes land-based activities. People should fish. Children should be sent on backpacking trips rather than to computer camps. Embedded in this advice is the logic of dualism. The past and nature are elevated as opposed to the present and technology. But what is ignored is the context of fishing in the present. Fishing involves technology, people, fish, and interactions among them. Fishing is based on intricate technological management, which is accomplished by a variety of experts and technologies. Fish habitat is manipulated and sometimes created by experts who determine stream flows and manipulate pools as well as create reservoirs and dams through technology and computer systems. Populations are engineered and manipulated by technical experts through poisoning of undesirable species, managed breeding of desirable species, stocking of lakes and streams, setting rules for where, how, and when people may fish, and managing natural interferences (such as shooting birds that eat young fish). On the other hand, the computer can be used to model the dynamic complexities of nature, in order that they might be better understood, so that industrial society might better care for nature (c.f. Botkin, 1990). The cyborg describes these kinds of interrelationships and enables environmental discourse that does not simplistically evade them. In a decolonized world in which the futures of particular instances of life are integrally related with not only other instances of life but also with technology and with a wide variety of cultural variations, environmental discourses need to adopt such timely reinventions.

A second metaphor is the Goddess imagery emphasized by some ecofeminists (Christ, 1982; Eisler, 1987; Sjoo & Mor, 1987; Starhawk, 1979). While some ecofeminists disavow such imagery in favor of more analytic approaches (c.f. Buege, 1994), others claim that the mistreatment of women and nature is not confined to structure, but rather is also accomplished through symbol use (Radford Reuther, 1993). Therefore, such imagery is central to the project of creating a transformative discourse (Yanni, 1993). This imagery is not a simple reversal that replaces male God imagery but rather a questioning of our dualistic category system of understanding the world and a working toward a reconstruction of our understanding of cosmology in a way that fosters an "ethics of awe" (Diamond, 1994, p. 53). Diamond, for example, reports the use of Hindu Goddess imagery to further a campaign for community survival that reconstitutes the managerial mentality of development.

Third, Krall (1994) has written of the ecotone, or the edge or marginal locations, as a rich metaphor for understanding. This encourages a view from locations where shifting borders or boundaries mutually evolve and change the landscape. From this perspective, she points out that deep ecology is clearly ecological in privileging nature, while social ecology is anthropological in privileging culture. She asserts that ecofeminism draws on the value of a position between nature and culture to avoid privileging one or the other. Instead, from this marginal position, ecofeminist analyses focus on the dynamic evolving relationships as they constantly reshape and transform both nature and culture (Merchant, 1980, 1989; Shiva, 1988).

Ecofeminist analyses of problems of human population are one example of the kinds of studies that can stem from this marginal position. Here I turn to a comparison of population analyses from the perspectives of deep ecology, social ecology, and ecofeminism.

Human Population

Human population is commonly considered as perhaps the most important eco-logical crisis. Both deep ecology and ecofeminism have considered it a central concern. Reducing the human population is one of eight points on the platform of deep ecology (Devall & Sessions, 1985). As noted earlier, overpopulation was central to the original 1974 formulation of ecofeminism (d'Eaubonne, 1994). Radical discourses, however, differ in their analyses of the population problem.

Deep ecology has centered on human population as a "cancer" on the Earth. Its analysis has evolved from Malthus's original claims that the Earth's carrying capacity is limited and as the human population grows, it outstrips the Earth's ability to provide food for the population. Deep ecology has relied on the slightly broader notion of carrying capacity, insisting that human population growth has outstripped carrying capacity. Dave Foreman's comments in an Earth First! news-letter that AIDS and third world famine are "necessary solutions" to the popu-lation problem illustrate the difficulties with this analysis. As deep ecology labels concern with social systems as anthropocentric, it is unable to examine social dy-namics of population. As both Bookchin (1994) and ecofeminist Cuomo (1994) note, such a unidimensional analysis is inadequate.

Both social ecology and ecofeminism have pointed out alarming similarities between deep ecology's premises and analysis of population, and nearly identical claims made by influential German Nazis (Bookchin, 1994; Zimmerman, 1994). The fascist implications of deep ecology surface in its analysis of population prob-lems. Bookchin has been outspoken in his criticism of the Malthusian tones of deep ecology's approach to population. He condemns this simplistic analysis as a "numbers game" (1994, p. 32) that ignores the complexity of the problem. Instead, he insists on understanding the problem's class and social roots. He locates his analysis, then, in a critique of hierarchy as manifested in both money economies and statist societies. He focuses on class differences to point out that all humans are not equally guilty and that solutions must stem from creating human freedom. Instead of adopting a position that blames victims by erasing social analyses, Bookchin proposes to solve the population problem through designing institu-tions to comply with humanity's natural capacity for cooperation. For example, famine in the Sudan, rather than being a simple demographic problem of humans exceeding the Earth's carrying capacity, is a consequence of historical exploitation by the British and more recent exploitation by the World Bank. Both coerced a large-scale agricultural change from growing food for subsistence to growing cotton for export. Such class-based patterns of domination and exploitation are the locus of population problems. As cultures are broken down and life expectan-cies decrease, people reproduce at higher rates. In arguing against the simplicity and resulting brutality of deep ecology, Bookchin reaffirms his commitment to reason as the most important voice. His solution includes encouraging humanity's rational voice rather than erasing humanity's unique role. He cites declines in some regional birth rates to argue that population is not a universal problem. When people are situated in conditions of more rather than less autonomy and freedom, they reduce their numbers.

Ecofeminist analyses do not quarrel with Bookchin so much as they focus on gender as more central. While Bookchin includes gender in his assumption that women's liberation has helped women to aspire to be "more than reproductive factories" (1994, p. 44), his treatment of gender is overly simplistic. He adopts the patriarchal assumption that reproduction, the body, and the physical are subordinate to the transcendent, the rational, and the mind. A serious treatment of gender adds complexity.

Overpopulation, from an ecofeminist view, emerged as a result of modernity's quest for control. Therefore, deep ecology's simplistic reversal of assuming natural control is inadequate. Instead, a more serious analysis of control is needed. Diamond (1994) points out that, by grounding an analysis in women's lives, we understand the issue in a more complex way. For example, while some women are forced to have children, others, particularly in the South (or third world) are forced into sterility. Cuomo (1994) points out several factors that ecofeminist analyses of population include, only one of which social ecology notes. Racism and classism, also considered by social ecology, must be viewed as contributing to population problems, as many impoverished people of the South need to have large families to assure their survival. Cuomo's analysis, however, is distinct from social ecology in its emphasis on several additional gender-related concerns. First, sexism is central to an understanding of population: women in many societies are unable to refuse men's demands for heterosex, as the glorification of male sexual prowess disempowers women. Similarly, definitions of women that are focused on motherhood serve to oppress women and encourage continued motherhood when alternative identity choices are limited. Cultural values, such as the nuclear family in Western culture and prohibitions of some birth control practices in some religions, regulate reproductive practices in ways that must be understood in order for population problems to be fully analyzed. Moreover, sexuality, in a number of cultures, takes on meaning for people that is severed from the population consequences for population. Another concern from an ecofeminist perspective is the alienation of women from their own bodies, as professionals have usurped control of women's health.

Diamond (1994) grounds her critique of the technology of reproductive control in the broader narrative of control. The image of a population "out of control" is a modern image, reproducing the narrative of control. As both Cuomo and Diamond point out, however, it is control that has produced the population crisis. An understanding of these oppressive gender dynamics, including the assumption that transcendence is valued over birth, is needed to address the population problem.

Ecofeminism: Transformation through Destabilization

Ecofeminist discourses hold radical potential that is different from dominant radical environmental discourses. As I have illustrated throughout this ecofeminist-based critique, radical environmental discourses that seek to transform modern Western discourse radically are themselves instances of the modern Western discourse they seek to transform. Discourses that unreflectively accept rather than critique the central modernist themes, such as the expanded Self, rational rights, war, the fall from grace, salvation through self-development, nature as separate, objectivity, and primary dualisms, are unable to begin alternative patterns. Rather, either they reproduce and extend the discourse they seek to problematize, as in the cases of rational rights, the self, the Western Judeo-Christian tradition, and objectivity, or they rely on a critique based on reversals and opposites, as in the case of the dualistic views of war and peace, modern society and primal cultures, technology and nature, West and East, and so on.

Ecofeminism, by considering gender and the feminist focus on "other" and difference as a central axis, has the potential to destabilize and repattern environmental discourse. This radical potential does not derive solely from including the category "women" as one more division of oppressed others and then elevating women, nature, and the wide variety of the oppressed to equal or superior status. Rather, by considering gender as central, ecofeminism is able to appropriate a wide variety of feminist and womanist discourses that have developed in diverse material and historical circumstances.

Ecofeminist discourse should make it more difficult to accept uncritically and to reproduce an environmental discourse based on universalizing, colonizing, Western themes such as those of deep ecology and, to some extent, social ecology. Ecofeminism has the potential to destabilize modern discourse by starting from different, feminist, narratives. Some of these alternatives are noted herein. Ecofeminist alternatives are based on feminist discourses that dominant "radical" environmental discourses have ignored. By seeking alternative starting points and patterns for discourses themselves, rather than reproducing extant discourses by extending them or reversing them, by embracing difference rather than assimilating difference, by treating relations between and among entities as figure rather than ground, by attending to the particular and local as well as the whole rather than privileging the whole at the expense of the particular, by seeking alternatives to dualisms rather than shifting from one pole to the other or transcending the dualisms, ecofeminism provides an alternative voice in radical environmental discourse.

References

Adams, C.J. (Ed.). (1993). *Ecofeminism and the sacred.* New York: Continuum.
———. (1994). *Neither man nor beast: Feminism and the defense of animals.* New York: Continuum.
Biehl, J. (1991). *Rethinking ecofeminist politics.* Boston: South End Press.
Bookchin, M. (1982). *The ecology of freedom.* Palo Alto, Calif.: Cheshire Books.
———. (1987). *The modern crisis.* 2nd ed., rev. Montreal: Black Rose Books.

————. (1989). *Remaking society.* Montreal: Black Rose Books.

————. (1994). *Which way for the ecology movement? Essays by Murray Bookchin.* Edinburgh: AK Press.

Botkin, D. (1990). *Discordant harmonies.* New York: Oxford University Press.

Buege, D.J. (1994). Rethinking again: A defense of ecofeminist philosophy. In Warren, 1994, pp. 42-63.

Bullis, C. (1990, November). Ecofeminism: An alternative approach to organizational communication. Paper presented at the annual meeting of the Speech Communication Association, Chicago.

Caldecott, L., & Leland, S. (Eds.). (1983). *Reclaim the Earth: Women speak out for life on Earth.* London: Women's Press.

Capra, F. (1982). *The turning point.* New York: Bantam Books.

Chase, S. (Ed.). (1991). *Defending the Earth: A dialogue between Murray Bookchin and Dave Foreman.* Boston: South End Press.

Cheney, J. (1987). Eco-feminism and deep ecology. *Environmental Ethics, 9,* pp. 115-45.

————. (1989). The neo-stoicism of radical environmentalism. *Environmental Ethics, 11,* pp. 293-325.

Chodorow, N. (1978). *The reproduction of mothering.* Berkeley: University of California Press.

Christ, C. (1982). Why women need the Goddess: Phenomenological, psychological, and political reasons. In C. Spretnak (Ed.), *The politics of women's spirituality: Essays on the rise of spiritual power within the feminist movement* (pp. 71-86). New York: Doubleday.

Cook, A., & Kirk, G. (1983). *Greenham women everywhere: Dreams, ideas and actions from the women's peace movement.* London: Pluto Press.

Cuomo, C. (1994). Ecofeminism, deep ecology, and human population. In Warren, 1994, pp. 88-105.

Dankelman, I., & Davidson, J. (1988). *Women and environment in the third world: Alliance for the future.* London: Earthscan.

d'Eaubonne, F. (1994). The time for ecofeminism. In Merchant, 1994, pp. 174-97.

Devall, B. (1988). *Simple in means, rich in ends: Practicing deep ecology.* Salt Lake City: Peregrine Smith.

Devall, B., & Sessions, G. (1985). *Deep ecology.* Salt Lake City: Gibbs Smith.

Diamond, I. (1994). *Fertile ground: Women, Earth, and the limits of control.* Boston: Beacon Press.

Diamond, I., & Orenstein, G.F. (Eds.). (1990). *Reweaving the world: The emergence of ecofeminism.* San Francisco: Sierra Club Books.

Dinnerstein, D. (1967). *The mermaid and the minotaur: Sexual arrangements and human malaise.* New York: Harper & Row.

Doubiago, . (1989). Mama coyote talks to the boys. In Plant, 1989, pp. 40-44.

Eckersley, R. (1992). *Environmentalism and political theory: Toward an ecocentric approach.* Albany: State University of New York Press.

Eisler, R. (1987). *The chalice and the blade.* San Francisco: Harper & Row.

Foreman, D. (1991). *Confessions of an eco-warrior.* New York: New Harmony.

Foreman, D., & Haywood, B. (Eds.). (1987). *Ecodefense: A field guide to monkeywrenching.* 2nd ed. Tucson, Ariz.: Ned Ludd.

Fox, W. (1989). The deep ecology–ecofeminism debate and its parallels. *Environmental Ethics, 11,* pp. 5-25.

————. (1990). *Towards a transpersonal ecology: Developing new foundations for environmentalism.* Boston: Shambhala.

Gaard, G. (Ed.). (1993). *Ecofeminism: Women, animals, nature.* Philadelphia: Temple University Press.

Garb, Y.J. (1990). Perspective or escape? Ecofeminist musings on contemporary Earth imagery. In Diamond & Orenstein, 1990, pp. 264-78.

Gilligan, C. (1982). *In a different voice: Psychological theory and women's development.* Cambridge, Mass.: Harvard University Press.

Griffin, S. (1978). *Woman and nature: The roaring inside her.* New York: Harper & Row.

Hamilton, C. (1990). Women, home, and community: The struggle in an urban environment. In Diamond & Orenstein, 1990, pp. 215-22.

Haraway, D. (1989). *Primate visions: Gender, race, and nature in the world of modern science.* New York: Routledge.

———. (1991). *Simians, cyborgs, and women.* New York: Routledge.

Harding, S. (1991). *Whose science? Whose knowledge?* Ithaca, N.Y.: Cornell University Press.

Jacobs, J.L. (1990). Women, ritual, and power. *Frontiers: A Journal of Women's Studies, 11,* 2-3, pp. 39-44.

Keller, C. (1990). Women against wasting the world: Notes on eschatology and ecology. In Diamond & Orenstein, 1990, pp. 249-63.

Kheel, M. (1990). Ecofeminism and deep ecology: Reflections on identity and difference. In Diamond & Orenstein, 1990, pp. 128-37.

King, Y. (1989). The ecology of feminism and the feminism of ecology. In Plant, 1989, pp. 18-28.

Krall, F.R. (1994). *Ecotone: Wayfaring on the margins.* Albany: State University of New York Press.

LaChapelle, D. (1978). *Earth wisdom.* Los Angeles: Guild of Tutors Press.

———. (1988). *Sacred land, sacred sex: Rapture of the deep.* Silverton, Colo.: Finn Hill Arts.

Leopold, A. (1949). *A Sand County almanac.* New York: Oxford University Press.

List, P.C. (Ed.). (1993). *Radical environmentalism: Philosophy and tactics.* Belmont, Calif.: Wadsworth Publishing Co.

Marshall, J. (1989). Re-visioning career concepts: A feminist invitation. In M. Arthur, D. Hall, & B. Lawrence (Eds.), *Handbook of career theory* (pp. 275-91). Cambridge: Cambridge University Press.

Merchant, C. (1980). *The death of nature: Women, ecology, and the scientific revolution.* New York: Harper & Row.

———. (1989). *Ecological revolutions: Nature, gender, and science in New England.* Chapel Hill: University of North Carolina Press.

———. (1992). *Radical ecology: The search for a livable world.* New York: Routledge.

——— (Ed.). (1994). *Ecology.* Atlantic Highlands, N.J.: Humanities Press.

Mies, M. (1993). The need for a new vision: The subsistence perspective. In Mies & Shiva, 1993, pp. 297-322.

Mies, M., & Shiva, V. (Eds.). (1993). *Ecofeminism.* London: Zed Books.

Naess, A. (1973). The shallow and the deep, long-range ecology movement. *Inquiry, 16,* pp. 95-100.

———. (1989). *Ecology, community and lifestyle: Outline of an ecosophy.* Cambridge: Cambridge University Press.

Nash, R.F. (1989). *The rights of nature.* Madison: University of Wisconsin Press.

Nelson, L. (1990). The place of women in polluted places. In Diamond & Orenstein, 1990, pp. 173-88.

Norwood, V. (1993). *Made from this Earth: American women and nature.* Chapel Hill: University of North Carolina Press.

Pepper, D. (1993). *Eco-socialism: From deep ecology to social justice.* London: Routledge.

Pister, E.P. (1987). A pilgrim's progress from group A to group B. In J.B. Callicott (Ed.), *Companion to "A Sand County almanac": Interpretive and critical essays* (pp. 221-32). Madison: University of Wisconsin Press.

———. (1990, November). Type A and type B managers. Lecture at the Utah Wildlife Symposium, Salt Lake City.

Plant, J. (Ed.). (1989). *Healing the wounds: The promise of ecofeminism.* Philadelphia: New Society Publishers.

Plumwood, V. (1991). Nature, self, and gender: Feminism, environmental philosophy, and the critique of rationalism. *Hypatia, 6,* pp. 3-27.

Radford Reuther, R. (1993). *Sexism and God-talk: Toward a feminist theology.* 10th anniversary ed. Boston: Beacon Press.

Razak, A. (1990). Toward a womanist analysis of birth. In Diamond & Orenstein, 1990, pp. 165-72.

Rifkin, J. (1992). *Biosphere politics: A cultural odyssey from the Middle Ages to the New Age.* San Francisco: HarperCollins.

Scarce, R. (1990). *Eco-warriors.* Chicago: Noble Press.

Seager, J. (1993). *Earth follies: Coming to feminist terms with the global environmental crisis.* New York: Routledge.

Sessions, G. (Ed.). (1995). *Deep ecology for the 21st century.* Boston: Shambhala.

Shiva, V. (1988). *Staying alive.* London: Zed Books.

———— (Ed.). (1994). *Close to home: Women reconnect ecology, health, and development worldwide.* Philadelphia: New Society Publishers.

Sjoo, M., & Mor, B. (1987). *The great cosmic Mother: Rediscovering the religion of the Earth.* San Francisco: Harper & Row.

Sontheimer, S. (Ed.). (1988). *Women and the environment: A reader.* New York: Monthly Review Press.

Starhawk. (1979). *The spiral dance: A rebirth of the ancient religion of the Great Goddess.* San Francisco: Harper & Row.

————. (1987). *Truth or dare: Encounters with power, authority, and mystery.* San Francisco: Harper & Row.

Tobias, M. (Ed.). (1985). *Deep ecology.* San Diego: Avant Books.

Waring, M. (1988). *If women counted: A new feminist economics.* New York: Harper & Row.

Warren, K. (1990). The power and the promise of ecological feminism. *Environmental Ethics, 12,* pp. 125-44.

———— (Ed.). (1994). *Ecological feminism.* New York: Routledge.

Yanni, D. (1993). Ecofeminism as transformative discourse. In J.G. Cantrill & M.J. Killingsworth (Eds.), *Proceedings of the conference on communication and our environment* (pp. 294-309). Big Sky, Mont.

Zimmerman, M.E. (1990). Deep ecology and ecofeminism: The emerging dialogue. In Diamond & Orenstein, 1990, pp. 138-54.

————. (1994). *Contesting Earth's future: Radical ecology and postmodernity.* Berkeley: University of California Press.

Part II

Case Studies in
Environmental Communication

7

"What to Do with the Mountain People?": The Darker Side of the Successful Campaign to Establish the Great Smoky Mountains National Park

Bruce J. Weaver

In an interview almost sixty-five years later, Zenith Whaley remembered that in 1925

> he and his young classmates stood outside Greenbrier School and watched a small plane circle the Great Smoky Mountains. They stared as it seemed to stop in mid-air for several seconds, then continue on. Most of the aircraft seen from this Tennessee mountain cove passed quickly over the churches, school, and businesses that were at the center of the community life for about a hundred families. This one looked as if it were suspended from the clouds, then it passed quickly out of sight. It would be more than a year before Whaley learned that the mysterious plane was making aerial surveys of his mountain home for a new national park. Within eight years, Whaley's father would be forced to sell his farm to Tennessee to transfer to the National Park Service (NPS), the cash he received would be lost in a bank failure, and his father would never own land again. Although Zenith Whaley remained in the Smokies area and found employment in tourism-related businesses, he never completely got over the loss of Greenbrier. "I nursed a grudge against the Park for a long time," he said. What made the pain so unforgetable [*sic*], was that he, like hundreds of others in surrounding coves, lost his home and community to a force as anonymous and remote as that airplane. [Brown, 1990, pp. 1–2]

The campaign leading to the dedication of the Great Smoky Mountains National Park on September 4, 1940, lasted almost two decades and involved hundreds of messages communicated through newspaper articles and feature stories, speeches, brochures, tours, broadsides, and lobbying activities. The campaign was unusual for its time because of its length, the diversity of media employed, the number of people involved, the complexity of issues discussed, and the modern public relations techniques used.

From the start, the campaign was decidedly a middle- and upper-class endeavor, with all of the leaders of the campaign being men of stature. These men of means appeared to have little concern for the likes of Zenith Whaley, either as a fellow campaign participant or as a potential receiver of persuasive messages. "Can we get them?" asked the writer of a 1925 article in *Outlook* magazine, referring to two proposed national parks: Shenandoah and the Great Smoky Mountains. "Ah!

There's the rub. We can if the two classes most interested will do their share: First, the class which will benefit financially by the tourist travel. Second, the class that travels, especially people of means" (Gregg, 1925, p. 667). Although this article used the word *class* as a synonym for *group*, it accurately described the people who successfully created a national park in the Smokies.[1]

Despite their elitist perspectives, Colonel David C. Chapman from Knoxville, often called the Father of the Park; Horace Kephart, famous author and outdoorsman; and other major communicators in this extensive campaign deserve some praise for their foresight, persistence, and ability to design and implement a dynamic and persuasive crusade promoting the Smoky Mountains as a national park. They had a product to sell, and they accomplished their goal with vigor by saying as many positive things about their product as often and through as many channels as possible. Using this basic communication strategy, they argued that the inclusion of the Great Smoky Mountains in the pantheon of national parks was inevitable. Their argument depended on the testimony of a wide variety of experts, including scientists, politicians, celebrities, artists, and common folk, who, in accord, painted a vivid picture of the Smoky Mountains as a biologically, geologically, and aesthetically unique area that promised exceptional recreational options. As an extension of this fundamental argument, they asserted that these uniquely "charming" mountains would revitalize the spirits of millions of urban dwellers living within driving distance of the proposed park. In turn, these tourists would bring millions of dollars in revenue to the area. The park promoters developed a persuasive campaign that had something for almost everyone.

Despite the thoroughness of the campaign, however, its organizers failed to achieve two important goals: they failed to persuade large numbers of people to contribute money for the purchase of land, and they failed to persuade the mountain people who occupied the proposed park area to leave their ancestral homes willingly. Instead of trying to reconcile the needs of the park and the people occupying it, promoters relied instead on negative stereotyping as a major argument for relocating hundreds of mountain families out of the proposed park area. In representing the mountain people as undeserving of the land they had occupied for generations, the park promoters cheapened a noble cause and rhetorically negated any possible compromise that would have allowed simultaneous park development and preservation of a unique indigenous community.

The campaign's first failure was overcome by a five-million-dollar contribution by John D. Rockefeller Jr. and by the federal government's eventual purchase of park land through a revised governmental policy under the administration of Franklin D. Roosevelt. The second failure of the campaign, however, was salvaged neither by a conservation-minded billionaire nor the federal government. Rather, the campaign failed to recognize the human cost of displacing the mountaineers, and this failure will be my focus in this chapter, as I explore the darker side of a successful environmental campaign. Through this exploration, useful lessons may emerge for contemporary environmentalists, whose rhetoric often frames the question of preserving and restoring the environment as a conflict between the needs of nature and the needs of people.

Upon closer examination of rhetorical arguments used in the campaign to establish the Great Smoky Mountains National Park, especially those based on Kephart's best-selling account of his adventures in the Smoky Mountains, *Our Southern Highlanders,* we discover that the campaign was successful only because it embraced a new form of exploitation, one that might be called benevolent capitalism. Benevolent capitalism favors making profits from tourist development rather than old-fashioned resource "hard" development such as lumbering and mining. As the park campaign developed, this new type of capitalism necessitated "civilizing" the wilderness of the Smoky Mountains and denigrating and eliminating the mountain people who threatened, by their very existence in the park, the vision of nature that dominated the leaders of the campaign and the influential people whom they wished to persuade. To remain consistent with their view of a benevolent capitalism, the campaign organizers eventually pictured the Smoky Mountains as feminine friendly mountains conducive to tourist-based development, while representing the mountaineers as an uncivilized threat, undeserving of the friendly and charming land they occupied. If left to their own devices, the argument continued, the mountaineers would destroy both the treasure that the promoters of the national park wished to preserve and a burgeoning tourist trade.

In this chapter, therefore, I investigate one of the most extensive environmental communication campaigns in American history and argue that the establishment of a national park in the Smoky Mountains was ultimately beneficial for the environment, the region, and the American people in general. A price was paid, however, not only in the harm inflicted upon the mountain people and their heritage, but also in the destruction of the purity of an environmental rhetorical position. Environmentalists like to believe they are on the side of the angels when arguing in favor of preservation and restoration of nature against the evil forces of development and exploitation. A thorough investigation of the campaign in the Smoky Mountains, though, leads to a more ambiguous, less reassuring view.

Selling the Park

In 1872 Yellowstone became the first national park in the world. Thereafter followed the establishment of other national parks, including Yosemite (1890), Mount Rainier (1899), Crater Lake (1902), Wind Cave (1903), Denali (1917), and Grand Canyon (1919). All these early parks had two things in common: they were formed from remote lands already owned by the federal government; and aside from some mining and ranching, they were undeveloped and unspoiled. In the 1920s, however, a movement developed in the nation's capital and in a number of local areas to establish more national parks in the East, close to major population centers, to accompany Acadia National Park in Maine, which was established in 1919. This movement introduced significant problems for park promoters. First, most of the land in the East was already privately owned, and more important, it was in various stages of development or abuse. For instance, one of the earliest scenic wonders considered for national park status, Niagara Falls, was deemed unsuitable because of complications of private ownership, development, and the potential of the Falls for energy generation. Second, the federal government had adopted a policy with the first national parks that no federal funds would be used to purchase park land. Because of these decisions, any group wishing to establish a park had to prove the proposed site's worthiness to be included in the system and had to raise funds for its purchase or persuade owners to donate their land for the general welfare.

The Great Smoky Mountains was an obvious candidate for national park status, being the nation's prime example of a hardwood forest and the highest mountains east of the Rockies. Despite the mountains' value as a natural area, however, designation as a national park posed major rhetorical problems for promoters. All of the land in the Smoky Mountains was privately owned, in more than sixty-six hundred separate tracts, and only one-third of the area still harbored first-growth forest. More than 85 percent of the area was owned by eighteen timber and pulpwood companies; the remaining 15 percent was divided among twelve hundred farms of various sizes plus five hundred summer homesites and lots. Also included were about one thousand tiny parcels, totaling only fifty-two acres, that had been sold in a promotional contest at a Knoxville cinema, in the hope that each winner would purchase at least two additional lots for summer cottages.

The park promoters' initial problem was compounded by the fact that considerable land in the Smokies was in terrible condition and continued to be harvested by lumber companies even as the negotiations for the national park progressed. A number of people in favor of lumbering argued for national forest designation instead of national park status. They claimed that the mountains, despite their obvious natural wonders, were undeserving of membership in the exclusive club with Yellowstone and the Grand Canyon because they were damaged goods, altered so significantly by development and exploitation as to offer only a faint reminder of past glories. In opposition to the image presented by the park promoters of endless lush mountain landscapes, those in favor of national forest designation and continued lumbering painted a very different picture. Instead of emphasizing mountain grandeur and biological diversity, they pictured devastation and paradise lost.

Rhetorical problems caused by private ownership and forest destruction were amplified somewhat by public ignorance of the area. While the necessity of preserving the natural wonders of the Smoky Mountains was obvious to park promoters and others who hoped to profit from increased tourism, it was not so obvious to many people in the area, and certainly not to most Americans, who in the 1920s probably could not even locate the mountains on a map. Although the area was within a day's drive of most eastern cities, very little was known about the Smokies, even by the locals, because of its exceedingly inhospitable terrain. No roads penetrated the area, and even the trails through the mountains were limited and in terrible repair. Some of the mountain locations, nicknamed "hells" because of the steepness of the mountains and the extensive growth of laurel, rhododendron, and other shrubs, were impassable. Few people even from Knoxville, only fifty miles away, had ever been to the mountains because of their inaccessibility and because of stories circulating about gun-toting mountaineers ready to ambush any outsiders stumbling upon bootleg stills. At the time tours of the area began as part of the park campaign, there were no detailed maps of the area, nor were many of the mountains and other distinct natural sites even named. Those that were named often had many and often conflicting toponyms.

The first major rhetorical task for the campaign's organizers was to shift the Smoky Mountains from being an unknown commodity to one instantly recognizable, and recognizable in a positive way. The campaign was especially successful in this task. Promoters spent most of their energy generating a continuous barrage of information of both the scientific and the personal type, praising the unique qualities of the Smokies. The purpose of this almost endless parade of articles and speeches about the Smokies was, in the early years of the campaign, to make the mountains an instantly recognizable entity and, later, after most Americans recognized the Smokies as a special place worthy of national park designation, to confirm the view that such designation was inevitable and beyond criticism. The strategy was to bombard the public with as many positive assessments as possible and to overpower any negative statements by sheer weight.[2]

Nationally prominent scientists were invited to write and speak on the botanical and biological significance of the area, emphasizing that the Smoky Mountains were more botanically diverse than any other sites in North America and had more species of trees than in all of Europe. For instance, in a speech given to the Southern Horticultural Association and covered extensively by local media, L. Bailey, a Cornell University botanist, discussed with much flourish a trip into the biologically rich Smokies, where he had discovered a number of plants that previously had not been named (*Knoxville Journal*, September 4, 1925).

Geologists wrote and spoke enthusiastically of the area's geologic diversity. The Smokies are among the oldest mountains on Earth, and they are as high as the Rockies minus the elevated plateaus from which the Rockies rise. World travelers in the 1920s emphasized the area's beauty by comparing the scenes in the Smoky Mountains with the grandeur of faraway places. For instance, Dr. H.C. Longwell of Princeton was reported to have said that after experiencing

> awe at the snow-capped Mount Olympus which the Greeks of old naturally believed the dwelling place of the Supreme Being; the glory of the Bay of Naples transformed into molten gold and silver by a sunset viewed from the top of Mt. Vesuvius; the beauties of Fujiyama and the "Alps" of Japan; the grandeur of Yellowstone, Yosemite and Grand Canyon . . . his experience on Mount LeConte will linger a memory as standing out unique, without parallel or duplication. The harmony of colors which I saw blended with the peculiar haze reminding me of a spiritual veil is absolutely indescribable. [*Knoxville Journal*, August 31, 1924]

Likewise, Mayor Robert J. Belton of Dawson, Yukon Territory, argued, after a stay in the Smokies, that the scenery he saw in Tennessee and North Carolina was more beautiful than any the Arctic region could offer. "Although not nearly so high as our mountains," he said, "the Smokies are certainly more beautiful for they are covered with flora and vegetation whereas our mountains are simply great masses of stone" (*Knoxville News-Sentinel,* September 21, 1930). The Smokies were compared by an almost endless parade of speakers and writers to Gothic cathedrals, the Alps, the towers of Bangkok, the lush growth of the Amazon, and other well-known sights in order to reinforce their value by making the unfamiliar familiar. Many communicators even justified using rather fanciful comparisons to distant wonders by noting that most people had a clearer vision of places such as European cathedrals and the Alps than of the mountains close at hand.[3]

To further emphasize the value and uniqueness of the mountains, representatives of the National Park Service compared the Smokies with other parks already in the system and painted a picture of the mountains as the superlative example of an environment not preserved elsewhere. "They are worthy of being included in our national park system along with the Yellowstone, the Yosemite, and the Grand Canyon," argued a representative of the Park Service. "Each is individual and unique and is unsurpassed in its particular type; neither surpasses the other for they are totally different that we cannot compare them" (*Knoxville Journal,* September 25, 1925).

Even celebrities were invited to respond to the beauties of the mountains, the most famous being the major participants in the Scopes "monkey trial" taking place in nearby Dayton, Tennessee. The first to come from Dayton, through an invitation by Chapman, was a large group of reporters assigned to cover the trial, most of whom represented newspapers in the nation's larger cities. Two of them wrote for London papers. The reporters were given a tour of the proposed park, and they wrote stories that reached national and international audiences. Harper Leach of the *Chicago Tribune,* for instance, named the mountain area "the Scotland of America," while arguing that the soon-to-be-completed road system would make the new national park an ideal vacation spot for Chicago residents (*Knoxville Journal,* July 23, 1925).

The coverage of celebrities in the Smokies became even more frenzied the next week, when Clarence Darrow, famous defense lawyer at the Scopes trial, and defendant John Scopes himself rode horseback in the mountain wonderland. Photographs of Darrow smoking a cigar and wearing a cowboy hat and young Scopes awed by the landscape adorned the front pages of newspapers throughout the country. Headlines announced that Darrow exclaimed, "There must have been a God," as he surveyed the majesty of Mount LeConte (*Knoxville Journal,* August 4, 1925). William Jennings Bryan, the prosecuting attorney, had also agreed to visit the proposed park, but he died while resting near Knoxville before he made his visit. Not to miss an opportunity to mention the name of the Great Smoky Mountains, reporters writing newspaper dispatches announcing Bryan's death pointed out that his final act before the fateful nap from which he did not awake was planning his intended visit to the mountains (Campbell, 1969, p. 41).

Even common travelers contributed to a popular syndicated series of articles entitled "What the Tourists Say about the Great Smoky Mountains," an amusing compilation of personal accounts praising the beauty and unusual qualities of the area as well as identifying places for tourists to stay and eat while they traveled.

Randomly selecting a week in 1925, I found more than fifty articles in each local newspaper concerning some aspect of the Great Smoky Mountains, the park, or individuals associated with the campaign. The strategy of bombarding both local and national media with stories from as many angles as possible about the proposed park helped reinforce the idea that, although obstacles slowed progress temporarily, nothing would stop the inevitability of the national park movement. This sentiment was reinforced through headlines that used terms such as *inevitable, only a matter of time, in the future, any day now, invariably,* and *certainly.* Editorial cartoons mirrored the sentiments of the stories. For instance, a 1925 cartoon pictured Uncle Sam and Colonel Tennessee hiking arm in arm toward the ultimate summit, National Park, with the caption reading, "Only a Step to the Summit" (*Knoxville Journal,* April 15, 1925).

The campaign had been bolstered somewhat in 1924, when the Southern Appalachian National Park Commission under the Secretary of the Interior concluded, after investigating numerous possible sites in the East, that two national parks, one in the Shenandoah Mountains of Virginia and one in the Smoky Mountains, deserved designation and that steps should be taken to guarantee their development. Despite the legitimation that this recommendation brought the local campaign, however, it pointed out clearly another major problem that the park leaders needed to face before the national park idea was concluded. The Southern Appalachian National Park Commission argued that the park in Virginia should be developed first because of accessible transportation and its proximity to the nation's capital, although they agreed that the Smokies were superior in splendor and biological importance. The report from the commission states:

> We have found many areas which could well be chosen, but the committee was charged with the responsibility of selecting the best, all things considered. Of these several possible sites the Great Smoky Mountains easily stand first because of the height of the mountains, depth of valleys, ruggedness of the area, and the unexampled variety of trees, shrubs, and plants. . . . The Great Smokies have some handicaps, however, which will make the development of them into a national park a matter of delay; their very ruggedness and height make road and other park developments a serious undertaking as to time and expense. The excessive rainfall also (not yet accurately determined) is an element for future study and investigation in relation both to the developmental work, subsequent administration, and recreational use as a national park. [U.S. Department of the Interior, 1931, pp. 7–8]

The rhetorical significance of the commission's report was that the very characteristics leading people to perceive the Smoky Mountains as the premier candidate for national park status—its rugged mountains, its diverse biology based on extensive rainfall, and its inaccessibility—were, in turn, hindering its acceptance and its development.

Indeed, the park promoters faced profound rhetorical problems: they needed to make an unknown commodity known in a highly favorable way. They did this, as we have seen, by characterizing the unique qualities of the area such as the height of the mountains and the steepness of the canyons. In doing so, however, they amplified another rhetorical problem. After all, one of the most important arguments in favor of the park was its proximity to urban areas and its ability to revitalize the urban masses who lived within driving distance. In presenting a strong case for the preservation of the Smokies, the communicators frequently identified the characteristics of the mountains that negated the argument stressing its proximity to urban areas. Of what value were mountains close by, if they were too wild and rugged to be used by the people? The campaign emphasized inaccessibility and mystery, qualities that made demands upon potential consumers such as strength and knowledge of wilderness, qualities not especially persuasive for most tourists and urban dwellers, who were necessary to bring economic prosperity to the area.

The park promoters solved this second rhetorical problem by adapting their representation of the Smokies and the mountain people to changing persuasive needs and to an evolving concept of benevolent capitalism that was consistent with the dominant view of their middle- and upper-class audience. To understand how the park promoters accomplished this difficult rhetorical task, it is useful to look at Claude Lévi-Strauss's discussion of binary oppositions and anomalous categories. These two concepts form the basis of the arguments formulated by the park promoters in both "selling" the significance of the park and eliminating the mountain people rhetorically and physically from the area.

Simply, Lévi-Strauss (1969, 1963, 1966) argues that human beings make sense out of the world in which they live by seeing and discussing things in binary oppositions. He believes this process is universal because it is a product of the physical structure of the human brain and is therefore specific to the species and not to any one culture or society. In the perfect binary opposition, everything is either in category A or category B, and by imposing such categories upon our world, we begin to make sense of it. Category A cannot exist on its own, but only in a structured relationship with category B. That is, category A makes sense only because it is not category B. So we have women and men, civilization and savagery, good and evil, order and anarchy. The meaning of one of these categories is significant only in relation to its opposite and vice versa. Order means nothing unless we define it in opposition to anarchy; male means only in relationship to female.

An anomalous category is one that does not fit the categories of the binary opposition but straddles them, blurring and confusing their clear boundaries. An anomalous category draws its characteristics from both binary opposites, and consequently, it has too much meaning and is conceptually too powerful. Because it has too much meaning and is too confusing, an anomalous category needs to be controlled, usually by designating it either sacred or profane. For instance, homosexuality threatens the clarity of gender binary opposites, and therefore, in societies that consider gender identity important, it is considered taboo, and reactions to it are highly emotional.

During the first few years of the campaign for the Great Smoky Mountains National Park, the promoters emphasized the unique qualities of the park. What emerged from their descriptions were huge and awe-inspiring mountains with a strong, masculine image. Such words as *awesome, powerful, steep, craggy, mysterious, savage, wild, distant,* and *massive* dominated their descriptions. These words, as well as comparisons with other powerful mountains such as the Rockies, focused people's attention on the wild qualities of the area, disputing the common belief that the East was completely settled, developed, and lacking in scenic grandeur. The very justification for the designation of national parks in the West, namely the wildness of the place, was used to substantiate designation of a park close to eastern urban areas.

The rhetorical picture of the mountains as awesome tree-covered versions of the Rockies was further reinforced by numerous messages arguing that most of the land recommended for the park was uninhabited. After all, if the mountains were to be seen as wilderness, they could not contain farmers whose families had lived in the area for more than a hundred years, no matter how backward or primitive their lives might be. When a congressman supporting the park was asked at a 1925 public hearing whether or not there were any "villages" in the proposed area, he answered, "I think none at all" (hearings before the Committee on Public Lands, January 1925, Great Smoky Mountains National Park Archives).

Writers of early promotional materials reinforced this claim. "Within the Park area there are whole tracts of forest where, even to this day, no human being has set foot, and only three or four living men have traversed the crest of the Smoky Mountains end to end," stated a booklet produced by the Great Smoky Mountains Conservation Association (n.d., p. 6). News articles labeled the Smokies "unknown mountains," describing them as "unvisited by man" and "uninhabited" (*Allentown [Pa.] News,* March 26, 1926). In places, one reporter wrote, "there are still gulfs and gorges which only the bear and wildcat have roamed" (*Nashville Tennessean,* April 4, 1926). It was recounted that "so rough is the park area that the only way of mapping it satisfactorily is by taking snapshots from the sky and patching them together." An army aviator flew for weeks about the mountains to take photographs for maps, and his exploits were reported almost daily by the press. In his many flights, the *Knoxville Journal* reported, "he did not see a house or a human being, except in a few spots where lumbermen were at work" (November 29, 1925).

A Washington, D.C., woman who had vacationed in the Smokies wrote the North Carolina publicity director to inquire how much land would be taken by the park and what impact the park would have on its inhabitants. He replied that the park would cover "400,000 acres" or "700 square miles" and asserted that "the Park area is not populated with the exception of some few clearings along its border" (Brown, 1990, p. 24). The "few clearings" probably contained a minimum of four thousand people. As late as the 1930s, promoters of the park continued their "empty wilderness" theme by stating in a booklet distributed by local merchants in Knoxville that "even today, there are gulfs in the Smokies that no man is known to have penetrated" (*Great Smoky Mountains National Park,* n.d., p. 1).

This claim is highly improbable. The Cherokees had hunted the Smoky Mountains for centuries, and mountaineers had lived in the area for 150 years. Certainly Horace Kephart, chairman of Great Smoky Mountains Incorporated, knew that the Smoky Mountains were neither "empty" nor primeval. A former librarian who had come to the mountains to escape city life and recover his health, Kephart had lived in the Smokies since 1904 and had written extensively about the area and its inhabitants. Yet he too, for persuasive reasons, repeated the "empty wilderness" claim in numerous articles and speeches. His justification for engaging in this rhetorical game was his firm belief that the Smoky Mountains had saved his life and that this mysterious environment had the power to perform such a miracle for others. Kephart stated, "I got my health back in these mountains and intend to stay here as long as I live . . . and I want them preserved that others may profit as I have" (Frome, 1966, p. 156). Although he understood that the mountains and the mountaineers would be changed by the influx of tourists, he believed that "millions of people hived" in "swarming industrial centers" needed "wing room every now and then," and that "small farmers and the like" would not be disturbed (Brown, 1990, p. 25).

Initially, the picture that emerged of the Smoky Mountains was one of a rough, unspoiled, breathtakingly beautiful but dangerous and unpeopled land, worthy of preservation because of its raw qualities and because it was the opposite of the rest of the East, which was polluted, used, and lacking in majesty. The Smokies were presented as primitive, and without any humanizing or civilizing influences—an absolute wilderness. The park promoters clearly described the Smoky Mountains as a binary opposite of the civilized eastern seaboard, the very area from which tourists would flock for rejuvenation. This image is a powerful one in American myth, and it gains meaning through its opposites: femininity, civilization, settlement, and defilement. There is a problem with this powerful image, however. It does not correspond with what visitors or reporters actually saw when they visited the area. Most important, it is questionable whether this image of the wilderness was persuasive to the capitalist audience that park promoters wished to persuade, because it was inconsistent with the basic arguments of proximity to eastern urban centers and the development of tourism. It was difficult to see how such a raw environment would be of use to rejuvenate the workers of eastern cities, a primary goal of park designation, without encouraging them to revert to being wild men in order to survive the experience. After all, Kephart and others did not want the masses of working people to come and live in the mountains, but rather periodically to visit and become refreshed by them, only to return home to their dirty industrial environments to continue working, renewed until their next vacation in the mountains. Nor did they want the masses to become "decivilized" and return to urban areas and cause problems. The last thing the experience in the Smokies should bring is a disorderly, savage workforce.

So by the mid-1920s, when the Southern Appalachian National Park Commission conducted its investigation of possible park sites in the East, most people in the country had learned about the Smoky Mountains. They could identify the mountains on a map, and as a result of the image of the mountains presented in the campaign, they had a vision of the peaks as a mysterious and awesome place that was worthy of preservation. However, the report of the commission clearly proved to the park promoters that a change in image was necessary if tourist demand was to continue as a powerful argument for park designation. From this point on, a clear change in the way the Smokies were represented in messages "selling" the park occurred. Instead of focusing on the masculine nature of the mountains—their massiveness, the steepness of their valleys, and their distant, inhospitable qualities—the promoters painted a more feminine image emphasizing "user-friendly" characteristics. Instead of being mysterious, unapproachable, and aloof, the mountains became "undulating," "rounded," "an unbroken vision of friendly crests," "welcoming," "quaint," "warm," and, most important, "charming." In an interview by Horace Kephart, Robert Yard, executive secretary of the National Park Association, stated, "Nobody can overrate the Smokies. They are supremely worthy. And they have one quality that is unique . . . Charm. The Smokies are natural wonders; but they are more than that. One can see a stupendous phenomenon of nature that awes one with its majesty; but when he has seen it once—well, he has seen it. But the Smoky Mountains have enduring charm. Having seen them once, they lure you back to them again and again. I love them. I am coming back" (Kephart, 1926, pp. 631–32). Kephart himself offered this reassuring and feminine description of the Smokies: "Pinnacles or serrated ridges are rare. There are few commanding peaks. From almost any summit in Carolina one looks out upon a sea of flowing curves and dome-shaped eminences undulating, with no great disparity of height, unto the horizon. Almost everywhere the contours are similar: steep sides gradually rounding to the tops, smooth-surfaced to the eye because of the endless verdure. Every ridge is separated from its sisters by deep and narrow ravines" (1913, p. 51).

To correspond with this shift in rhetorical image, the promoters "discovered" that people lived in the mountains and began to talk about them in a way complementary to the feminine image of the "charming" mountains. "They are fascinating people these mountaineers," stated a booklet that romanticized the newly discovered mountaineers.

> Descendants of settlers of Revolutionary days still live within the Park and around its borders. White wood smoke curls lazily from the chimneys of log cabins where Daniel Boone and Davy Crockett would feel perfectly at home. In pioneer cabins within the Park a pioneer way of life goes on, and, if you listen sharply to mountaineer conversations you will be rewarded with fragments of early English speech. The coves know well the echoes of old Scottish ballads, not written, but handed down from generation to generation. Mountain men still scrape their fiddles in lively old tunes, and pioneer crafts of spinning, weaving, pottery and basket making still survive. [*Great Smoky Mountains National Park*, n.d., pp. 2–3]

To further the image of the Smokies as a charming location, park promoters presented the mountain people as living museum pieces. "As inhabitants of the Park, these picturesque southern highlanders will be an asset, and so will their ancient log cabins, their foot-logs bridging streams, and their astonishing huge water wheels," claimed Tennessee promoters (Great Smoky Mountains Conservation Association, n.d., p. 12). "Several typical mountain communities remain intact, and many constitute valuable outdoor exhibits in a 'museum of mountain culture.' Already large collections of household goods, tools, farm equipment, weapons, chiefly primitive and hand-wrought, have been assembled" (*Great Smoky Mountains National Park,* n.d., p. 14). White mountaineers were not the only ones used to sell the new image of the Smokies. "Like the mountaineers, our Indians will retain possession of their abodes within the Park, and—perhaps— enjoy their new dignity being objects of interest to million of tourists" (Great Smoky Mountains Conservation Association, n.d., pp. 12–13).

Journalists who swarmed into the region at the request of the promoters saw an opportunity to sell newspapers by perpetuating the image of "quaint" and "charming" people. Accompanying descriptions of mountaineers, who always had "families of ten, twelve, and fifteen children" (James, 1928, p. 373) and who lived in "crude cabins built of unchinked logs," was a parade of photographs that reinforced the "charm" of the area. Old men with great beards smoked their pipes while rocking on ancient chairs, old women in long dresses toiled at spinning wheels, and only the most primitive dwellings were photographed, usually set in front of a steep cliff or a rushing cascade. One writer borrowed romantic hyperbole from an earlier age: "At the very top of LeConte there is a boy living alone in a cabin made of slabs. The writer saw in that cabin a single volume, namely Thoreau's *Walden*" (Bohn, 1926, p. 7).

Instead of ignoring the mountaineers, as they had done earlier in the campaign, the park promoters used them, following the commission report, to sell the new rhetorical vision of the Smokies as being friendly, welcoming mountains rather than threatening and distant. By consciously changing the Great Smoky Mountains from masculine mountains to feminine mountains, and thereby from forbidding to friendly, volatile to comforting, the park promoters effectively maintained the consensus they had established earlier when they promoted the unique qualities of the park. Also, they successfully set the image of the park for future generations. All publications about the Smoky Mountains since the campaign emphasize their friendliness and charm and de-emphasize ruggedness and mystery. The Great Smoky Mountains National Park is always represented as the family park, the natural area that has something for everyone, even those unaccustomed to the outdoors. The park can be enjoyed by the rugged mountain climber and the inexperienced day hiker alike; it beckons all people to explore it without danger.

The mountains continue to be represented as feminine rounded peaks that are inviting, almost seductive, in contrast to the stark Rockies, which continue to be described as formidable and requiring active participation. The image of the Smoky Mountains as charming and quaint became so dominant by the time the park was officially opened in 1940 that the image even influenced decisions concerning preservation of mountaineer dwellings in the park. All modern structures, including numerous homes of frame construction, were obliterated. The single guiding principle was that anything that remotely suggested progress or advancement beyond the most primitive stages should be destroyed. A sort of pioneer primitivism alone survived in the cove structures left standing. A visit to Cades Cove today, where the majority of preserved farms exist, leaves the visitor with a single impression: This is a charming and quaint place.

What to Do with the Mountaineers?

Michael Frome, in the most popular history of the Great Smoky Mountains (1966), depicts the campaign leading to the establishment of the park as a battle between the forces of good and evil: unselfish lovers of nature are pitted against rapacious capitalists. The urban upper-class promoters are represented as courageously fighting for the preservation of trees and wilderness, while their primary adversaries, logging and mining interests, are pictured as greedy exploiters who hope to cut over the land and then manage it for a future harvest as a national reserve or forest. For instance, Frome quotes extensively from Kephart's fiery attacks against logging companies to reinforce the purity of motive of park promoters. "'Why should this last stand of splendid, irreplaceable trees be sacrificed to the greedy maw of the sawmill?'" asked Kephart. "'Why should future generations be robbed of all chance to see with their own eyes what a real forest, a real wildwood, a real unimproved work of God, is like. . . . There is no use, then, in talking about conserving the Smoky forest by turning it into a national forest after the lumbermen get through with it. The question, the only question, is: Shall the Smoky Mountains be made a national park or a desert?'" (Frome, 1966, p. 188).

Reading the Knoxville and Asheville newspapers of the time, however, makes it difficult to characterize the movement to create the park in the Smoky Mountains as a battle, or even a fight. It is impossible to conclude that conventional capitalism was the villain in this campaign. Rather, the campaign could be described more accurately as a consensus-building crusade based on a reformulated definition of capitalism, a definition that stresses benevolence, not exploitation, one that, contrary to the capitalism of old that abused the land and its people, focuses on tourism in the guise of preservation and rejuvenation. Certainly, throughout the campaign, individuals or groups periodically expressed reservations about the park and what it might mean for the area, and the lumber industry for a time did try to stop park development with expensive advertising and smear tactics against park leaders. But despite these efforts, no serious opposition ever materialized to oppose the park, and it certainly does not appear as if most people in Knoxville and Asheville ever wavered from their support of the venture. In fact, most middle- and upper-class people, who were best positioned to take advantage of the opportunities for tourist development, were so sold on the park from the start that it was practically impossible to hear a contrary word about the issue in polite society. "Asheville has come out flat-footed for the Park and one of the most ardent boosters is Mrs. Vanderbilt," wrote Kephart in the *Knoxville Journal* (August 17, 1925).

Convincing mountain residents and the small mountain middle class that thousands of acres should be donated to the federal government for a scenic park, thereby removing the land from tax rolls, eliminating possible lumbering jobs, and bringing in thousands of urban tourists, proved more challenging than persuading the wealthy and more urban middle class that money could be made while they indulged their civic-minded inclinations by preserving nature and aiding the country. "Our hardest job was to convince our own folks in and around Bryson City that the Park would be worthwhile. They could not see what the coming of hundreds of tourists daily through the park in Swain County would mean for Bryson City," Kephart wrote (*Knoxville Journal,* August 17, 1925). Often those who earn a living from the land are not sympathetic to preservationist or environmental arguments. They see preservation as an elitist activity that deprives them of access or restricts actions essential for survival while the wealthy benefit. They discount environmental arguments as only speaking to the needs of others. Such views seemed to dominate the reactions of the local mountain residents in the Smokies in the 1920s, and the promotional campaign did little to dislodge these impressions.

Therefore, almost from the start the major audience for the campaign was the group of people already convinced that the park would economically benefit the area. Those opposed to the park or even those expressing some dissatisfaction with certain policies soon were ignored by the promoters or, in the case of the mountain people, became the target of vilification for the larger rhetorical goal of reinforcing people of influence in the belief that the park was an inevitability.

All of the local and most national newspapers favored the park, and through these important channels of communication, popular sentiment remained dominant. When a prominent Knoxville judge supported the people who might be dispossessed, he asked for anonymity (*Knoxville News-Sentinel,* May 10, 1926). When his identity was uncovered, he qualified his statement by noting that he was not really against the park (*Knoxville News-Sentinel,* April 23, 1927). Likewise, a columnist who stated that "on a small scale we are re-enacting the famous hegiras of history—the exodus of the Israelites, the migration of the Acadians, and the flight of the Tartar tribes"—qualified these remarks with three paragraphs of support for the park (Brown, 1990, p. 49).

The mountain residents organized at least two petitions against selling their lands, but there was no violence and little formal resistance.[4] Zenith Whaley, the former resident of Greenbrier, was asked to explain why, if the mountain people did not want to leave their homes, more of them did not resist. "Some of 'em did, but the bigger portion knew that they was fighting a losing battle," he said. "As those [that resisted] folded, why then, the little ones, the small landowners followed. They knew, with the big dog gone, they couldn't handle the bear" (Brown, 1990, pp. 10–11).

Despite the promoters' apparent success in shifting the focus of the campaign from one binary opposite to another when describing the mountains, they were not as successful when dealing rhetorically with the mountain people, no matter how hard they tried to present highlanders as quaint extensions of the newly formulated feminine mountains and no matter how influential this view came to be for future visitors to the park. One fundamental reason led to their difficulty in connecting the mountaineers with the charming mountains: neither the park promoters nor the National Park Service ever intended to allow the mountain dwellers to stay in the park. Kephart and Chapman both fought to save the mountains, not to save or even to preserve as a museum a way of life that grew out of them.

While the public ignorance of the Great Smoky Mountains allowed its promoters to represent the area rhetorically in whatever way they wished, a different kind of public ignorance affected the topic of the mountaineers. Because so few people had any clear idea of what the mountains looked like or what biological and aesthetic wonders they possessed, promoters could call them massive and formidable at one time, charming and welcoming the next, without causing considerable problems. A different type of ignorance came into play, however, when the promoters tried to fit the mountain people into their newly devised vision of the mountains. That is, despite the lack of concrete knowledge about the inhabitants of the area, people throughout the country and especially in the immediate vicinity of the park had a clear idea of what to expect when encountering mountain people.

The nature of the people of Appalachia, including their physical and mental traits, behavior, history, and culture, had provided material for an American fable for generations. The people of Appalachia, so the fable argued, constituted "a culturally homogeneous group, most often said to be Scotch-Irish and English, mostly Presbyterian, independent, fatalistic, and culturally and geographically isolated" (Perdue & Martin-Perdue, 1979–80, p. 85). These characteristics were part of a generalized complex of stereotypes that had defined the so-called hillbilly and had been applied by different writers at various times to the people of the Great Smoky Mountains. They were supposedly cultural primitives who purposely isolated themselves for undesirable reasons, tended to farm the smallest amount of land capable of producing the barest subsistence, possessed diminished abilities to feel love, affection, loyalty, and responsibility, and had exaggerated propensities for violence and immoral behavior (pp. 85–86). Thus, when the park promoters began to define the mountain people as "quaint" to correspond with their vision of the Smoky Mountains as charming, a conflict emerged between the fable of the hillbilly and the vision espoused by park promoters. The mountain people became an anomalous category, both barbaric and civilized, quaint yet repulsive, creative and dim-witted, possessing simultaneous characteristics of both the masculine mountains of the early campaign—wild, ruthless, untamed, and unknown—and the feminine mountains described later in the campaign—caring, family-oriented, historical, tradition-bound, and friendly.

The conflict inherent in this rhetorical dilemma came to a head later in the campaign, when it was obvious that promises made to the mountain people by Chapman and others concerning guarantees of homesteads and jobs were not to be kept, when it was clear that the mountaineers without exception were to be moved from their homes, and when the park movement was granted the legal means to dislodge them. The conflict inherent in this confusing view of the mountaineer was expressed, for instance, when two assistant editors of the *National*

Geographic in the spring of 1930 argued that the Park Service should let the people of the Shenandoah and the Great Smokies stay on the land because "the mountaineers, not the scenery, were the real tourist attraction." The assistant director of the Park Service, Arno Cammerer, replied that the editors were "all wet." Cammerer's assessment was that "the worthy mountaineers . . . would leave; the only ones anxious to stay were those anxious to make money from the tourists." He claimed further, "There is no person so canny as certain types of mountaineers, and none so disreputable" (Perdue & Martin-Perdue, 1979–80, p. 89). It is ironic that the same interest in tourism that motivated businessmen and politicians to praise the park was seen as disreputable if it motivated mountaineers.

Park promoters decided early in the campaign to ignore the issue of the mountain people as much as possible. Limited messages were designed to appeal to or inform them of the reasons for the park or what was going to happen to them once the park was established. When they thought about the native people at all, the park's founding fathers, like the timber barons who preceded them, imagined that the changes they brought would be in the best interests of the local people. Such a view sometimes forced men like Chapman to make promises they could not or did not intend to keep, just to gain support for the cause. In 1925, for instance, when a petition was framed protesting the taking of land from the mountain people, Chapman strongly criticized rumors of possible expulsions by announcing that mountain families would be permitted to remain in their homes and that some folks would even gain employment caring for the park (*Knoxville Journal*, December 6, 1925).

In 1926, just a year before Tennessee and North Carolina received permission to use the power of eminent domain to acquire land, Chapman made an even more powerful announcement: "Let's have an end to this constant talk about dispossession of the mountain people. They are the ones who should be—and many of them are—eager for consummation of a project that is going to mean to many of them real independence for the first time in their lives." He then promised lifetime free housing and employment in the park (*Knoxville News-Sentinel*, February 16, 1926).

The promoters claimed that tourism would help the people of east Tennessee "modernize." They located a "native mountaineer" in Mascot, a small town outside Knoxville, to voice this opinion. "There are mountain folks in that section today living in rude cabins," stated the Mascot resident, "who will have comfortable homes in 10 years just like city people if they take advantage of what the park will do for them." He went on to say that he was "surprised by a lot of opposition to the park among some of the people up there. It has all emanated from some city people, whose financial interest will be hurt, spreading false ideas" (Brown, 1990, p. 32).

The sentiment that the park would help the people in the area was pushed hard by Horace Kephart, who believed that only economic revolution would introduce the mountaineers to a "finer more liberal social life" (1913, p. 451). He also hoped that development would force the mountain people to examine their tendency to be "high-strung and sensitive to criticism." Kephart wrote, "Of late years they are growing conscious of their own belatedness and that touches a

tender spot. . . . Since they do not see how anyone can find beauty or historic interest in ways of life that the rest of the world has set aside, so they resent every
exposure of their peculiarities as if that were holding them up to ridicule or
blame. . . . Hence it is next to impossible for anyone to write much about these
people without offending them" (pp. 280–81).

It is easy, however, to understand the mountain people's apprehensions concerning scrutiny. All one needs to do is to look at how the mountaineers had been
represented consistently by popular writers and had been misrepresented by park
promoters. Charles W. Todd, for instance, used Montvale Springs, Tennessee, as
the setting for *Woodville; or, Anchoret Reclaimed*, published in 1832. Todd painted
glorious pictures of the mountain scenery, while representing mountaineers as
lazy yet combative. Also, Sidney Lanier stereotyped mountain inhabitants in
Tiger-Lilies, published in 1867, as people totally out of touch with life outside their
immediate area. Mary Noailles Murfree, under the pseudonym Charles Egbert
Craddock, achieved some fame in the 1880s for her short stories and novels set in
the Cumberland Mountains of middle Tennessee and in the Great Smoky Mountains. Despite her advantage of having spent considerable time in these areas, she
lacked the power to understand either the folk culture or individual motivation on
any but the most superficial level, and so her characters are static, unrealistic
manikins who bear no likeness to real mountain people. Likewise, her own Victo-
rian code forced her to moralize about what she saw in the mountains and to
censor the speech and actions of her characters in such a way as to portray them in
a condescending and distorted fashion (Dunn 1988, p. 161).

With these authors as predecessors, it is little wonder that Horace Kephart,
through his book *Our Southern Highlanders* (1913), was immediately seen by
many as an honest and precise chronicler of the people who lived in the Smoky
Mountains. Kephart had associated extensively with mountaineers for years. His
amusing and lucid style gave the impression that his observations were objective
and lacked any dominant ideological underpinnings. Also, he frequently praised
the mountain people for their ability to endure great hardships, to eke out livings
from a harsh environment, and to carry on traditions that the rest of society was
abandoning. It appears as if Kephart held the mountain people in high esteem
and respected them, and therefore he did not purposely distort their actions or
represent them in stereotypic ways for effect. Upon closer investigation, however,
these initial reactions to Kephart's work soon disappear, especially when one looks
at the role Kephart played later in the campaign to establish the Great Smoky
Mountains National Park and how his vision of the highlander was used to legitimate the expulsion of the mountaineers from the area.

Our Southern Highlanders

What to do with the mountain people? It was a question that plagued the campaign. Ignoring them did not work. They had some champions and increased
attention from the media. Presenting them as quaint throwbacks to a picturesque
time did not work either, because the image failed to correspond completely with
the commonly held view of them. Nor did this view seem consistent with the

government and park promoters' move toward a park devoid of any human in-habitants or of their praise of modern enlightened sensibilities. If the highlanders were so quaint and so obviously a national treasure, one might ask, why force them out of the park and, therefore, destroy their uniqueness? This argument became even more persuasive as the promoters continued to espouse a benevolent capital-ism that would benefit the area without exploiting people or the environment. How could they justify relocation of some of the nation's poorest peoples from areas where they might increase their standard of living from tourism, while they championed local development? How could they talk about a benevolent system concerned with the renewal of the spirit for hundreds of urban dwellers, while they destroyed the spirit of hundreds of local people? Such were the rhetorical problems that faced the park promoters in the latter part of the campaign.

In order to reconcile the rhetorical problems caused by the highlanders, the campaign identified and amplified the conflicts inherent in the anomalous cate-gory "mountaineers" to justify the necessity of eliminating them from the park. By asserting that the mountaineers possessed characteristics of both the wilderness and civilization, the park promoters argued that the mountain people, as is true of most anomalous categories, were profane and should be expunged from the land. While surely the mountaineers were holdovers from an ancient time, deemed "quaint" by some, they were an unruly wild people who were prone to base, primi-tive instincts, as evidenced in their feuding, bootlegging, and general lawlessness. Instead of being quaint, they were dangerous; and while they were primitive and wild, they were also lazy and shiftless. Although they were trusting and naive, they were also devilishly and cruelly shrewd. Although they appeared childlike and in-nocent, they were planning to steal your wallet. While they complained like children about the lack of game in the forest, they indiscriminately shot all the game they could, including young-bearing females. In other words, the mountain people had inherited the bad qualities of both their wilderness home and the civi-lization that surrounded them. They were shiftless and destructive. They lived in the wilderness and littered it with debris; they were greedy but neither clever nor hard-working enough to satisfy their greed without reverting to dishonesty.

So too, we find two strong images emerging of the mountaineer that were used by the park promoters to justify expulsion from the area in the name of hu-manitarianism for both the mountain people themselves and for the survival of the land. On the one hand, we have the active wily animal of the wilderness, dan-gerous, clever, and ever alert for the chance to make a strike; on the other hand, we have the dull, lazy, slothful, and unwashed degenerate who lives in squalor and lacks either the ambition or the wit to improve his plight. Both of these stereo-types emerged from the conflicting tendencies of the anomalous category and were used to argue that expulsion from their ancestral homes was good for both the mountain people themselves—because it would save them from their natural tendencies toward savagery and sloth—and for the land—because the mountain people, in their ignorance and failure to develop a sophisticated social conscience and an enlightened capitalism, would destroy the treasure others were desperately trying to save. All of these arguments can be traced back to the work of Horace Kephart in his classic book on the mountain people, *Our Southern Highlanders.*

Kephart, in chronic bad health, abandoned his wife and children and moved from St. Louis into the wilderness of western North Carolina at age forty-two, looking for "a free life in the open air, the thrill of exploring new ground, the joys of the chase, and the man's game of matching my woodcraft against the forces of nature" (1913, p. 30). Aside from extended drunken periods, he spent much of the rest of his life in robust health, in the mountains, enjoying long spans of isolation. He recorded his experiences, including an exploration of mountaineer behavior and speech patterns, in two books and numerous articles. These highly readable works made him a local celebrity and were viewed by many as an accurate picture of the area and the mountaineers. Recent scholarship, however, has questioned Kephart's objectivity and concluded that his observations more clearly represented his own agenda and personal biases than anything close to how mountain people actually spoke or acted.[5]

The real heroes of Kephart's *Our Southern Highlanders* are not the mountaineers for whom the book is titled but rather the glorious Smoky Mountains and himself. In Kephart's work, the mountains are described as wondrous, mysterious, all-giving, and powerful enough to rejuvenate his tainted body and soul. He represents himself as a stranger to the mountains. Similar to a white hunter in Africa or a British soldier in India, he is spiritually enriched and comes to understand the "meaning" of the environment better than the people who have always inhabited it. In many ways, Kephart's *Our Southern Highlanders* conforms to the characteristics of a conversion narrative (Griffin, 1990). The purpose of Kephart's writing is to give meaning to his own conversion from a sickly, spiritually corrupt city slicker to an enlightened person at one with the natural environment. Likewise, the campaign in favor of the national park appealed to those who were in a position to reach enlightenment, namely the middle and upper class who understood the importance of preservation and benevolent capitalism. The highlanders, trapped in their primitive unenlightened state, had to be sacrificed for the park because they were not capable of conversion. As W.C. Spengemann and L.R. Lundquist argue, the primary goal of a conversion autobiography is to establish a believable identity for the author in the form of "an integrated, continuous personality which transcends the limitations and irregularities of time and space and unites . . . experiences into an identifiable whole" (1965, p. 516). Often, autobiographies become mythic tales, rather than true descriptions of events and people.

Such is the case with *Our Southern Highlanders.* In order to establish clearly the importance and credibility of his own conversion, Kephart had to emphasize the mystery and remoteness of the land and to show that not all people are chosen, unlike himself, the convert. Some people are not ready for spiritual renewal, although the opportunity is given to them. Even those who have lived their entire lives in an environment can be alienated from it and become immune to the powers available to the convert, a theme explored in a number of stories vividly detailed in the book. For instance, when Kephart accompanies a small group of mountain men on a bear hunt, he remains aloof from the action, observing the tragicomic local hunters as they are abused by the natural world and, in turn, victimize it through their crass killing of bears and their mistreatment of hunting dogs. The characters, as presented by Kephart, are superstitious, inept, unsympathetic to their own circumstances, undernourished, somewhat dim-witted, and, most important, incapable of understanding how their isolation has made them grotesque and their actions ludicrous. Only Kephart sees the killing of the bear as an act beyond mere food gathering, and therefore, he is spiritually up-lifted by the experience (1913, pp. 75–109).

Charles Griffin argues that it may be inappropriate to evaluate the "truth" in autobiographical conversion stories by using the criteria appropriate to other types of communication. Instead of using validity tests, we should see if the de-scribed events reinforce the myth created in the narrative (1990, p. 161). Using this means of evaluation, Kephart's descriptions of the mountain folk may be deemed "true" if the communicative purpose is to establish the spiritual powers of the land. The important question is, however, What happens when "truths" that are deemed appropriate to a specific form of communication become the basis for other forms of communication, such as practical persuasion that concerns issues like the selling of property and the relocation of people? This transfer of "truth" took place as Kephart's mythical view formed the basis of the discussion of the mountain people in the Smoky Mountains.

What images of the mountain people emerged from Kephart's work? First, they were castaways in time. Kephart painted this picture by emphasizing the iso-lation and inaccessibility of the Smokies, a view that proved so useful for the park promoters in the early campaign. As a result of the mountaineers' natural isola-tion, Kephart argued that their social development was arrested in the eighteenth century, an interpretation used to justify both the argument that the highlanders were "quaint" and that they posed a social danger to the rest of civilization and therefore needed to be eliminated. In preparation for his initial trip to the Smok-ies, for instance, Kephart described his inability to find materials concerning either the area or its inhabitants. "Had I been going to Teneriffe or Timbuctu, the libraries would have furnished information a-plenty; but about this housetop of eastern America they were strangely silent; it was 'terra incognita'" (1913, p. 13). More was known about the mountains than about the "odd" people living in them, who, according to Kephart, viewed even themselves as unique:

> The mountaineers of the South are marked apart from all other folks by dialect, by customs, by character, by self-conscious isolation. So true is this that they call all out-siders "furriners." ... No one can understand the attitude of our highlanders toward the rest of the earth until he realizes their amazing isolation from all that lies beyond the blue, hazy skyline of their mountains. Conceive a shipload of emigrants cast away on some unknown island, far from the regular tract of vessels, and left there for five or six generations, unaided and untroubled by the growth of civilization. ... They are creatures of the environment, enmeshed in a labyrinth that has deflected and repelled the march of our nation for three hundred years. [1913, pp. 16–17, 19]

This glorious isolation, according to Kephart, saved an Anglo-Saxon enclave intact. Although many of the mountaineers actually had German and Italian back-grounds, Kephart and promoters who used his image of the mountain people for the campaign ignored this fact and emphasized a Scotch-Irish ancestry instead. Kephart wrote, "There is something intrinsically, stubbornly English in the nature of the mountaineer: he will assimilate nothing foreign" (1913, p. 364). One pro-motional article declared that there were "no foreign-born and no negroes" living in the Smokies, implying that it was a desirable vacation spot for racist American tourists (*Washington [D.C.] Star*, April 1, 1928). The racial purity of the mountain people resulting from their isolation was a recurrent theme in Kephart's vision, and it was expressed, for instance, through a story in which mountain people visited a railroad for the first time. "Nearing the way station, a girl in advance came upon the first negro she ever saw in her life, and ran screaming back: 'My goddamighty, Mam, thar's the boogerman—I done seed him!'" (1913, p. 24).

Kephart further proved mountaineer isolation by describing the many "old-fashioned terms preserved in Appalachia that sounded delightfully quaint to strangers who never met them outside of books" (1913, p. 364). The highlander, he argues, "often speaks in Elizabethan or Chaucerian or even pre-Chaucerian terms" (p. 361). This characterization became a dominant theme for the cam-paign, as seen in one of many articles describing the mountain people. "Sharp ears are rewarded," said the article, "with fragments of early English speech, un-changed from the days of Chaucer's pilgrims. The mountain stock is pure Anglo-Saxon" (*Knoxville News-Sentinel*, March 3, 1940).

The second image of the mountain man that emerges from Kephart's work is that he is uneducated, unkempt, gun-toting, crude, violent, and bootlegging. Although there is considerable evidence to indicate that the God-fearing people of the Smoky Mountains were as concerned about education, tidiness, and law and order as the rest of rural America, they did not appear this way in Kephart's work. Considering the fact that six chapters of his book were spent on bootlegging, an-other one on family feuds, and another on violence in the mountains, while only one chapter discussed family life or religious beliefs, it can be concluded that Kephart's view of the highlanders was obviously influenced by his own interests and that he probably held the mountaineers in low regard, while praising them. He described the mountaineers as a "fierce and uncouth race of men" who com-mitted so many murders that every adult he knew had been directly interested in some murder case, "either as principal, officer, witness, kinsman, or friend" (1913, p. 266). To expand his negative image, Kephart painted all of the characters he met as dirty, semimoronic, undernourished medieval grotesques who, lost in the past, operated by a moral code that contemporary people could term quaint but certainly would not wish to follow or have as the basis for the rest of society. As Kephart said, "The mountain code of conduct is a curious mixture of savagery and civility. One man will kill another over a pig or a panel of fence (not for the property's sake, but because of hot words ensuing) and he will 'come clear' in court because every fellow on the jury feels he would have done the same thing himself under similar provocation; yet these very men, vengeful and cruel though they are, regard hospitality as a sacred duty towards wayfarers of any degree" (p. 267). What we had, according to Kephart, was a flawed social system, one that had been preserved in quaint isolation, one reminiscent of a less civilized time, but one that bred violence and poverty and therefore demanded change.

Two other related characteristics of the mountaineers rounded out Kephart's negative image. First, he painted them as cunning, although uneducated. They were excellent "deal" makers who enjoyed tricking "furriners" and figuring out the way to beat people at their own game. This characteristic became important for the campaign, as promoters argued that mountain people did not need to be treated in any special way. Ben W. Hooper, purchasing agent for the park, warned at a public meeting against "sickly sentimentality" toward the mountaineers in the proposed park area. He argued that "you fellows don't have to worry about them. They're shrewd traders. They skinned the yankees from snout to tail" (*Knoxville News-Sentinel*, March 6, 1929).

The final characteristic in this negative litany is that the mountain people were destroyers of the land. Kephart spent considerable space in his book describing the sad farms and even sadder condition of women and children, burdened by excessive work and poor food, in what he named "The Land of Do Without." The Smoky Mountains were not responsible for this poverty; the people themselves were, because they had exploited the land by using outdated farming techniques that had robbed the land of its richness, by overhunting so that few deer and other game were left, and finally by working for lumbering companies, the final exterminators of a glorious landscape. Although Kephart admired their ability to survive in a hostile environment, he presented mountain folk as backward ravagers of the land who failed to use the resources provided them by the mountains, because of a lack of understanding, modern sensitivity, and wit; an excess of stubbornness and laziness; and an inability to develop an environmental ethic. Highlanders as destroyers and modern-day Visigoths became an important part of the rhetorical image later in the campaign, when park promoters felt a need to respond to increased criticism of their policy to eliminate all people from the park area.

Lessons from the Mountains

In this chapter I have discussed somewhat briefly a complex, multifaceted, and lengthy rhetorical campaign. Throughout this discussion, however, one significant point has become clear. The campaign to save the Smoky Mountains from destruction and preserve them as a national park was noble in intent and displayed a high level of communicative sophistication, at least when the promoters were "selling" the value of the area and persuading people that establishing the park was both economically and intrinsically valuable. It failed significantly, however, when it attempted to deal with the complicated issue of the people who had made their homes in the mountains for generations. While the campaign successfully created consensus among the more successful members of society, it failed to persuade those people who were not in a position to see the direct value of the park to their lives. As a result of their class prejudices and the negative stereotypical image of the mountain people perpetuated by popular writings, especially those of Horace Kephart, the park promoters initially tended to treat the mountain people in a condescending way, and eventually they abused them rhetorically as they became more "troublesome."

Instead of focusing on legitimate grievances, campaigners reinforced stereotypes by emphasizing the flawed nature of mountaineer society, calling outspoken mountain people liars and troublemakers, arguing that their only reasons for disagreeing were to inflate selling prices of land, meanness, stupidity, or a failure to understand the importance of the park project for a modern world. For instance, in response to direct questions concerning relocation at a Knoxville town meeting, pro-park speakers said, "If they insist on living in the mountains, there are plenty of other mountains around," displaying an insensitivity rarely seen in the rest of the campaign (*Knoxville Journal,* April 4, 1929).

This abrupt behavior resulted somewhat from class differences but, most important, also from a low opinion of the mountaineers. The campaign organizers assumed they knew what was best for everyone. Nowhere in the information produced by the campaign, however, is there any clear evidence that the leaders ever genuinely understood the problems raised by moving hundreds of people out of their homes. This failure did not seem to affect negatively the success of their campaign, however. There are probably two explanations for this: the mountain people, being rather unsophisticated in the area of communication, were unable to generate enough interest in their issue; and most people outside the mountains, even those much poorer than Chapman and Kephart, accepted the vision of the mountain people held by campaign organizers as underfed rustics who really did not know what was good for them and who certainly did not deserve much consideration when an issue as significant as a national park and millions of dollars of tourist money were at stake.

We can only hope that contemporary environmentalists do not fall prey to the same misconceptions that forged the words and deeds of the promoters of the Great Smoky Mountains National Park. As the world becomes more crowded and the need for wilderness increases, finding land that is not inhabited or is not connected somehow with human interaction becomes almost impossible. Virtually no land exists that can be rhetorically separated from human habitation. Therefore, it becomes imperative for those concerned with the preservation and rejuvenation of wild environments to frame arguments that incorporate human needs. As Alfred Runte points out in his book *National Parks: The American Experience* (2nd ed., 1987), Americans have rarely found compelling either the argument that scenery is worthy of preservation for its own sake or that preservation of the environment is meaningful separate from human interaction. The Smoky Mountains campaign shows how in the process of arguing in favor of one type of human use, namely tourism, we can simultaneously destroy another compelling argument: that the interaction of human history with an environment makes a place more meaningful and even more worthy of preservation.

Notes

1. Wealthy and successful city dwellers, the park promoters argued for economic growth and development; and like the timber barons who had controlled the region for thirty years, the new developers who believed in tourism equated progress with increased personal wealth. While some historians represented them as Progressives, others argued that their failure to embrace important social reforms associated with progressivism, such

as programs for the poor, clearly showed their conservative leanings (McDonald & Wheeler, 1983, p. 42).

2. Of course, this strategy may have ultimately backfired for the promoters, as the campaign failed to produce the funds necessary to purchase land for the park. In order to be persuaded to donate money, especially in hard economic times, people need to feel that something important may actually be lost or fail to materialize if they do not personally take action. Perhaps the park promoters, in emphasizing the inevitability of the park, replaced the urgency necessary for personal action with complacency and lessened the need people felt to donate.

3. It was a well-used convention in the nineteenth and early twentieth centuries for writers, when describing American landscapes, to compare the unknown environments of the New World to more easily recognizable ones, especially those familiar to well-to-do travelers to Europe. For a thorough examination of this practice, see Runte, 1987. Note Runte's discussion of early descriptions of Yosemite and Yellowstone; the natural areas were described in such a way as to help define a historical culture for the United States. According to early chroniclers, a limestone formation "bore a strong resemblance to an old castle," whose "rampart and bulwark were slowly yielding to the ravages of time." Indeed, the explorers could almost imagine "that it was the stronghold of some baron of feudal times, and that we were his retainers returning laden with the spoils of a successful foray" (Runte, 1987, p. 35).

4. In his study of miners in Yellow Creek, Tennessee, John Gaventa argues that their apparent willingness to be exploited was a result of "the effective wielding of power in all its dimensions to achieve consensus," rather than an inherent problem with the people (1980, p. 75). Likewise, other historians conclude that subordinate groups, such as the mountain people, share a "kind of half-conscious complicity in their own victimization" because they "participate in maintaining a symbolic universe [that of the promoters, for example], even if it serves to legitimate their domination" (Lears, 1985, p. 573).

5. Durwood Dunn, a critic of Kephart, attacks him for portraying "Southern Appalachian dialect in such a distorted and grotesque fashion he completely misinterpreted regional speech patterns" (1988, p. 156). Another critic argues that Kephart "seems to have been impressed particularly by what would look like good dialect on paper, and his notes and published writings scarcely do justice to the speech which he seeks to represent" (Hall, 1942, p. 4). Critical reaction to Kephart's representation of the behavior and social life of mountain people is even more damning, as he is accused of focusing on people and behavior unrepresentative of the majority living in the mountains and of reinforcing stereotypes of hillbillies and bootleggers. "Despite these enduring stereotypes," argues Dunn, "the historical record clearly shows that . . . family life in Cades Cove at the turn of the century was largely indistinguishable from that of other rural Tennesseans" (1988, p. 200).

References

Manuscript Collections

Great Smoky Mountains National Park Archives, Gatlinburg, Tenn.
Zebulon Weaver Papers, Western Carolina University Library, Cullowhee, N.C.

Newspapers

Allentown (Pa.) News, March 26, 1926
Asheville Citizen, January 1, 1921–December 31, 1941
Knoxville Journal, January 1, 1921–December 31, 1941
Knoxville News-Sentinel, January 1, 1921–December 31, 1941

Nashville Tennessean, January 1, 1926–December 31, 1928
Washington (D.C.) Star, April 1, 1928

Books and Articles

Bohn, F. (1926, January 25). A new national park. *New York Times,* late ed., p. 7.

Brown, M.L. (1990). Power, privilege, and tourism: A revision of the Great Smoky Mountains National Park story. Unpublished M.A. thesis, University of Kentucky.

Campbell, C. (1969). *Birth of a national park in the Great Smoky Mountains.* Knoxville: University of Tennessee Press.

Dunn, D. (1988). *Cades Cove: The life and death of a southern Appalachian community, 1818-1937.* Knoxville: University of Tennessee Press.

Frome, M. (1966). *Strangers in high places.* Garden City, N.Y.: Doubleday.

Gaventa, J. (1980). *Power and powerlessness: Acquiescence and rebellion in an Appalachian valley.* Urbana: University of Illinois Press.

Great Smoky Mountains Conservation Association. (N.d.) *Great Smoky Mountains.* Knoxville: Conservation Association.

Great Smoky Mountains National Park. (N.d.) Knoxville.

Gregg, W.C. (1925, December). Two new national parks? *Outlook,* pp. 662-67.

Griffin, C.J.G. (1990). The rhetoric of form in conversion narratives. *Quarterly Journal of Speech, 76,* pp. 152-63.

Hall, J.S. (1942). *Phonetics of Great Smoky Mountain speech.* New York: King's Crown Press.

James, H. (1928, October). The Great Smokies: Site of a proposed national park. *Review of Reviews,* pp. 373-77.

Kephart, H. (1913). *Our southern Highlanders.* New York: Outing Publishing.

———. (1926). The last of the eastern wilderness. *World's World,* pp. 617-30. Rpt. in *Tarheel Banker* (1929), pp. 43-50.

Lears, T.J.J. (1985, June). The concept of cultural hegemony: Problems and possibilities. *American Historical Review,* pp. 567-93.

Lévi-Strauss, C. (1963). *Structural anthropology.* C. Jacobson & B. Grundfest Schoef (Trans.). New York: Basic Books.

———. (1966). *The savage mind.* Chicago: University of Chicago Press.

———. (1969). *The raw and the cooked.* J. Weightman & D. Weightman (Trans.). New York: Harper & Row.

McDonald, M., & Wheeler, W.B. (1983). *Knoxville, Tennessee: Continuity and change in an Appalachian city.* Knoxville: University of Tennessee Press.

Perdue, C.L., & Martin-Perdue, N.J. (1979-80). Appalachian fables and facts: A case study of the Shenandoah National Park removals. *Appalachian Journal,* pp. 84-104.

Runte, A. (1987). *National parks: The American experience.* 2nd ed., rev. Lincoln: University of Nebraska Press.

Spengemann, W.C., & Lundquist, L.R. (1965). Autobiography and the American myth. *American Quarterly, 17,* pp. 501-19.

U.S. Department of the Interior. (1931). *Final report of the Southern Appalachian National Park Commission.* 1931. Washington, D.C.: GPO.

8

Plastics as a "Natural Resource": Perspective by Incongruity for an Industry in Crisis

Patricia Paystrup

The photograph frames the ocher-colored angles of two pyramids against the contrasting background of a deep blue Egyptian sky. The print ad from the Environmental Challenge Fund juxtaposes this striking image of timelessness and human ingenuity with what some see as the modern world's symbol of engineered durability—the polystyrene "clamshell." The headline claims: "Your cheeseburger box will be around even longer." The copy block begins: "Most things made on this planet last a few centuries. But styrofoam is forever. It will never decompose. Never disintegrate. Never go away. And neither will the garbage problems it creates, unless we find solutions."

The pyramid–cheeseburger box ad began running in *Time* in November 1990—about the same time the president of McDonald's USA, Edward A. Rensi, announced the company's response to boycotts by environmentally concerned consumers: "Although some scientific studies indicate that foam packaging is environmentally sound, our customers just don't feel good about it. So we're changing" (Hume, 1991). The fast-food chain's change spelled victory for grade schoolers in the eight hundred chapters of Kids Against Pollution. In addition to a standard letter-writing campaign, these young activists also mounted a Send-It-Back effort, where they packed up greasy polystyrene containers and mailed them to local McDonald's stores or to the company's headquarters in Illinois (Castro, 1990). Kids Against Pollution celebrated, but overall reaction to the announcement was mixed.[1]

An executive with the polystyrene plastics division of Dow Chemical saw the McDonald's action as a caving in to customer pressures based on "perceptions and without understanding the myths and realities involved" (Sternberg, 1990). A representative of Huntsman Chemical, the company headed by the man who pioneered the polystyrene clamshell in the early 1970s, noted that McDonald's "has responded to the *perception* of polystyrene foam rather than to the *reality*" (Leaversuch, 1990). But there was more to the McDonald's polystyrene phaseout than bowing to public perceptions: McDonald's senior vice president, Shelby Yarrow, claimed that the company was disappointed with the polystyrene industry's slowness in building a viable recycling infrastructure and felt that the time and money invested with the National Polystyrene Recycling Company "had created a financial drain on the company" (Wolf & Feldman, 1991, p. xii; Sternberg, 1990; Leaversuch, 1990).

McDonald's admission that the decision to eliminate polystyrene was based on more than environmental boycotts illustrates how the plastics industry's recycling advocacy advertisements and supporting public relations campaigns are sabotaged by the harsh reality of an unstable and inadequate plastics recycling infrastructure, further weakened by market forces that make virgin plastic resin much cheaper to use than recycled resins. By 1993—after some five years of pushing public awareness of plastics recycling—only 2.2 percent of the plastics packaging produced made it to recycling facilities. When compared with aluminum's 70 percent rate, the roles played by an established infrastructure and favorable market forces become more obvious in understanding the irony of high-tech plastic's eventual fate as valueless waste. "Of all the junk that constitutes America's household wastes, plastics are the most recycling resistant. The material comes in so many varieties, produced from such different polymers, that just separating it in the recycling process is cumbersome and expensive" (Van Voorst, 1993, pp. 79–80).

In this chapter I will examine the communication strategies used in advocacy advertisements developed by the plastics industry to address its "image" and "public opinion" problems during two periods—the first in the late 1980s and the second in the early 1990s. I will also explore the ironies surrounding the industry's efforts to push plastics recycling in the first wave as a way to combat antiplastics legislation and then to work just as hard to defeat measures that would create market forces needed to support plastics recycling in the second phase of the industry's advocacy efforts. For an industry in crisis, advocacy advertising—as a rhetorical activity—is "a mode of altering reality, not by direct application of energy to objects, but by the creation of discourse which changes reality through the mediation of thought and action" (Bitzer, 1968, p. 3). These two very different campaigns both set out to alter what the plastics industry sees as the threatening realities of harsh legislation and widespread public misperceptions of plastics.

The first set of advocacy advertisements used in this analysis ran in major magazines during 1989 and 1990. These ads were launched in an attempt to defeat plastics bans and to change what the industry called misunderstandings and inaccurate perceptions of the "myths" and "realities" of plastics' contribution to the solid waste crisis. A number of these advertisements employ a rhetorical device Kenneth Burke calls "perspective by incongruity"—placing two previously unconnected things together in ways that transfer meanings and revise perceptions—to recategorize plastics as both a recyclable material and the new "natural resource." The "planned incongruity" in these advertisements recycles synthetic plastics to mimic the closed system found in nature. Postconsumer plastics become symbolically connected to natural life cycles: they are "born again" with a "second life" and then finally slough off their negative synthetic identity as they are positively transformed into the new "natural resource." The second set of advocacy messages—the initial "Take another look at plastic" campaign—began running in late 1992 with media buys scheduled until May 1993. While the first series was primarily a defensive or reactive response formulated to fight legislative plastics bans, the second campaign was launched as a proactive or positive attempt to improve plastics' overall poor image by showing how high-technology plastics contribute to modern living and also, incidently, help "save the planet."

When these two campaigns are compared and contrasted, they contradict the accepted wisdom in public relations and issues management circles that says that proactive communication programs are inherently better than reactive responses. In this case, the reactive campaign has a focus and clarity that is missing in the proactive campaign. The first campaign's use of perspective by incongruity powerfully challenges readers radically to revisualize and revise their perception of plastics. By contrast, the positive "Take another look at plastic" campaign merely presents a quick, superficial glance at a glossy surface. The reactive campaign tells a fascinating—almost archetypal—story of transformation and rebirth, as it reclassifies plastic as recyclable. The proactive campaign, borrowing the patter of a stand-up comic, throws out one-liners in an attempt to appeal to everyone with an upbeat "plastic is your friend" type of image message.

The major differences between the two campaigns are grounded in an understanding of the larger "rhetorical situation." The first campaign responds to a pressing "exigence"—plastics bans—and addresses one set of "constraints"—the belief that plastics cannot be recycled. The "Take another look at plastic" ads do not answer any specific exigence-generated questions. Nor do they demonstrate an understanding of the target audience's primary concerns or the broader cultural "constraints" operating in the situation. These cultural constraints include what one anthropologist describes as the schizophrenic love-hate relationship we have with plastic as the basic substance of our modern material culture (Gutin, 1992).

Understanding how advocacy advertising campaigns—when used to issue manage public policy initiatives or to reshape images—need to respond to the demands of a larger "rhetorical situation" points to the importance of answering a concerned audience's questions. To examine how these two major advertising campaigns address the components of the rhetorical situation as they fit into the larger issue advocacy campaigns waged by the plastics industry, this chapter will explore how the plastics and solid waste disposal issues developed in the late 1980s, by placing them into an issues management context; discuss the industry's decision to use advocacy advertising as an issues management strategy; analyze how the first recycling advocacy campaign employed perspective by incongruity to show an industry transforming used plastics into a "natural resource"; discuss how changing public policy issues in the early 1990s met with contradictory issues management tactics; show how the "Take another look at plastic" campaign fits into the industry's advocacy efforts; and analyze how this second "positive" campaign may actually further damage the industry's image by failing to address the real issue on consumers' minds—a still woefully inadequate plastics recycling infrastructure.

Turning Up the Heat on Plastics

In 1959, Procter and Gamble's new packaging for Ivory Liquid placed the first plastic bottles on supermarket shelves. Thirty years later, the offspring of those original shatterproof bottles faced some tough legislation that threatened to break the plastics industry. By the end of 1989, industry analysts estimated, at least eight hundred solid waste bills involving plastics were on the dockets of state and local policy-making bodies in at least thirty-five states. A few years earlier there might have been ten at the very most (Stuller, 1990). With environmental activists pressing for outright bans on plastics, the industry was forced to enter into a high-stakes game to influence policy makers and to persuade the public on solid waste disposal issues.

Faced with increasing public policy pressures in recent decades, business and industry turned to a more powerful form of public relations called issues management. Issues management is a strategic planning approach aimed at combating the public policy agendas of activist groups. Originally designed by American Can executive Howard Chase to defend the business sector in a hostile "court of public opinion," issues management would become the way industry could exercise "every moral and legal right to participate in *formation* of public policy—not merely to react, or be responsive, to policies designed by government" (Jones & Chase, 1979, p. 7).

The issues management process model of Barrie Jones and Howard Chase gave industry a tool for monitoring developing issues, based on systems theory. Coming from a background in rhetorical theory, Richard Crable and Steven Vibbert (1985) added another perspective to the growing field of issues management literature. The Jones and Chase model neglects issue development. Crable and Vibbert—and subsequently Brad Hainsworth (1990)—address the need to understand the life cycles of issues. According to Crable and Vibbert, issues go through five basic stages: "potential," where there is latent interest; "imminent," where the interest builds and legitimation occurs, as people begin to see linkages; "current," where people participate in discussions of the issue and the media cover it; "critical," when the issue demands a decision; and "dormant," when the issue is resolved or forgotten. Often a "trigger event"—like the 1987 media coverage of the Islip, New York, garbage barge sailing up and down the east coasts of North and South America looking for a willing dumping ground—will catapult an issue into the center of the current or critical stage.

When Jones and Chase (1979) identified an issues management model of the public policy process, they defined three basic strategy options for issue change—reactive, adaptive, and dynamic. Crable and Vibbert (1985) reinterpreted the Jones and Chase model to add a fourth option—catalytic. The type of change strategy option an organization may decide to use to respond to challenges or opportunities is also related to where the issue—defined as a contestable question of fact, value, or policy—is in the issue status life cycle.

The four change option strategies suggest a sort of hierarchy of time and effort: long-term, mindful "management" versus short-term, defensive "firefighting." The catalytic strategy begins as early as the potential stage in the form of "agenda stimulation." The other strategies, Crable and Vibbert note, primarily respond to agendas set by others. An organization choosing the dynamic stance anticipates action and may initiate projects and policies as early as the imminent stage, but, as explained by Jones and Chase and critiqued by Crable and Vibbert, it usually waits until the issue is current before acting. The adaptive strategy adjusts to change through compromise or proposed alternatives. Adaptive strategies often begin in the current stage, but they are also adopted after the issue has reached the critical stage. When organizations employ reactive strategies, they usually face issues only in the critical stage, when a crisis demands policy decisions (Crable & Vibbert, 1985). Robert Heath and Richard Nelson observe: "Issue management is doomed to fail if it is not based on the stark reality that opinion is the foundation of laws, and that considerable media discussion transpires prior to an issue reaching legislative chambers" (1986, p. 141). Catalytic and dynamic strategies involve more time and effort in the early stages in order to favorably "manage" the status of the issue. Adaptive and reactive strategies are mainly defensive tactics, which demand attention at the current or critical stage, where others have set the agenda and have defined the issues.

At first, no single group or sector seemed to be actively "issue managing" the solid waste problem. By the late 1980s, however, proposed plastics bans multiplied, as one reaction to larger concerns over solid waste disposal. Nationwide, the number of available landfill sites declined from eighteen thousand in 1985 to nine thousand in 1989 (Wood, 1990). In New York State, fourteen landfills closed between 1980 and 1989, and experts estimated that more than half the landfills serving the East Coast would be out of room by 1990 (EarthWorks Group, 1989). The developing solid waste disposal issue grabbed public attention and entered the "current" stage in the summer of 1987, "when the sorry image of the wandering garbage barge towed by the tug *Break of Dawn* was seared onto the collective retina" through ongoing news coverage (Gutin, 1992).

The ill-fated Islip, New York, garbage barge made trash the new cause célèbre, as Americans began to take a closer look at what went into the "waste stream" (Gutin, 1992; Callari, 1989). As a result of the renewed scrutiny, plastics—the fastest-growing segment of the packaging industry—received critical reviews in just about every popular discussion of environmental and solid waste issues (Lawren, 1990). The public's growing antiplastics sentiments fed on statistics found in environmental best-sellers like *50 Simple Things You Can Do to Save the Earth:* "Each American uses about 190 pounds of plastic per year—and about 60 pounds of it is packaging which we discard as soon as the package is opened. . . . Americans go through 2.5 million plastic bottles every hour" (EarthWorks Group, 1989, p. 66).

With the public's eye now on solid waste disposal, plastics' original virtues—high-performance qualities like strength, durability, and light weight—became their new vices. Another irony emerged: the new definition of plastics as a major part of the pressing national environmental crisis also symbolized how successful the plastics industry had been in developing new product applications. The plastics industry saw itself as the most direct expression of ultimate product utility and convenience in packaging; environmentalists saw the plastics industry as the foremost expression of product wastefulness and social and economic dysfunction (Blumberg & Gottlieb, 1989). As the proposed plastics packaging bans added up, a plastics industry journalist wrote, "Obviously, this entire controversy amounts to more than a hill of garbage. The future of an entire industry—or a least certain key components of it—is at stake" (Callari, 1989).

The plastics industry first adopted a reactive stance when the solid waste issue was catapulted onto the critical stage, because the industry just was not ready. Not ready, that is, to be labeled a "villain," "scapegoat," or "target," as the industry lamented in its own publications (Callari, 1989). Ironically, Louis Blumberg and Robert Gottlieb point out, industry claims that it was singled out as the "symbol" of the waste crisis were true, true because the bans resulted from the activist public's perception of one of the plastics industry's most enduring and characteristic features—its continuing drive to expand existing markets, establish new ones, and produce more and more plastic products destined to become part of the burgeoning solid waste stream (1989, p. 273). With its advantages having become disadvantages, plastic was caught in another ironic cosmic joke, as the versatility that enabled it to fill so many packaging niches also made it difficult to recycle. Glass is glass, and aluminum is aluminum, but plastic has a thousand faces (Gutin, 1992). Since plastic comes in so many varieties, separating it is so cumbersome and expensive that plastics become the most recycling-resistant part of our household trash (Van Voorst, 1993, p. 79).

Firing Back

By 1989 the plastics industry was at a crossroads. Would it "reactively" continue to battle legislation piece by piece? Or would it commit to an "adaptive" issues management strategy more consistent with the environmental concerns of a changed business and political climate? Experts like Susan Selke, a professor at Michigan State University's School of Packaging, advised: "The plastics packaging industry can no longer proceed with business as usual because legislators will no longer allow it. Through the use of taxes, bans, deposits, and other tools, restrictions will be placed on the ability of businesses to use plastics packaging. The most effective way to combat restrictions is for the industry to promote recycling, because recycling is politically popular" (Stuller, 1990). From the plastics industry's perspective, the decision to push recycling as the "way out" of the solid waste crisis signaled a shift from a predominantly reactive issues management style to more adaptive strategies. But there were two serious problems associated with pushing recycling as the new adaptive strategy. First, a postconsumer plastics recycling infrastructure was not yet in place, and second, consumers retained the widely and deeply held belief that plastics are not recyclable. Ironically, the first problem—the fact that virtually no plastics were being collected for recycling—was probably the biggest factor in the public's perception of plastics as the nonrecyclable "villain" in the waste stream.

Industry groups scrambled to begin building a postuse recycling system. Resin producers and manufacturers of soft drink bottles formed the National Association for Plastic Container Recovery and set the goal that by 1992, 50 percent of the polyethylene terephthalate (PETE) produced would be recycled. In June 1989 the seven polystyrene (PS) giants—Amoco, ARCO, Dow, Fina, Huntsman,

Mobil, and Polysar—chipped in two million dollars each to start the National Polystyrene Recycling Company (NPRC). Seven major resin suppliers formed a new division of the Society of the Plastics Industry (SPI) in June 1988. Called the National Council for Solid Waste Solutions (CSWS), the group planned "to put money, brain power, and political clout behind a push into plastics' waste reuse" ("Resin Suppliers," 1988).

With a recycling infrastructure slowly taking shape, the plastics industry was finally ready to begin its national recycling advocacy campaign by late 1989, two years after the Islip garbage barge sailed the solid waste issue into the current-critical stage of the policy agenda. But getting the campaign together was no easy task. The overall cautiousness of the SPI was reflected in the announcement of its plans for a new image campaign in the spring of 1989: "SPI has chosen this moment to start what it calls a 'special communications program' that 'seeks to educate external audiences.'" SPI's communications vice president was quoted as saying that the current wave of antiplastics sentiment and legislation was great enough to warrant spending on "proactive" as well as "defensive" actions: "We've reached a point where a positive image program is no longer a luxury" ("SPI Creates," 1989).[2]

Some members of the SPI's new Council for Solid Waste Solutions pushed for a higher-profile, more aggressive issue and image campaign.[3] At the December 1989 meeting of the CSWS, Dow's call for the SPI to mount a "media blitz" and a three-year, $150 million campaign to improve plastics' image "ignited a fierce debate within the industry." Some voiced a desire to see that kind of money spent on developing new technologies and building more recycling programs, not on advertising. This proposal split the association into opposing camps and torpedoed plans for more CSWS advertisements like the twelve-page insert that kicked off the industry's plastics-are-the-most-recyclable-of-recyclables drive ("Dow Calls," 1990).[4]

Natural Plastics

The plastics industry's inability to choose between the more conservative, reactionary change strategy options and more aggressive, adaptive change strategies or styles is apparent in an analysis of the series of advertisements produced by the CSWS and the individual resin companies. The same pattern of tensions between more reactive strategy messages versus messages with a more adaptive stance continues throughout the series of ads. In fact, the copy for many ads mixes elements of both. The "reactive" elements are mundane arguments about "the facts" of the solid waste issue. The better-executed "adaptive" messages, however, transcend the mundane through Burke's notion of perspective by incongruity or the "comic corrective." This use of perspective by incongruity reclassifies plastics as recyclable and further connects plastics to the life cycles of the natural plant and animal worlds, so that plastics too are "born again" into a "second life," with new identities as useful long-lived products. In this way, postconsumer plastics packaging changes from "trash" to a new "natural resource."

Perspective by incongruity gives the plastics industry a way to attempt to re-define and recategorize a product the public sees as inherently *unnatural* and environmentally antagonistic in three major ways: it is a nonrenewable petroleum product, it is nonbiodegradable, it is not recyclable. In a case like this, Burke would argue, the "comic frame" or perspective by incongruity becomes a method for gauging situations and changing meanings through verbal "atom cracking": "That is, a word belongs by custom to a certain category—and by rational plan-ning you wrench it loose and metaphorically apply it to a different category" (Burke, 1984a, p. 308). By linking previously unlinked words, concepts, and cate-gories, perspective by incongruity impiously challenges our established sense of what properly goes with what and transfers meaning from one setting to another (Burke, 1984a, p. 309; 1984b, p. 90).

Burke sees perspective by incongruity as more than just a literary conceit. It is a serious method for holding traditional views up to their opposites, to over-come the limitations a classification places on us: "The metaphorical extension of perspective by incongruity involves casuistic stretching since it interprets new situations by removing words from their 'constitutional' setting. It is not 'demor-alizing,' however, since it is done by the 'transcendence' of a new start. It is not negative smuggling, but positive cards-face-up-on-the-table. It is designed to 're-moralize' by accurately naming a situation already demoralized by inaccuracy" (Burke, 1984a, p. 309). The "transcendence" used in perspective by incongruity turns "problems" into "assets": "Thus we 'win' by subtly changing the rules of the game—and by a mere trick of bookkeeping, like the accountants of big utility cor-porations, we make 'assets' out of liabilities" (p. 308). Acceptance of the new perspective or categorization offered by the comic frame completes the transla-tion or revision of the situation (p. 173). The conceptual revisions offered by the comic frame occur when the material aspect or "essence" combines with the tran-scendent, as A and B cease to be opposites and a new categorical identity forms. As Burke explains: "Or, viewing the matter in terms of ecological balance, one might say of the comic frame: It also makes us sensitive to the point at which one of these ingredients becomes hypertrophied, with the corresponding atrophy of the other. A well-balanced ecology requires the symbiosis of the two" (p. 166).

The material—used plastic packaging—and transcendent—useful new durable products—join in the comic frame in Amoco's two-page spread, head-lined "We'd like to recycle the thinking that plastics can't be recycled." The verso page of the spread shows typical plastic containers—a PETE two-liter soda bottle, a milk jug, a PS clamshell and cup, and so on—and has "Before" as a subhead. The artwork on the recto page includes a parka-type jacket, a park bench, a roll of carpet, a paintbrush, tulips growing in flowerpots, and an ultramodern house. The subhead reads, "After." The copy further juxtaposes the material with the transcendent: "Contrary to public opinion, plastics are among the easiest materi-als to recycle. In South Carolina, one company is recycling 100 million pounds of used plastic soft drink bottles a year into carpet yarn, flower pots, and fiberfill for ski parkas. In Chicago, another company is recycling 2 million plastic milk jugs a year into 'plastic lumber' for decks. In Tennessee, another company is recycling plastic beverage containers into bathtubs and shower stalls." After this series of examples of the new, second lives of recycled plastics, the copy builds to a final transcendent leap: "The recycling of plastics is rapidly catching on. Recycling is transforming used plastics into a 'natural resource' that can be used to produce many new products." Placed side by side, the plastics pictured on the left side of the spread are no longer garbage, as they become, on the right side, "a 'natural resource.'" With this visual as well as verbal "atom cracking," the plastics industry attempts to reclassify plastics as "among the easiest materials to recycle" and thus, as Burke wrote, "'remoralize' by accurately naming a situation already demoralized by inaccuracy" (1984a, p. 308).

Perspective by incongruity also helps the reader look at the photograph of a polystyrene clamshell and cup in a Huntsman Chemical ad and, as the headline says, "Think of them as your new home. Think *recycle*." The copy begins: "Look closely. What do you see? We see a convenient and economical means of packaging—a resilient resource that can be brought back to life in any number of forms. Like insulation board in the walls of your home. Or products for your office. Or playground equipment. That's because polystyrene is recyclable." The Huntsman ad gives more examples of how this "recyclable" and "resilient resource" is transformed "into a variety of durable consumer products, from video cassette boxes to cafeteria trays to trash cans." Burke's observation that perspective by incongruity turns liabilities into assets through transcendent renaming explains how these "convenient and economical means of packaging" are no longer part of the solid waste disposal crisis but are instead "a resilient resource that can be brought back to life."

The comic frame is at work when Dow Plastics shows a strikingly designed chair and tells readers: "You're looking at 64 milk bottles and 2 shampoo containers." The material and the transcendent join again in Fina's rebus-styled ad with a whimsical headline and artwork combination that simply declares: "How to recycle"—with a picture of a polystyrene clamshell and cup—"into"—then a series of pictures with commas between them: a tape dispenser, a plastic spray bottle, a brush and dustpan, a watering can, and poker chips. These two simple ads illustrate Burke's observation that the comic frame "also makes us sensitive to the point at which one of these ingredients becomes hypertrophied, with the corresponding atrophy of the other" (1984a, p. 166). When the emphasis is on the wide range of new products made from recycled plastics, the transcendent hypertrophies while the material—postconsumer waste—atrophies. Plastics are no longer trash. They are chairs and office supplies.

When the plastics industry's advocacy ads do not focus on transforming recycled plastics into the new "natural resource," they refute "the facts" about plastics packaging and solid waste disposal. Because these ads do not employ perspective by incongruity to correct or rename the situation, they contrast sharply with those ads that use the comic frame to invite reclassifications and transcendent translations. One example of the ads designed to present the industry's side of "the facts" is an Amoco ad with a left-page art piece suggesting a shower in progress as a glass shampoo bottle shatters on the floor—with female toes perilously close to the shards—and the right-page headline, "Do we really want to return to those good, old-fashioned days before plastics?" The ad's copy reads: "Some people believe that banning plastics and substituting other materials will solve the problem. We don't think they have all the facts. If plastics were banned, we'd lose safety and convenience features such as closures for foods and medicines, shatter-resistant bottles, freezer-to-microwave packages, and wrappers that preserve food freshness." Continuing in the defense of plastics, the next paragraph adds: "A 1987 study shows that if paper and other packaging materials were to replace plastics, the energy needed to produce the packaging would double, the weight of the packaging would increase four-fold, the packaging costs would double, and the volume of waste collected would increase about 2 $^1/_2$ times."

The ad's straight talk continues with "the facts" to refute charges that plastics' volume takes up one-third of the space in a landfill: "According to a recent study, plastics make up about 18% of the volume of solid waste in our landfills; paper and paperboard about 38%; metals, 14%; glass, 2%; and other wastes, 28%." In another Amoco ad headlined, "Let's dig a little deeper into the notion that much of our garbage is made up of plastics," the solid waste issue is framed by weight, not volume: "A lot of well-intentioned solutions are being offered. One is that foam plastics, plastic bottles and plastic packaging should be banned. The fact is that plastics make up less than 8%, by weight, of our nation's waste. Paper and paperboard make up about 36%, glass and metal about 9% each, all by weight. The rest is anything from yard wastes to lumber to rubber tires." Either way, according to the solid waste studies that the plastics industry chooses to cite, plastics are not that large—or heavy—a problem. After citing what the company is doing to make plastics recycling a reality, the ad closes with this: "At Amoco Chemical, we believe we're only beginning to see the benefits of recycling. In the not-too-distant future, it can turn solid waste from a national problem into a national resource."

Nonetheless, these ads do not recycle our thinking about plastics. Without perspective by incongruity, the plastics in these ads remain throwaway packaging and trash. As such, they are still part of the solid waste problem. When refuting the solid waste "facts" by weight or volume, the issue focuses again on plastics as garbage. Without the powerful transformations offered by perspective by incongruity, plastics end up headed for the landfill. But with the "atom-cracking" reorientations offered by the comic frame, our perceptions of postconsumer plastics change from the material—nonrecyclable, nonbiodegradable garbage—to the transcendent—a recyclable resource that becomes durable products like plastic lumber park benches.

Shifting from Bans to Recycled-Content Laws

The advocacy advertisements the individual resin companies ran in 1989 and 1990 using perspective by incongruity to show how recycling technologies could turn "solid waste from a national problem into a national resource" increased awareness of plastics' recyclability yet also created new criticisms and expectations. Now that state and local decision makers and the consumer public knew that plastics are recyclable, they began demanding plastics recycling in their communities. While the plastics industry had been working to build recycling plants, the recycling infrastructure was still so inadequate that the Minnesota state attorney general, Hubert H. Humphrey III, asked a poignant question: "Is it really fair to tell Minnesota consumers that plastic foam cups are 'recyclable' when there isn't a single recycling plant between Minneapolis and Brooklyn, N.Y.?" (Humphrey, 1990). The Environmental Protection Agency advised: "Plastics are technically recyclable—that is, they can be remelted and formed into other items. The term recyclable, however, means that there is a way to collect and separate the materials. . . . The recyclability of plastics can therefore vary over time and location" (Javna 1991).

With the undeveloped recycling infrastructure in mind, environmentalist John Javna took the industry to task for confusing consumers with "misleading public relations campaigns" that are "more likely to trumpet information about limited recycling programs than to discuss how widespread plastics recycling really is. As a result, consumers think that plastics recycling is imminently available at local recycling centers, when in fact it may be a long way off." But Javna found one positive aspect: "The good news is that, having given us the impression that their products are recyclable, plastics manufacturers now have to meet our expectations. So they're working hard to create the infrastructure that will make more plastics recycling practical" (1991).

While environmentalists like Javna expressed hope that a plastics recycling infrastructure would soon take shape, leading members of the plastics industry worked to sabotage plastics recycling. As shifting pressures in the public policy arena added to the plastics industry's need to clarify issues management strategies, the industry instead turned to confusingly mixed and contradictory tactics that obfuscated the issues. On the policy front, the industry publicly touted recyclability while at the same time reactively fighting to defeat the types of recycled-content packaging laws that would have made plastics recycling a practical reality, not just a promise. In the public relations area, the industry's willingness to create new organizations with large budgets and an "outreach" mandate suggested that industry leaders recognized the importance of communicating with various publics and pursuing adaptive strategies. The programs launched by these new groups, on the other hand, reveal that the industry's definition of *communication* was stuck in the old, linear "Let us tell our side of the story" model. The advocacy messages demonstrate no clear understanding of the relevant audiences' interests or concerns. In an attempt to point out the "positive" points about plastics, the industry's advocacy program sidesteps the real issues on the public agenda.

The best examples of how the industry's recycling policy efforts worked at cross-purposes are found in its reactions as recycled-content packaging bills replaced the earlier outright plastics bans. With recycled-content packaging laws, lawmakers seemed to be saying: "OK, if your products are so recyclable, let's see you recycle." The industry, led by Dow and Mobil, fought the laws on the national, state, and local levels by manufacturing arguments that recycled-content packaging laws would increase consumer costs and ultimately limit consumer choices because smaller companies would be unable to compete if forced to use recycled-content packages.[5]

While the industry fought measures that would make recycling economically feasible, consumers became increasingly frustrated with the fact that they could not recycle their supposedly recyclable plastic trash. Citizens in eleven states expressed their outrage on Earth Day 1993 with "Take the Wrap" demonstrations. Take the Wrap participants collected plastics refused by their local recycling programs and mailed them to the SPI. One representative of the recycling industry, who organized the demonstration in San Diego, remarked: "Most plastics are not recyclable today, and yet, if you turn over any plastic container, you see the recycling symbol with the familiar chasing arrows with a number inside of it." This is misleading and confuses consumers, the recycling executive claimed, as he called for "the plastics industry to remove the recycling symbol until a national system for recycling those plastics has been established" (La Rue, 1993).

Attacked by environmentalists and feeling misunderstood by consumers, the industry continued its efforts to defend itself by telling the recycling story. Dow's 1990 television spots—one with an earnest backpacker telling his hiking companions about recycled plastics and another set at a Dow company picnic with a plastics recycling engineer's young son pitching the plastics recycling story against the backdrop of a softball game—were designed as "an attempt to bring plastics up to parity in the public's perception with other well-known recyclable products" ("Dow Ads," 1990). But these ads apparently had little impact on the public's overall assessment of plastics' environmental friendliness.[6]

By mid-1991 industry publications mentioned that "a massive public outreach campaign is being quietly revived by top leaders of the U.S. plastics industry." Unlike the earlier, divisive Dow proposal, this "new effort is being planned with care" through "low-key deliberations" at "high decision-making levels." According to a representative, "The idea is to avoid a Band-Aid approach" by addressing both plastics' role in solid waste issues and plastics' positive contributions to modern life. The group's efforts would offer a "balance" between fostering recycling and taking the industry's case to the public ("Education on Solid Waste," 1991).

Tensions continued between those in the industry who favored aggressive campaigns to confront plastics' foes and correct "misconceptions," and those who more cautiously advocated changing the situation before advertising about recycling successes. The industry disbanded the CSWS and replaced it with the new Partnership for Plastics Progress (PPP). In announcing the new group, leaders emphasized one industry problem: the perceived failure to communicate with the general consumer public. One industry executive claimed: "We have done too little too late to communicate the facts and to correct misinformation so that practical and lasting solutions can be implemented." Another echoed these remarks: "The challenge now is at the consumer level, where the industry's message has yet to be heard" ("Education on Solid Waste," 1991). With annual funding of around sixty million dollars—more than three times greater than the CSWS ever enjoyed—the PPP, plastics industry insiders believed, would "give our industry direction and momentum" and "change the ways in which it responds to its environmental critics" ("Industry Group," 1992; "Powerful New Group," 1991).

The PPP's major mission—taking the industry's story to the consumer public—took shape in its first major advertising and public relations campaign, launched under the slogan "Take another look at plastic." As with the earlier costly "image" advertising campaign proposed by Dow, some industry members feared that this new campaign might backfire. One environmental officer at a major resin company argued that "a majority of Americans are extremely skeptical (even cynical) when evaluating messages on the environment presented by industry. The industry can't talk its way out of the solid waste crisis." Others agreed and claimed that "a publicity campaign should await more substantive industry solutions and implementation of practical (albeit often painful and slow) achievements" ("Industry Group," 1992). One industry consultant had noted earlier: "The public is more likely to believe plastics are recyclable when they themselves are involved in a curbside program" ("Education on Solid Waste," 1991).[7]

Disagreements over executing an advertising campaign constituted some of the "travails and doubts" that went into forming the PPP coalition ("Industry Group," 1992). The PPP was renamed the American Plastics Council right before the initial eighteen-million-dollar campaign kicked off in late 1992. Timed to lead into the new 1993 congressional and state legislative sessions, the campaign included four television spots, supporting radio spots, and print advertisements that sought "to restore plastics' image before a skeptical public" (Gardner, 1992).

Take Another Look

The broad basic purpose of the "Take another look at plastic" campaign—to polish plastics' tarnished image—suggests why these ads, when compared with the earlier ads using the powerful reframing device of planned perspective by incongruity, present a fuzzily myopic and confusingly unplanned or unconsciously incongruous view of plastic. Lloyd Bitzer's definition of the "rhetorical situation" (1968)—as adapted by public relations theorist Ron Pearson (1987)—points to how the first campaign's compelling "exigence" or question led to a clear response: "Yes, plastics are recyclable." This second campaign, on the other hand, addresses vague self-centered industry concerns about an overall negative image with vacuous self-congratulation. With the "Take another look at plastic" campaign, there is no compelling exigence, no clear response, and no need for the reader to accept the plastics industry's invitation to do even a cursory double-take. While a number of exigencies or problems may need to be addressed in a rhetorical situation, a controlling exigence becomes the organizing principle. The controlling exigence, Bitzer argues, specifies the audience—those who can act as "mediators of change"—and the change to be effected. The controlling exigence, as Pearson notes, also dictates the message and the central theme or thesis of an advocacy campaign. Bitzer warns about the trickiness of clearly perceiving the essential exigence—especially as it relates to the audience's interests.

The two advocacy campaigns mounted by the plastics industry serve as prime examples of Bitzer's description of key factors that the rhetor must understand when faced with the challenge of clearly perceiving the factors in a rhetorical situation. In the first series, the controlling exigence, its accompanying audience interests, and important operating cultural constraints are addressed and used to the industry's advantage. In the "Take another look at plastic" campaign, the industry misreads the audience's interests and overlooks cultural constraints. The campaign's fuzzy view of plastics' environmental attributes—it saves water, energy, and topsoil—misses what the public sees as the issue—a still woefully inadequate recycling infrastructure. While the industry may have set a new environmental issue agenda in the attempt to polish plastics' image, the public holds tight to its own perceptions of what is at stake. Ironically, as noted in an article in *Time,* the solid waste "crisis" seemed to have eased by October 1993, as more landfill space became available and more waste incineration plants began operating: "But Americans don't want to send their garbage into the earth or into the air. They want to recycle" (Van Voorst, 1993, p. 78). While the industry fought recycled-content laws that would create a market for recycled resins, the public pushed for more plastics recycling in their communities, as fewer and fewer programs collected plastics.

Pearson's Bitzer-based guidelines for gauging campaign acuity suggest that this effort misreads its target audience's interests and concerns—solid waste reduction, not energy savings. The industry and the ads' creators also fail to see that certain audience-centered cultural "constraints"—like our ambivalence toward plastic—are not addressed. Actually, the attempt to present plastics' positive contributions to modern living faces several cultural constraints that work against the industry. Bitzer (1968) identifies the constraints operating in a rhetorical situation as the audience-centered attitudes, beliefs, images, interests, facts, traditions, and so on, that have the power to constrain decision and action needed to modify the exigence. But the savvy rhetor can make positive use of constraints—as illustrated in the ads using perspective by incongruity to reclassify plastics as the new "natural resource." By working within the constraints of current cultural beliefs that said "natural" is good, the plastics industry co-opted the natural life cycle—as mimicked by the born-again plastics. Since the natural life cycle is good, "natural resource" plastics are good. While this did not necessarily become the new widespread definition of plastics, it did tap into positive current beliefs and images.

There are more deeply embedded cultural definitions of "plastic" that a campaign focused on the ways plastic permeates our lives verbalizes and depicts. Unfortunately for the plastics industry, these associative meanings speak to truths about our culture that we might be reluctant to admit. As one anthropologist noted: "The fact is that plastic will soon be to modern Americans what the walrus was to the Aleut or the buffalo was to the Sioux: nothing less than the basis of an entire material culture." As such, "in some form or other, we wear it, eat with it, write with it, cover our floors with it, insulate our houses with it—the list is prac- tically endless." But we are ambivalent about the substance of our material culture: "Being less spiritually evolved than the Aleut or the Sioux, though, we loath plastic instead of respect it; we're addicted to it, but make fun of it." In this sense, "plastic is not just a substance, but a code word for a way of life that all right-thinking people despise." As code word, "plastic has gone in thirty years from a symbol for high technology and inventiveness to a symbol of rampant consumerism, from the space program to Cup O'Noodles." For an anthropologist, "it is this schizophrenic attitude that makes the current struggle over what to do with plastic trash so interesting to watch" (Gutin, 1992).

The major problem with the "Take another look at plastic" campaign is that in attempting to appeal to both aspects of our love-hate relationship with plastic, it addresses neither. Instead of showing high-tech problem-solving inventiveness, it pictures plastic-encased hot dogs and styrofoam egg cartons. While it could show plastic "lumber" or a vast array of durable products made with postcon- sumer recycled plastics, it pictures a section of plastic-wrapped watermelon. By trying to combine the two levels of the meaning of *plastic* as code word without using a powerful device like perspective by incongruity to place the two side by side and then transform them, little happens to plastic packaging's image prob- lem. Ironically, there is an unconscious, unplanned—even natural—incongruity to the very concept of *plastic* as code word for both the best and the worst that high technology offers.

Without a clear understanding of the cultural constraints of *plastic* as code word, the "Take another look at plastic" campaign ignores its intended audience's persistent issue concern—uncollected "recyclable" plastic containers filling their trash cans—and consequently misidentifies the controlling exigence. Before the campaign even began, critics pointed out that these ads sidestep real consumer concerns in their attempt to tell the industry's positive story. After viewing the campaign's four television spots, Bob Garfield, resident television spot critic and columnist at *Advertising Age,* found that

> it all seems so Pollyanna, so obvious, so . . . plastic. D'Arcy Masius Benton and Bowles—which also gave us the unbelievably insipid "Dow lets you do great things"—tells the plastics story as if in a vacuum, as if there were scant consumer suspicion to overcome, as if viewers were just salivating to hear *more great news about polymers!!!* They aren't. We're talking about a hostile audience here, many of whom think foam coffee cups will destroy the earth, possibly by Thursday. How do you suppose they regard the motives of those who tell us to "Take another look at plastic"? Answer: dismissively. [1992]

Several scientists working with environmental groups reviewed the "Take another look at plastic" campaign and noted that consumers already know the advantages of plastics packaging and products. Recycling is their major concern. "People don't dislike plastics; they just don't like the fact that they don't know what to do with them when they're done. An advertising campaign that focuses on perceptions, on plastics' advantages, is not going to work. You've got to change the underlying reality." Another scientist agreed: "I can't see that the public will be fooled by this. Women in the home know what they can put in their recycling bins" (Gardner, 1992).

While the television spots may not be able to "change the underlying reality" or fool the public, there is an air of magic tricks—a sleight of hand—as plastic products become illusory shape shifters. Two of the four television spots emphasize that plastics are recyclable, with special effects designed to create metamorphosing visual images. With this technique the spots employ a visual application of perspective by incongruity, as one object transmutes into another. In one spot, the promoter tosses plastic detergent bottles, which are then transformed into plastic trash cans that seem to descend from the sky. But the wonder of this magic act is interrupted by a disclaimer at the bottom of the screen telling viewers to check to see if plastics recycling is available in their communities.

Of the two ads illustrating the metamorphosis of plastics into new forms, Garfield prefers the spot that opens with small boys playing with toy cars on a carpeted floor. One toy car turns into a real car. The voice-over says the car's lightweight plastic side panels save energy and resist denting, as a runaway shopping cart filled with plastic two-liter soda bottles slams into its door. One bottle tumbles out of the cart and onto the ground, as the voice-over says, "You probably don't think plastic can be recycled, but it can," and the bottle turns into the toy car being pushed on the carpet. Both the toy car and the carpet are now revealed as products made from recycled plastics. With this spot, Garfield writes, "we see how plastic fits into the full circle of our lives" (1992).

The other two television spots also use visual shifts as they address the safety and energy-saving features of plastics. In these commercials the background and foreground vantage points change as the promoters tout plastics' advantages. Garfield re-creates the spot that opens at a grocery checkout stand as the clerk spies a host of plastic containers and says:

> "When you look at plastic, you know how it helps things stay fresh and safe and unbreakable and strong and easy to carry. But take another look. [The whole checkout is passed by a moving car, and then itself seems to pass in front of a suburban home.] Plastic also saves energy, because it helps make cars lighter and saves gas. And plastic insulation helps save energy at home. [Now the checkout stand moves off-camera, and the clerk grabs two plastic bags full of groceries.] Even these strong plastic bags, because they take less energy to make than other grocery bags." And so on, sponsored by the American Plastics Council, urging you to "Take another look at plastic." [1992]

The other plastics' spot features a father and his young son—plastic safety helmets in place—bicycling through the storytelling backgrounds and foregrounds.

While the television spots use special effects for swiftly changing images to expand on the notion of taking another look at plastics, the print ads are visually static—and conceptually confusing. Three ads appear in both two-page spreads and full-page versions. The spreads' verso page is devoted to a photograph, and the headline and the copy appear on the recto page. The single-page ads use the same photograph and headline with shortened copy. These ads must operate on the assumption that the reader already knows certain things, since the most important connecting information—relating the headline to the picture—is often delayed or suppressed. If informing the assumedly unaware public is the major purpose of the "Take another look at plastic" campaign, it is curious that the two print ads dealing with recycling and environmental issues require the readers to fill in gaps with information they may not have. Even if they stick with the ad copy—which makes little sense—and read all the way to the end of a copy block, the connecting information is only alluded to, not given. While the print ads' headlines make provocative claims, the photographic art shows mundane plastic packaging, and the copy block falls short of that transcendent leap that would take plastic packaging to the heroic planet-saving heights—or floor-covering plains—hinted at in the headlines.

The best example of how these print ads can be more confusing than clarifying to all but the careful or patient reader is found in the ad that features a photograph of the inside of a refrigerator, featuring items such as two-liter soda bottles, two egg cartons, a one-gallon plastic milk jug, and a maple syrup bottle. The headline reads: "Your new carpeting may already be in your refrigerator." If readers do not already know that the PETE plastic in two-liter soda bottles is commonly recycled into carpeting fibers, then the headline, the photograph, and the copy block will not make much sense unless they stick with it. The final paragraph merely hints at the connection between the contents of the refrigerator and the future new carpet. The longer version of the copy begins:

> If you're looking for ways to help make the earth a better place to live, one solution is in your refrigerator right now. Next to the pillow. Behind the picnic table.
> It's called recycling. And when you look at plastic and recycling, you'll see it's turning into some pretty remarkable things.
> Look at plastic bottles turning into toys, pillows, garbage cans, sailboat sails, even plastic "lumber." Not to mention back into new bottles. And polystyrene foam dishes and cups recycled into building insulation, office accessories, and VCR tape cassettes. . . .
> But if you think recyclability is the only earth-friendly benefit of plastic, take another look. Plastic saves energy by insulating homes to save fuel. And helps reduce pollution by making cars lighter to save gas.
> Plastic even helps save the earth, literally. Plastic geotextile fabrics protect beachfront land from erosion and encourage plant growth.

After urging readers to call a toll-free number for a free booklet, the ad copy ends: "And just think. Someday that soda bottle in your fridge may be a beautiful addition to your living room floor." But the copy failed to mention that carpeting is one of the major products using recycled plastics. That is a huge gap the reader has to fill in, especially if the original assumption of the campaign is that consumers are not aware of all the everyday products using recycled plastics.

Changing the Subject to Sidestep the Issues

The industry's "Take another look at plastic" campaign to remind the American consumer how plastics makes life safer and better rightfully boasts about one plastics application in the print ad headlined "Some benefits of plastic last for only half a second." Its picture shows both driver- and passenger-side airbags inflated to protect a mother and son in a collision. The ad's copy also points out other "lifesaving benefits": "Look at the protective clothing firemen wear. The bullet-resistant vests policemen wear. And the bicycle helmets you and your family wear. They are all made from plastic."

If we do accept the American Plastics Council's invitation to "take another look at plastic," we might admit that in many ways we do appreciate plastic, even in its mundane applications. With safety and convenience in the front of our minds, we reach for the shatterproof plastic juice bottles instead of the glass. When asked which type of bag we want to use to tote our groceries home safely and conveniently, we respond, "Plastic." As dissenting voices within the industry who argued against the eighteen-million-dollar image campaign feared, deep down we already know that we love plastic—and that we hate it. An image campaign is not going to do much to change the intensity of either emotion in our culture-induced schizophrenic relationship with plastic.

When it comes to plastics, it is not automobile air bags, protective clothing for public safety personnel, or safety helmets that Americans want to talk about. Many Americans are still hanging on to the solid waste issue and want to know why the only thing they can do to lessen the plastics load at the local landfill is stomp on those two-liter PETE soda bottles and HDPE milk jugs and detergent containers, at least to compress their volume before consigning them to the garbage can. When will plastics recycling reach their community? Rather than answer those concerns, the industry would rather change the subject by talking about the "positive" aspects of plastics safety or its energy-saving features. Just as the "Take another look at plastic" campaign tries to tout plastics' advantages without clearly showing them, it also attempts to allay our solid waste anxieties without fully addressing plastics recycling issues. It reassures us that those used one-gallon milk jugs and two-liter soda bottles will find new life as pillows, toys, or even garbage cans, but it does not show or explain the transformations.[8]

While the advertisements in the "Take another look at plastic" campaign are upbeat about the wonders of plastics recycling, industry publications find plastics executives sounding themes that counter the industry's pro-recycling messages. The spokesperson for the industry's "product stewardship" task force advances the need to push for plastics' new, environmentally friendly "transformation"— incineration or the more euphemistic "waste-to-energy recovery" or "waste-to-energy conversion"—while the "advocacy" task force creates a network of industry-employed "activists" to fight against recycled-content legislation. Industry members point to the realities of the marketplace to explain why plastics recycling is not cost-effective and in the same breath decry any type of government intervention, claiming that the industry must retain control so that the recycled resins market will develop freely (Carbone, 1992; "Education on Solid Waste," 1991; "Industry Group," 1992).

While two of the industry's three task forces are basically undermining plastics recycling, the plastics industry's third task force—"outreach"—put together the eighteen-million-dollar "Take another look at plastic" campaign as a public education effort to "correct misconceptions" and "make the case that plastics have benefited consumers and are recyclable" (Gardner, 1992). While the campaign's two major messages seem to address the two sides of our love-hate relationship with what is indeed the basis of our entire material culture, these advertisements are so lacking in informative "substance" that we have little to cause us to intensify our infatuation or to soften our disdain. Little happens to *plastic* as the code word for features of our culture—hollow or phony consumerism—we often despise.

Throughout time, cultural artifacts remain to speak of the material basis of a way of life—as anthropologist JoAnn Gutin points to those connecting the Sioux with the buffalo and the Aleut with the walrus. The final plastics recycling irony may be predicted in the last paragraph of the Environmental Challenge Fund advertisement that juxtaposed the polystyrene clamshell cheeseburger box with the pyramids of ancient Egypt. The ad closes with this observation of what may be the most telling artifact of our civilization if we do not pursue answers to the solid waste problem with the same vigor that we pursue new packaging applications: "The most enduring monument left on earth by our civilization may be a mountain of trash."

Notes

1. The announcement surprised members of the National Polystyrene Recycling Company (NPRC) as they prepared for a joint news conference on plans to expand the NPRC and McDonald's polystyrene recycling program from its pilot stage to a nationwide effort involving eighty-five hundred restaurants (Leaversuch, 1990). One advertising executive working with the NPRC accused McDonald's of "getting out for the wrong reason" because its leaders "just didn't want to fight the battle anymore." "They made a business decision, not an environmental one" (Colford, 1991). The executive director of the Polystyrene Packaging Coalition lamented, "McDonald's decision tarnishes polystyrene in all its applications" (Leaversuch, 1990). McDonald's vice president for environmental affairs, Michael Roberts, pointed to opposition from environmentalists committed to recycling and responded, "Whenever you move and act decisively, you'll be criticized. No question there have been many different opinions on the decisions we've reached" (Hume, 1991). Nancy Wolf and Ellen Feldman, of the Environmental Action Coalition, wrote: "Ironically, McDonald's decision to switch to bleached paper/polyethylene wrappers led the company to abandon a recyclable package in favor of one that is neither recyclable nor compostable" (1991, p. xii). The irony deepened: The switch reversed what McDonald's founder Ray Kroc believed was the environmentally sound choice based on the Stanford Research Institute's environmental impact study, which concluded that when all stages of production were considered—from manufacturing through disposal—polystyrene was superior to paper. The 1976 study told Kroc: "Polystyrene uses less energy than paper in its production, conserves natural re-

sources, represents less weight and volume in landfills, and is recyclable" (Hume, 1991).

2. SPI's seeming reluctance to enter the fray may have stemmed from three decades of self-imposed silence. The last time the plastics industry carried its case on a controversial issue directly to the U.S. public was in 1959, when it refuted claims that polyethylene bags were to blame in infant suffocation deaths ("Resin Companies' New Tactic," 1989).

3. An article in the November 1989 issue of *Modern Plastics* discussed advertising campaigns planned by individual resin producers—Amoco, Mobil, and Huntsman—and suggested developing tensions between CSWS members. After a description of how the companies had targeted "influentials" with the goal "to convince opinion leaders that plastics (especially PS) are recyclable," the article ended with the question of whether one-company ads are the best way to get the desired message across or whether a unified approach by the CSWS would be more effective. Since all twenty-one members of the CSWS had to reach unanimous agreement before actions were taken, Huntsman Chemical vice president Don Olsen was quoted as saying, "It is difficult to get a resin company consensus on the issues at stake in planning an advertising campaign." Amoco's director of marketing communication noted that the company's commitment of one million dollars for the campaign was based on the feeling that the company could act quicker to "meet a need it defines as urgent" by acting alone ("Resin Companies' New Tactic," 1989).

4. SPI president Larry L. Thomas agreed with Dow. In a confidential letter to members about the proposal's discussion in the coming January 15, 1990, meeting, he cited the advice of opinion poll experts that the public's view of plastics had plummeted so fast that it approached "the point of no return." These experts predicted that the industry would be swamped by "a fast-moving tidal wave of growing negative public perception." An opponent of Dow's proposal leaked Thomas's letter to columnist Jack Anderson, who ridiculed it as a "propaganda drive" in a January column. The head of a public relations firm representing one of the major resin producers suggested that approving Dow's proposal would be like waving a flag under the nose of environmentalists: "It would be pure Hollywood and it will provoke a huge backlash" ("Dow Calls," 1990, p. 11).

5. For example, the Dow-Mobil lead industry coalition defeated a 1992 Massachusetts recycled-content ballot initiative—the most far-reaching to date—by using these arguments to cast the law as essentially antibusiness and anticompetition. The irony of the industry's adopting this reactive and defensive public policy stance is that plastics recycling has been hampered by market forces that make virgin resin cheaper than recycled resin. Recycled-content laws would create the needed market demand for recycled resin—and the accompanying higher prices needed to make recycling cost-effective. With the industry both trumpeting its self-proclaimed recycling successes and at the same time maneuvering to defeat measures that would make recycling a solid reality, one scientist working for an environmental interest group accused "Dow and other plastics producers of lacking sincerity" when they oppose legislation mandating recycled material in packaging (Reisch, 1992).

6. Opinion polls showed that the public's perceptions of plastics further deteriorated despite the industry's advocacy efforts. By late 1990, 51 percent rated plastics "unfavorable"—up from 34 percent in 1989. Sixty-one percent believed the risks associated with plastics outweighed plastics' benefits—up from 57 percent in 1988 ("Education on Solid Waste," 1991).

7. One critic notes that the industry has spent more money on advertising and public relations than actually goes into funding research or pilot projects that might begin to address the plastic waste problem (Gutin, 1992). Plastics industry members who thought the eighteen million dollars spent on the image advertising campaign would be better spent on further developing the recycling infrastructure—to change the reality—must also face the dark reality of the ironies that make successful plastics recycling programs scarce. The first set of ironies are found in the marketplace and in the industry's attempts to defeat legislation that would change market forces to favor recycled resins. As the by-products of petroleum refining, virgin resins will be cheap and abundant—as long as the oil supply lasts.

Recycled resins, on the other hand, are currently scarce and expensive—due in part to plastics' advantages turning to disadvantages as the wide range of resins and resin mixes used to fill a limitless number of packaging niches becomes the very reason why, from the perspective of cost, plastics recycling is "difficult and nearly pointless" (Gutin, 1992). When the plastics industry marshals its forces to fight recycled-content packaging legislation that would create a demand for recycled resins, it defeats market forces that would make recycled resins more valuable.

8. While both print ads aimed at environmental concerns brag that "650 *million* pounds of plastic packaging was recycled last year," environmentalists counter that with this campaign, "The plastics industry is putting a happy face on plastics recycling." The 650 million pounds may sound impressive, but only 2.2 percent of plastics—mainly PETE soda bottles—were "reborn" in 1990. While the industry claims significant progress in increasing recycling programs, the amount of plastics used grows each year, as more and more manufacturers adopt plastic packaging because of its advantages. A 52 percent increase from 4.4 million tons of plastic packaging discarded in 1985 to 6.7 million tons in 1990 indicates the growing solid waste problem that recycling 650 million pounds of plastics barely begins to reduce (Stipp, 1992). Recycling rates for plastic held steady at a mere 2.2 percent in 1993, while the rates for aluminum reached nearly 70 percent, paper hovered at 50 percent, and glass rose to 33 percent (Van Voorst, 1993). It may be that the plastics industry—focused on its "Take another look at plastic" campaign—hopes that the public will overlook the way it slights solid waste and recycling issues.

References

Advertisements
American Plastics Council. (1992a). How to save the planet and the picnic at the same time.
———. (1992b). Some benefits of plastic last for only half a second.
———. (1992c). Your new carpeting may already be in your refrigerator.
Amoco. (1989a). Do we really want to return to those good, old-fashioned days before plastics?
———. (1989b). Let's dig a little deeper into the notion that much of our garbage is made up of plastics.
———. (1989c). We'd like to uncover a hidden natural resource.
———. (1990). We'd like to recycle the thinking that plastics can't be recycled.
Dow Plastics. (1989). You're looking at 64 milk bottles and 2 shampoo containers.
Du Pont. (1990). We've got to stop treating our garbage like garbage.
Environmental Challenge Fund. (1990). Your cheeseburger box will be around even longer.
Fina. (1990). How to recycle . . . into. . . .
GE Plastics. (1990). Life after death: A recycling strategy to stop burying technology alive.
Huntsman Chemical. (1990). Think of them as your new home. Think *recycle.*
National Council for Solid Waste Solutions. (1989). The urgent need to recycle.

Books and Articles
Bitzer, L. (1968). The rhetorical situation. *Philosophy and Rhetoric, 1,* pp. 1-14.
Blumberg, L., & Gottlieb, R. (1989). *War on waste: Can American win its battle with garbage?* Washington, D.C.: Island Press.
Burke, K. (1984a). *Attitudes toward history.* 3rd ed. Berkeley: University of California Press.
———. (1984b). *Permanence and change.* 3rd ed. Berkeley: University of California Press.
Callari, J. (1989, September). How did plastics become the target? *Plastics World,* pp. 12-18.

Carbone, A. (1992, March). Industry and the environment: Making business part of the solution. *USA Today: The Magazine of the American Scene,* pp. 32-34.

Castro, J. (1990, June 25). One Big Mac, hold the box! *Time,* p. 44.

Colford, S.W. (1991, February 25). Polystyrene ad fights bad image. *Advertising Age,* p. 16.

Crable, R.E., & Vibbert, S.L. (1985). Managing issues and influencing public policy. *Public Relations Review, 11,* pp. 3-15.

Dow ads boost recycling. (1990, January 15). *Advertising Age,* p. 28.

Dow calls for media blitz to stem anti-plastics tide. (1990, March). *Modern Plastics,* pp. 10-12.

EarthWorks Group. (1989). *50 simple things you can do to save the Earth.* Berkeley, Calif.: EarthWorks Press.

Education on solid waste is goal of new campaign. (1991, June). *Modern Plastics,* pp. 42-44.

Gardner, J. (1992, November 9). Plastics seek $18M image boost. *Advertising Age,* p. 12.

Garfield, B. (1992, December 23). Plastics industry molds wrong approach in ads. *Advertising Age,* p. 18.

Gutin, J. (1992, March-April). Plastics-a-go-go. *Mother Jones,* pp. 56-59.

Hainsworth, B.E. (1990). The distribution of advantages and disadvantages. *Public Relations Review, 16,* pp. 33-39.

Heath, R.L., & Nelson, R.A. (1986). *Issues management: Corporate public policymaking in an information society.* Beverly Hills, Calif.: Sage.

Hume, S. (1991, January 29). McDonalds: Case study. *Advertising Age,* p. 32.

Humphrey, H.H., III. (1990, August). Let's keep "green" clean! *Progressive Grocer,* pp. 130-31.

Industry group remaps stand on solid waste. (1992, July). *Modern Plastics,* p. 35.

Javna, J. (1991, May 29). Recycling plastics a tricky issue. *Lafayette, (Ind.) Journal and Courier,* p. C9.

Jones, B.L., & Chase, W.H. (1979). Managing public issues. *Public Relations Review, 5,* pp. 3-23.

La Rue, S. (1993, April 23). Plastics get activists' stamp of disapproval. *San Diego Union Tribune,* p. B1.

Lawren, B. (1990, October-November). Plastic rapt. *National Wildlife,* pp. 11-19.

Leaversuch, R.D. (1990, December). Will McDonald's switch have a ripple effect? *Modern Plastics,* pp. 42-45.

Pearson, R. (1987). Public relations writing methods by objectives. *Public Relations Review, 13,* pp. 14-26.

Powerful new group will take over from CSWS. (1991, December). *Modern Plastics,* pp. 36-38.

Reisch, M. (1992, June 22). Dow sets big plastics recycling program. *Chemical and Engineering News,* p. 16.

Resin companies' new tactic in solid waste: Consumer ads. (1989, November). *Modern Plastics,* pp. 11-12.

Resin suppliers organize for solid waste battles. (1988, July). *Modern Plastics,* p. 14.

SPI creates an image program for industry. (1989, May). *Modern Plastics,* pp. 24-25.

Sternberg, K. (1990, November 14). McDonald's polystyrene pullout draws mixed reviews. *Chemical Week,* p. 22.

Stipp, D. (1992, September 21). Lag in plastics recycling sparks heated debate. *Wall Street Journal,* natl. ed., p. B1.

Stuller, J. (1990, January-February). The politics of packaging. *Across the Board,* pp. 41-50.

Van Voorst, B. (1993, October 18). Recycling: Stalled at curbside. *Time,* pp. 78-80.

Wolf, N., & Feldman, E. (1991). *Plastics: America's packaging dilemma.* Washington, D.C.: Island Press.

Wood, A. (1990, May 2). Plastics: Can more be made of less? *Chemical Week,* pp. 36-38.

9

Valuation Analysis in Environmental Policy Making: How Economic Models Limit Possibilities for Environmental Advocacy

Tarla Rai Peterson & Markus J. Peterson

At 12:04 A.M. on March 24, 1989, the *Exxon Valdez* ran aground on Alaska's Bligh Reef. Rather than backing off the reef, the captain ordered the engines to be run at full speed forward. Despite his chief mate's advice to the contrary, he held the throttle forward on sea speed for fifteen minutes before admitting defeat. At 12:27 he finally radioed the local Coast Guard's traffic controllers, informing them that the tanker was leaking oil. At 12:30 the Coast Guard notified the Alyeska night shift superintendent of the spill.[1] Alyeska's contingency plan stated that disaster equipment would arrive at the scene of any oil spill no more than five and a half hours after the spill had occurred. Larry Shier, Alyeska's marine supervisor, assured both the Coast Guard and Alaska's Department of Environmental Conservation (DEC) that he would activate the contingency plan; then he notified the night crews of the spill and went back to sleep. At 3:00 A.M. a Coast Guard lieutenant and local DEC representative reached the *Exxon Valdez* aboard a Coast Guard pilot boat. Both called Alyeska again to request that equipment to transfer the remaining oil from the *Exxon Valdez* be sent immediately. Shier told them that the equipment was on its way and would arrive at first light. At 5:00 A.M. Alyeska employees arrived at the terminal, but the needed equipment had been misplaced. By 7:30 A.M. union members were arriving at the union hall of Cordova District Fishermen United. They assembled thirty fishing boats and tried to contact Alyeska. Around 9:00 A.M. Alyeska responded to their offer to help contain the oil: "We'll get back to you." At noon the union members called to report that they now had seventy-five boats ready. Again, they were told that someone would get back to them. Alyeska's barge, equipped to transfer the oil from the *Valdez* to another vessel and to skim the spilled oil from the sea, arrived at Bligh Reef at 2:30 P.M. In the meantime, Alyeska had contacted Exxon Headquarters in Houston. By 8:36 A.M. equipment and supplies from up and down the West Coast were being diverted to the disaster, and two Exxon executives were on their way to Valdez. At 5:37 P.M. Exxon's corporate jet landed in Valdez, and Exxon took over from Alyeska, assuming responsibility for the spill (Davidson, 1990, pp. 3–36).[2]

The *Exxon Valdez* spilled approximately ten million gallons of crude oil into the previously pristine waters of Prince William Sound (U.S. General Accounting Office, 1989). The spill confirmed the fears of area citizens who had joined environmentalists' campaigns during the 1970s, attempting to prevent the transport of oil through the sound. Their allegations of the ecological dangers posed by pipeline and tanker transport had delayed pipeline construction for several years. Industry claims, however, that the expense of an alternative route (building the pipeline overland through Canada and into the midwestern United States) would price Alaskan oil out of market range ultimately prevailed. At the time of the spill, one-fourth of U.S. domestic oil production moved through Prince William Sound (Came, Quinn, & Lowther, 1989; Laycock et al., 1989).

The events surrounding the *Exxon Valdez* oil spill illustrate an important trend in natural resource management policy and also clarify potential problems with this tendency. First, the conceptual foundation of the contemporary move toward privatization has influenced environmental management policy. Because Alyeska—as well as the individual companies constituting the consortium—was responsible for damaging resources that belonged to someone else, it was required to pay the owners (the public) for the loss and degradation of those resources. Damage assessments were determined by valuation methods based on capitalist economic dogma that has become central in U.S. natural resource policy decisions. Second, the widespread dissatisfaction with Exxon's handling of the spill illustrates the limitations of using even widely accepted economic dogma as the sole means for determining, and then justifying, environmental policy. Although some degree of economic restitution was expected, its significance was diminished by images of oil-soaked birds, blackened beaches, and dead sea otters. Public response to these images, combined with identification of the Alaska spill as the nation's worst environmental disaster, fueled demands for more than short-term economic restitution (Came, Quinn, & Lowther, 1989; Hackett et al., 1989). The economic valuation models upon which oil spill policies, and particularly compensation claims, were based, however, offered no such options. Because economic valuation does not recognize noneconomic motives, it cannot respond to them.

Debates among social theorists reflect an ongoing controversy over the relative significance of symbolic (communicative) action and material existence in structuring social situations. There is general agreement, however, that each influences and is influenced by the other (Giddens, 1979, 1984). Kenneth Burke (1984) argues that, while the human experience is grounded in material existence, materiality is insufficient to explain social motivation. Rather, more complete explanation of social reality can be achieved by exploring rhetorical action, or language used to provoke cooperation among human beings. He explains that, "once you have a word-using animal, you can properly look for the linguistic motive as a possible strand of motivation in all its behavior" (Burke, 1966, p. 456). He argues further that whatever rhetoric one uses serves as a "terministic screen," emphasizing some aspects of reality while de-emphasizing others.

In this chapter we rely on Burke's notion of rhetorical selectivity, grounded in the framework developed in Peter Berger and Thomas Luckmann's *Social Construction of Reality* (1966), to argue that because economic valuation models deflect attention from all but economic aspects of wildlife, they provide an inappropriate foundation for decision making. We follow a brief history of economic valuation as it has been applied to natural resources with an example that epitomizes the underlying assumptions of several models designed to assign monetary value to natural resources that have little direct market value. The problem is not that economic models are limited to economic perspectives, for any model of reality suffers from limitations. Rather, the problem is that these models have been applied without explicit consciousness of their limitations. Niklas Luhmann's theory of modern function systems (1989) provides a conceptual pattern for analyzing the limitations of these models within the context of contemporary U.S. culture. We then use the response to the *Exxon Valdez* disaster as an example of how exclusive reliance on these tools can constrain policy. We explain how, by excluding other social spheres, the language of economic valuation produces a terministic screen that limits policy makers' vision of appropriate environmental advocacy. Finally, we argue that privileging private economic motives over others presents significant legitimation dilemmas for democracies that rely on broad public participation in political decision making.

Economic Valuation of U.S. Natural Resources

The U.S. Public Trust Doctrine assigns ownership of the natural resources found on vast tracts of land to the public. In contrast to the policies of most European countries, this doctrine also assigns ownership of indigenous wildlife, whether on public or private land, to the citizenry. State and federal government agencies, as trustees, have been given the responsibility for managing these resources for the benefit of the public, symbolizing a recognition that natural resources have "value" to the populace. Wild animals and plants, however, need habitats in which to live. Humans increasingly compete with other creatures for space by turning prairies and wetlands into cropland, forests into urban sprawl, canyons into reservoirs, and deserts into golf courses. Activities such as mining and oil exploration and extraction have caused additional change. Such landscape alteration has led to the extinction of some species and the expansion of other populations, which society may perceive as either beneficial or detrimental. For example, human activity led to the near extinction of bison (*Bison bison*) in North America, which many consider a tragedy, yet human-induced environmental change also led to greatly increased numbers of white-tailed deer (*Odocoileus virginianus*), which many of the same people consider beneficial.

A corollary to the Public Trust Doctrine is that, because wilderness has "value," this value must be weighed against that of other natural resources such as minerals, timber, water, and space. Industries dependent on natural resources, such as mining, farming, hydrologic exploitation, and urban development, remain critical to the U.S. economy. Thus, when industry wants to explore for oil in critical wilderness areas, government wildlife management agencies are pressed to justify, in terms of benefit to the public, why the exploration should not be allowed. Simply put, they must explain why wilderness is worth more money than oil. If exploration and subsequent oil extraction take place, then agencies are asked to calculate, in dollars, a mitigation value for wilderness loss. Although a growing number of people agree that wilderness has value, the task of quantifying it in dollars and achieving numbers comparable to the billions a new oil field on Alaska's North Slope is worth, for example, is difficult at best.

Such quantification—which we call "value" throughout this section of the chapter—was initially based on valuation methods used for marketed natural resources (commodities). To determine the value of a stand of timber, for example, one can ascertain what price the logs would fetch at a sawmill, add to this the economic benefits the logging and milling operations contribute to the community, and subtract the cost of building roads into the area, felling and bucking the trees, and hauling them to the mill. What is left is the value, or "net economic surplus." The same process is used for valuating mineral or grazing resources and also has been used for valuating wild plants and animals. One could argue, for example, that creeled trout, processed venison, or harvested huckleberries are worth some number of dollars per pound, as replacements for purchased fish, beef, or blueberries.

It was clear to economists, however, that the recreational value of consumptive public use of natural resources was far greater than the market value of related commodities taken from the area. Most sport hunters and fishers spent far more on their avocations than they gained in the form of meat and fish. Thus, since the experience had no net economic surplus, or value, there was no apparent incentive for people to hunt, fish, harvest mushrooms, or pick wild fruit. Additionally, economists reasoned that relatively nonconsumptive uses of wildlands, such as bird watching, mountaineering, backpacking, photography, and sightseeing also must have some value, or people would not spend money (and time) performing these activities.

Nonmarket Economic Models

Economists now consider some aspects of recreation to be "nonmarket commodities," because, since they are not traded in the market, their value cannot be derived from the competitive market structure. Despite its nonmarket status, however, leisure is still defined as a commodity. In an attempt to resolve the paradox embodied in a nonmarket commodity, an evolutionary series of models designed to place dollar values on recreation have been proposed and tested. Recreational value models based upon this research were then adapted to place price tags on wildlife and wilderness.

Current models date back to 1947, when the Prewitt Commission was attempting to determine the economic value of a recreational site (Stoll, 1983, 1986). Because recreationists must travel from where they live to the recreation site, Hotelling proposed that incurred travel costs could be used to impute a value to recreation at a given location. Thus, the travel cost method (TCM) of valuating recreation was born. This model enjoyed only brief popularity until economists rediscovered it in the early 1960s. Since then the TCM has been modified to incorporate site quality, the availability of substitute recreation sites, the cost of both travel and on-site time, and cost to nonparticipants.

Problems with the TCM recognized by economists (Stoll, 1983, 1986) include difficulty in separately valuating specific recreational components at a site, incorporating substitute recreational experiences for unique sites (such as the

Grand Canyon), calculating the value for recreational experiences where travel is unnecessary (such as urban forests), calculating travel cost when the recreationist visits more than one site on a single trip, and determining the proper allocation of costs when one takes a multipurpose trip (combining business with recreation). Several modifications of the TCM are used to address these shortcomings, yet all are based on observing and then drawing inferences from recreationist behavior. Many economists find this a major drawback, because there must be some form of related market behavior from which inferences can be drawn. Despite problems, variations of the TCM are still used to valuate wildlife and wildlife-based recreation (Lyon & Keith, 1984; Miller & Hay, 1984; Hvenegaard, Butler, & Krystofiak, 1989).

Economists who maintain that the TCM's shortcomings render it ineffective have turned to another economic valuation model. The contingent valuation method (CVM) is widely used as an alternative technique, which economists maintain addresses many of the TCM's shortcomings. The CVM is defined as "any approach to valuation that relies upon individual responses to contingent circumstances posited in an artificially structured market" (Stoll, 1983, p. 120). There are many variations to this approach, but most rely on either iterative or noniterative bidding, which are further subdivided into open-ended and closed-ended questioning techniques (including dichotomous-choice models).

Briefly, in the iterative bidding approach, respondents are asked if they would continue a given recreational experience if the cost were increased to some higher value (Stoll, 1983, 1986). If the respondent says yes, the value is then iteratively raised until the respondent says no. The highest yes amount is then assumed to measure the value of the recreational experience associated with that resource. To be administered effectively, this method requires a personal interview. Noniterative bidding circumvents this necessity. The open-ended question format is as follows: "I would not continue backpacking in Yellowstone National Park (or some other activity) if the fees (or some other factor) cost _____ annually." The respondent is instructed to write in the highest applicable value. In the closed-ended format, the dollar value is specified. Questionnaires with varying dollar amounts can be sent, and the results analyzed statistically, to estimate demand curves.

The amount of money one is willing to spend on a leisure experience involving wilderness, however, may not represent the "total" value of that wilderness. For example, people have expressed willingness to pay for wildlife-oriented experiences that transcend individual spaces and times (nonuse value), such as having the option of using the resource later (an "insurance premium"), the knowledge that the resource exists somewhere, and the ability to bequeath wildlife, or other natural resources, to future generations (Brookshire, Eubanks, & Randall, 1983; Loomis, Peterson, & Sorg, 1984; Walsh, Loomis, & Gillman, 1984). In other words, people have declared their willingness to provide financial support for conservation, to preserve resources they may never directly experience.

Admitting motives that transcend space and time into the recreational value equation led to Randall and Stoll's (1983) total value paradigm (TVP). This valuation model combines "all" on- and off-site use and nonuse values for a given location, and it has been used to the satisfaction of several economists and managers for valuating wildlife and other natural resources (Loomis, Peterson, & Sorg, 1984; Bowker & Stoll, 1988; Bergstrom et al., 1990). The TVP relies on previous economic valuation models, using the artificially structured market of the CVM, to sum the dollar values of nonuse items and, in combination with the value of recreation (use), to calculate the "total" value of the resource.

Nonmarket Valuation of Whooping Cranes

One of the earliest attempts to use the TVP for valuating a population of wildlife having no consumptive value involved whooping cranes. The only wild self-perpetuating flock (fewer than one hundred individuals) winters in and around the Aransas National Wildlife Refuge on the Texas Gulf Coast and breeds and rears it young in Wood Buffalo National Park, which straddles the Alberta–Northwest Territories border in Canada. Because of the whooper's grand size and appearance, haunting call, long migration route, and rarity, this species has come to symbolize the significance of the U.S. endangered species program (U.S. Fish and Wildlife Service, 1986; Binkley & Miller, 1980).

Management policies associated with any endangered species have become increasingly volatile, as they have threatened to limit the financial returns available to established interests, such as the lumber and oil industries. Whooping crane management is further complicated by the fact that the birds winter in one country and summer in another. Human activity at Aransas, such as oyster dredging and construction of the Gulf Intracoastal Waterway through the refuge, with its associated traffic and dredging, has adversely affected the whoopers in their wintering grounds (U.S. Fish and Wildlife Service, 1986). Environmental pollution, including agricultural chemicals and oil spills from barges passing through the refuge, poses both direct and indirect hazards to whooping cranes through possible oil fouling and elimination or contamination of forage. Although human activity in and around Wood Buffalo National Park has been less intrusive, timber harvest along park borders (as well as within the park), proposed "ecotourism" development, as well as both existing and proposed hydrologic projects are likely to affect the park adversely (Environmental Assessment Panel, 1990). It was the government-subsidized cattle grazing near park boundaries, however, that focused attention on the area. Because many of the park's bison are infected with bovine brucellosis and tuberculosis (diseases that can be transmitted to cattle, reducing industry profits), the Northern Diseased Bison Environmental Assessment Panel recommended that all bison in and around the park be killed and eventually replaced with disease-free, genetically "pure" wood bison (*B. b. athabascae*) (Environmental Assessment Panel, 1990). Killing such a large number of bison and propagating thousands of replacement animals could dramatically change the greater Wood Buffalo National Park ecosystem.

Stoll and Johnson (1984) used the valuation of the whooping crane "resource" to provide an example of how one can assign a monetary value to a "priceless" resource. They used a closed-ended, noniterative bidding form of contingent valuation (dichotomous choice) to determine the value of the whooping crane resource based on current use (refuge entry fee), anticipated future use (option), and nonuse (existence). Stoll and Johnson made no attempt to quantify what people would be willing to pay for related travel expenses, so they would not consider this valuation "total." If, however, this element had been included, the valuation would be interpreted as complete. The questionnaire prepared by Stoll and Johnson was given to visitors entering Aransas National Wildlife Refuge and mailed to a cross-section of Texas residents and persons living in three large out-of-state U.S. cities. After analyzing survey responses, they estimated the total combined option and existence value of the whooping crane resource in the United States to be $1.58 billion, if the bids represented individuals, and $573 million if they represented households. Not surprisingly, Bowker and Stoll's reevaluation of these data (1988) determined that whooping crane value was greater for those respondents having higher incomes and/or memberships in wildlife-oriented organizations.

Even though the value of the whooping crane resource may be between $573 million and $1.58 billion, this is a paltry sum compared with the commercial value of the Gulf Intracoastal Waterway, the U.S. petrochemical industry, or western Canada's proposed hydrologic system. If economic measures are assumed to provide "total" gauges of wilderness values, then wilderness will be found insignificant, and human alteration of habitat essential to the existence of wild plants and animals will continue unabated. Additionally, it appears that, rather than using nonmarket values to determine dollar value, these models simply use hypothetical market values. Surely many people value wilderness either more than they can afford to pay or in ways that money does not address. We argue that "total" valuation of wildlife and other publicly owned natural resources must include factors that do not lend themselves to measurement on a scale of dollars. Hence "willingness to pay" is at least inadequate, and perhaps inappropriate, for measuring the "total" value with which people regard publicly owned natural resources such as wilderness. The language used to describe economic valuation conceals several problematic assumptions required by these models.

Limitations of Economic Valuation Models

The decision to base natural resource policy on results of cost-benefit analyses is rhetorical in its selectivity. Despite economists' claims that they are simply creating tools to measure existing reality, the tools warrant the construction of one reality rather than another. As within any social reality, privileged experience becomes information, whereas the reality of dispreferred experience is denied. Within the economic valuation paradigm, experience becomes preferred or dispreferred by satisfying (or failing to satisfy) two basic presumptions. First, the basic justification for cost-benefit analysis is the assumption that actions should be undertaken only to maximize benefits. Kelman (1990) claims that for many public decisions, the question of whether benefits outweigh costs is insufficient at best, and inappropriate at worst. He argues that "in areas of environmental . . .

regulation there may be many instances where a certain decision might be right even though its benefits do not outweigh its costs" (1990, p. 132). Perhaps this is why, in the midst of global privatization, so many countries have made environmental protection the business of national, and even international, governing bodies, rather than relying on private initiative.

The second assumption develops out of the first. In order to determine when an act's benefits outweigh its costs, all factors must be expressed in a common denomination so they can be compared against others. For economic valuation in the United States, that measure is dollars, and the possibility that some things cannot be expressed accurately in dollar terms does not exist. There is some difficulty, however, in determining the dollar value (using either market or nonmarket methods) of such "commodities" as life, peace and quiet, or fresh air. Economists have responded to the challenge of imputing a dollar value to nonmarketed goods by determining their value as it relates to marketed goods, thus creating an economically based nonmarket value. For example, while peace and quiet is not marketed, houses are sold in both noisy and quiet locations. Therefore, the value of peace and quiet is the difference between the purchase price of two homes (otherwise alike, and in the same real estate market) with varying levels of noise. Although this example drastically oversimplifies the technical aspects of cost-benefit analysis, it does not oversimplify the fundamental assumptions upon which it rests.

Modernity and Social Fragmentation

Luhmann's functionalist social theory (1989) both clarifies the extent to which the fundamental assumptions underlying economic valuation models provide a narrowly deterministic basis for public policy decisions and also provides a framework within which we can critique potential repercussions of basing management decisions about publicly owned natural resources on this perspective. Luhmann proposes a radicalized functionalism as a theoretical perspective toward society and its environment. Rather than viewing functional relations as causal, he characterizes "cause" as a special, and singularly opaque, case of function. Functional relations exist between a problem and a range of possible responses, and problems that do not acquiesce to such a range are not social problems.

Luhmann uses the concept of autopoiesis to model society as simultaneously closed (organizationally) and open (structurally). He defines human society as an "all encompassing social system of mutually referring communications . . . [that] originates through communicative acts alone and differentiates itself from an environment of other kinds of systems through the continual reproduction of communication by communication" (1989, p. 7). The theory of autopoiesis relies on the powerful notion that all systems examine themselves and regulate their own functioning through a process analogous to cognition. Autopoiesis suggests that the most basic communicative operation is that of categorization (Ulrich & Probst, 1984). Any unity, including human society, can be differentiated into its constituent parts, by drawing further distinctions. One can differentiate elements of the *Valdez* spill as the discussion between the captain and the chief mate over appropriate procedures once the ship ran aground; the captain's radio conversation with the Coast Guard; the Coast Guard's telephone conversation with Alyeska, and so on. Alternately, one can distinguish between the social system within which the oil spill occurred and its environment, thus emphasizing differentiations within the environment. For example, the social system includes neither the temperature nor the contour of the bay. The whole process of differentiating entities from their background is based upon this simple cognitive process, which specifies the organization of a system.

While this interpretation recognizes that society has an environment, it presumes that social relations with the environment are internally driven responses to, rather than interactions with, the environment. In other words, the fouling of Prince William Sound by oil from the *Exxon Valdez* was not an interaction between society and its environment so much as it was an internally driven response to environmental conditions. Factors such as the captain's inebriated condition, an organizational reward system that encouraged ships' captains to value speed over safety, and an organizational culture that discouraged other crew members from acting on their own best judgment combined to invite the accident. The spill and subsequent cleanup operation were the culmination of elements internal to the social world that constrained both awareness of and responses to environmental conditions. Instead of asserting that the system adapts to its environment or that the environment selects the system that survives, autopoiesis emphasizes the way the social system shapes its own future.

The organizational closure assumed by autopoiesis means that society can react to its environment only according to its own mode of operation. Society is seen as an autonomous, closed system because it strives to maintain an identity by subordinating all changes to the maintenance of its own organization as a given set of relations. It does so by engaging in circular patterns of interaction (within itself) whereby change in one element of the system is coupled with changes elsewhere, setting up continuous patterns of interaction that are always self-referential. Both the decision to leave port on schedule despite the captain's physical condition and Alyeska's subsequent inability to use the assistance offered by local fishers the morning after the spill illustrate system closure at the institutional level. Further, although Alyeska had already demonstrated that its contingency plan was dysfunctional, when Exxon took over the cleanup, it focused more energy on making the plan work than on cleaning up the oil. In all three of these examples, a system's self-referential nature prevented it from acting in ways that were not specified in the pattern of relations that defined its organization. Thus, as an autonomous system, society's supposed interaction with its environment is really a reflection and part of its internal organization. It responds to the environment in ways that facilitate its own self-production.

In describing society as closed and autonomous, Luhmann is not characterizing it as completely isolated. The closure and autonomy to which he refers is merely organizational. Society closes in on itself to maintain stable patterns of relations, and this process ultimately distinguishes society as a system. There is no beginning and no end to the system because it is a closed loop of interaction. In other words, society is seen as a system that possesses a logic of its own, rather than as a network of separate parts. Luhmann's rejection of the input-output model distinguishes his approach to social theory from that used by most systems theorists. Because the system envisioned by Luhmann cannot escape the closed loop, it makes no sense to say that society interacts with its external environment. Rather, apparent transactions between society and its environment are really transactions within the system that have been prompted by resonance between society's function systems.

Although society is organizationally closed, it remains structurally open. Systems maintain stability by sustaining processes of negative feedback that allow them to detect and correct deviations from operating norms, and systems can evolve by developing capacities for modifying these norms to take account of new circumstances. The source of change, then, is located in random variations occurring within the system. This structural openness allows seemingly unrelated aspects of a system to interact with each other. For example, environmentalists had long opposed transporting oil through Prince William Sound because they feared contamination of the pristine region. They also had complained repeatedly about the habitual drunkenness of tanker captains. When Captain Joseph Hazelwood stepped onto the *Exxon Valdez,* even when he reported its grounding on Bligh Reef, however, few would have predicted that he soon would be accused by New York State Supreme Court justice Kenneth Rohl of perpetrating "man-made destruction that probably has not been equaled since Hiroshima" (Hackett et al., 1989, p. 17). The random combination of events that led to the long-feared spill enabled Senator Joseph Lieberman and fellow opponents of a bill to permit exploratory drilling in the Arctic National Wildlife Refuge to "put the brakes on it and put it into reverse" (Hackett et al., 1989, p. 18). It provided support for proponents of the proposed requirement that oil tankers be equipped with double hulls. The structural openness specified by the theory of autopoiesis encourages us to understand these and other transformations of society as the result of internally generated change, rather than as adaptation to external forces.

Chaos theory, which began developing in the 1960s, suggests that random changes in a system can lead to new patterns of order and stability (Crutchfield et al., 1986). Random variation within society, then, generates possibilities for emergence and evolution of new system identities. Of course, possibilities do not necessarily translate into practices, and the attendant potential for importing negative entropy does not always result in the importation of negative entropy. Erratic changes can trigger interactions that reverberate through the system, the final consequences being determined by whether the current identity of the system dampens the effects of the disturbance through compensatory changes elsewhere, or whether it allows a new configuration of relations to emerge. In the case of the *Exxon Valdez,* the decision to defer drilling in the Arctic National Wildlife Refuge, and implementation of the double hull requirement, dampened the potential for fundamental changes in relations between the petroleum industry and the U.S. government.

Luhmann interprets the aspects of social systems that enable these transactions to occur as communicative interactions rather than as individual elements. Society, thus, is structured by self-referential operations (communication) that are produced within society's subsystems. Luhmann characterizes these operations as communicative acts, which are the sole means for differentiating society from its environment (1989, p. 7). Communication, which refers to "the common actualization of meaning," rather than to information transfer, provides society's mode of operation, and the environment comprises everything that does not operate communicatively (p. x).

Luhmann describes the society wherein these communicative transactions take place as a centerless set of "function systems" and insists that both what can be communicated and how it is communicated are constrained by these subsystems. He argues that because each subsystem fulfills only one primary function (hence the name *function system*), it cannot substitute for another, as was the case within traditional societies that were differentiated through stratification. In medieval Europe, for example, the authority of the pope, who occupied the pinnacle of the social hierarchy, could be brought to bear on any sort of problem. Whether the issue was economics, politics, or education, the same ultimate authority ruled. Modern society, however, recognizes no single authority figure that can cut across different social functions.

These function systems—the most important are economy, law, science, politics, religion, and education—sort all experience that is allowed to become information according to a binary code, wherein negation secures system closure by assuring that every value refers exclusively to its countervalue. Binary codes reproduce system closure by resolving tautologies and paradoxes, and by limiting further possibilities. For example, within the function system of science, a claim that is not true is false, and a claim that is not false is true. Members of society are spared both the tautology that "truth is truth" and the paradox that "one cannot truthfully maintain that one is not truthful." The principle of negation imputes binary codes with universal validity, because something that is not identified by one term must be identified by the other. Thus, the binary code of truth/falsity precludes the consideration of alternative criteria when evaluating a scientific event. Binary codes operate similarly in each function system. While the principle of negation (as materialized in the binary code) ensures organizational closure, it also ensures structural openness by inducing society to examine the possibility of that which does not exist.

Each system's programs, which refer to its binary code yet are not terms of the code, further retain the system's openness. At the same time that they operationalize the system's binary code, they must remain variable, because determining the relative suitability of one or the other binary value when appraising an experience requires information from outside the system. Programs, then, refer to the conditions necessary to determine the selection of one binary term over the other. For example, decisions regarding whether to perform experiments designed to determine the truth or falsity of scientific claims regarding the damage done by the *Exxon Valdez* spill relied on the binary codes of the legal system (legal/illegal) and the economic system (ability/inability to pay). Structural openness, then, allows social systems to use terms from within other function systems without losing their previously determined identities.

Luhmann argues that functional differentiation limits society's potential responses to environmental disturbances, for responses can be formulated only in terms of function systems. Whenever society is unable to ignore environmental disturbances, the resulting "resonance" between society and its environment is channeled into a function system and treated in accordance with that system's binary code. Experience that cannot be translated into the binary code of a function system never becomes information. Even though function systems screen society from its environment by sharply reducing what counts as information, they make up for this by producing resonance at the internal boundaries of society— where communication across function systems defines society. Additionally, function systems form each others' environments, for the world is not constituted so that events fit neatly within the framework of one function alone. For example, "*scientific* research has made the construction of nuclear plants *economically* possible through a *political* decision about *legal* liability limitations" (Luhmann, 1989, p. 49).

In the case of the *Valdez* oil spill, an environmental disaster was amplified when a *political* decision provided Alyeska with *legal* justification for an *economic* decision to postpone technical changes suggested by *science* (to station more oil spill equipment in Valdez). Despite overlap between function systems, the systems lack integration to the degree that a positive valuation in one system does not automatically entail a positive valuation in the other systems. For example, even if Exxon's two-billion-dollar cleanup operation in Prince William Sound entailed a positive economic valuation of the spill and resulting cleanup, because it infused the local economy with financial capital or increased the local ability to make payments, the Alaska Oil Spill Commission determined that the entity responsible for causing a major oil spill should never control the cleanup effort (Laycock, 1990, p. 110). In this case, the binary codes for both the legal and economic function systems dictated that Exxon should pay for, and manage, the oil spill cleanup. However, structural openness produced resonance between several function systems, which enabled the investigating commission to base its recommendation on conditions relating more closely to the function systems of science and education. The commission's recommendation reflected the belief that basing all decisions of environmental value on the economic function system was inappropriate. Even in cases such as the *Valdez* oil spill, wherein function systems did not produce coordinated responses, their communicative interdependency ensures that operations can switch quickly from the code of one function system (the economic system) to the code of another (the political system).

Luhmann cautions against defining other function systems solely in terms of their relationship with the economic sphere. Although he admits that "among society's many function systems the economy deserves first consideration," he claims that the attempt to derive the near totality of other sociological phenomenon from any one sphere is hopelessly reductionist (1989, p. 51). This is the problem with using cost-benefit analyses to determine "total" worth of publicly owned natural resources. Contrary to some economists' claims, utilitarian considerations do not provide a "total" picture of any social dilemma, and to assert that they do is to deny the information value we gain from other function systems. When the criteria and programming of one system are privileged over all others, the number and variety of experiences that count as information in a society are sharply reduced. Because society's ability to find resonance with its environment is almost completely dependent on the secondary resonance that develops between its function systems, this boundary activity is essential to the perception of environmental disturbances. Society can, however, compensate for this limitation. Through recognizing the limitations of any function system, society can benefit from the internal complexity of an integrated system. For example, although one cannot, theoretically, purchase indigenous wildlife in the U.S. *economic* system, one can manage it through the *political* system and safeguard it through the *legal* system.

The rhetorical situation (Bitzer, 1968) encountered when operating within each function system's binary code differs markedly from the rhetorical situation encountered when operating within another function system's code. This variation occurs because each function system's binary code provides specific constraints. For example, the economy refers to all operations transacted through the payment of money, and only to such operations. Economic valuation models must, therefore, determine "total value" in money. Luhmann argues that economic programs provide the means for cycling the capacity and incapacity to make payments from one segment of society to another (1989, pp. 51–62). The economy's binary code of payment/nonpayment limits economic valuation models to conceiving of value according to how much money people have spent, or how much they report willingness (and/or ability) to spend, on the resource in question. Therefore, social resonance with wilderness, for example, is possible only after wilderness is reinterpreted according to how human wilderness experiences fit in the cycle of the capacity and incapacity for making payments. According to Kelman, assuming that values expressed in market transactions should drive public policy "denudes politics of any independent role in society, reducing it to a mechanistic, mimicking recalculation" (1990, p. 134). As Luhmann (1989) points out, each function system experiences the environment through its own programs and codes. Terministic screens that privilege economic modes of experiencing over all others threaten to distort the social experience. When observation of environmental issues is interpreted in light of the ability to pay, the economic system can only observe social experiences with wilderness after arbitrarily decontextualizing the wilderness in question from its noneconomic milieu. Additionally, it can communicate about the relationship of society to the environment in question only through economic theories already in existence, thereby choosing which experience will become information without external (noneconomic) means of rationalizing the selection.

How Economic Valuation Distorts Environmental Advocacy

Economically determined valuation models distort analysis of ecological problems by trivializing other social functions such as education, politics, or law. For example, despite the relative accuracy of various economic valuation models in estimating the money lost by the Alaskan fishing industry because of the *Exxon Valdez* spill, the cultural damage to area residents cannot be appropriately measured in dollars. Whatever amount of money Exxon spends reimbursing those whose livelihoods were threatened by the spill, it will not repair damage dealt to the local culture (Scott, 1991). Descriptions of the slump that Alaska experienced in the tourist industry provide only peripheral characterizations of the cultural impact of the spill.

In March 1990, following a year of studies attempting to quantify the damage in economic terms, the U.S. Department of Justice and the State of Alaska announced lawsuits seeking millions of dollars in compensation (Dayton, 1990). Based on its assessment of Exxon's cleanup, the Alaska Oil Spill Commission (appointed by Alaska's governor, Steve Cowper) recommended: "Never again, should the spiller be in charge of a major oil spill." The study found that "privatization and self-regulation" contributed to both the spill and questionable cleanup techniques (Laycock, 1990, p. 110). The report suggested that neither the oil industry nor other private corporations were appropriate guardians for publicly owned natural resources. Rather, this was a responsibility of state and federal governments, as they represented their citizens. This view previously had been expressed by the U.S. General Accounting Office, in a report to Congressional Requesters regarding the *Exxon Valdez* spill. After stating that the preparation for and response to the *Valdez* oil spill was "clearly inadequate" and that a major "reason for this state of national unpreparedness is that there is no single designated leader or authority to ensure that preparations are adequate," the report proposed that "the federal government should perform this leadership role" (U.S. General Accounting Office, 1989, p. 12). Exxon's chairman, Lawrence Rawl, responded angrily to the lawsuits, justifying corporate handling of the spill by claiming, "We took responsibility, we spent over $2 billion, and we gave Alaska fishermen $200 million on no more than their showing us a fishing license and last year's tax return" (Behar, 1990, p. 62).

Statements made by those whose lives were most deeply touched by the spill, however, expand on the government report's claim that Exxon's billions were "clearly inadequate" by pointing to the inadequacy of the economic model upon which it was based. Armin Koernig had established a pink salmon hatchery at nearby Sawmill Bay. The hatchery was ready to release 117 million pink salmon fingerlings at the time of the spill. In addition to jeopardizing the fingerlings' food supply, the leading edge of oil was moving toward the hatchery itself. The local fishers, disgusted with corporate inactivity, decided they had to save their hatcheries. So they formed the Mosquito Fleet and started an independent recovery operation. Although there was some question of legal jurisdiction, because of the general confusion they were able to save the Koernig Hatchery, as well as several others, before anyone could pin them down. Koernig explained why they had to save the hatcheries, rather than waiting for reimbursement from Exxon: "We're not just producing income. We're producing food, supporting a lifestyle. And we're proud of it. A check from Exxon won't work. It would hurt our hearts. . . .

We're not interested in having Exxon just pay us off on a straight dollar-and-cent loss. And we know these same oil companies have safer operations elsewhere in the world. Here in Alaska they have tried to buy us with bullshit, expensive ad campaigns" (Davidson, 1990, p. 103). Tom Copeland, another fisher, got tired of waiting for Exxon to clean up the oil. He bought a pump, gas, and buckets, and then he and his crew went looking for oil. The first day they scooped up fifteen hundred gallons, and on their best day they got twenty-five hundred. Exxon's most productive skimmer collected only twelve hundred gallons of oil a day (p. 109).

These fishers saw beaches around the sound covered with dead and dying scoters, auklets, cormorants, and oystercatchers. Some of the birds were too mutilated to identify. They watched harbor seals ingest oil until they slowly died. John Thomas, a fisher who volunteered to assist in the search for otters, described the condition of the sea otters he brought in. Many had been blinded or had sustained severe damage to the central nervous system. Others died of oil damage to their lungs, kidneys, and liver. Their digestive tracts were ulcerated from ingesting oil. "After a while," said Thomas, "you don't get angry. Anger is way in the back. You have moved far beyond being angry, because everything around you is dead" (Davidson, 1990, p. 153). Fisher Linda Herrington headed the bird rescue fleet. Exxon ordered her not to pick up the oily debris with the birds and attempted to force her resignation when she refused to comply. She stayed on, claiming that she couldn't "leave that stuff floating around to get another bird" (p. 145).

Fishers, carpenters, schoolteachers, loggers, and other residents of the area were dismayed when oil-soaked globs barely recognizable as birds began washing ashore. Sea otters crawled ashore to get out of the oil and died of exposure. Bears coming out of hibernation scavenged on oil-soaked birds and otters. Eagles fed on contaminated carrion (Davidson, 1990, p. 57). The native villagers were warned that they should not eat the birds, shellfish, seals, and other food they took from the sea. Exxon sent them cases of chicken. Upon learning that chicken was not the same as fish, they offered to fly fish to the villages. Exxon employees who had been flown from Texas to manage the operation were offended when their offer was refused (p. 291). From an economic perspective, Exxon, which was only one Alyeska partner and had accepted the responsibility for the spill, was deemed generous. The spill, however, damaged much more than the region's economic potential.

If economic reasoning alone does not provide an adequate means for explaining and directing responses to the *Valdez* spill, perhaps it also is inappropriate as a means for determining management policy for other publicly owned natural resources. For example, traffic on the Gulf Intracoastal Waterway, which provides a relatively inexpensive shipping route for petrochemicals and other toxic materials, continues at high levels despite potential adverse effects on the endangered whooping crane population. Based on Stoll and Johnson's TVP (1984), the value of the cranes ranges anywhere from $573 million to $1.58 billion. The ability to transport toxic materials through the refuge is worth considerably more than $1.58 billion to those industries who directly or indirectly use the waterway. Further, to the extent that they control the means for cycling the ability or inability to make payments, industry representatives also control the economic function system. If policy decisions regarding Aransas National Wildlife Refuge rely on results obtained through total valuation models, the refuge will be managed to facilitate industrial transportation rather than to preserve whooping crane habitat.

Advocates of cost-benefit analysis argue that all human decisions are implicitly based on utilitarianism and that economic valuation models simply ensure rational conclusions by making this process explicit. Even if economists are correct in assuming that utilitarianism is implicated in all decisions, market values are not necessarily given causal roles. At most they may reflect, rather than precipitate, final decisions. In the models described in this essay, however, market equivalencies are established in advance, and they provide the raw materials for calculating a natural resource's value.

Additionally, there is some question as to whether utilitarian calculations control decision making. Luhmann (1989) argues that social decisions are deliberately opaque, for in order to create the illusion of a natural response, all decision structures conceal their own contingencies. Any instrument for acquiring or organizing knowledge is merely a form of simplifying the observation of self-observations. Further, its institutionalization releases it from the restraints imposed by unfettered critique. As John Stuart Mill argued, doctrines that "make the deepest impression upon the mind may remain in it as dead beliefs, without being ever realized in the imagination, the feelings, or understanding" ([1859] 1947, p. 40). By privileging the economic function system and valuation models driven by it, current methodologies naturalize the notion that those who are at an economic advantage not only do, but should, control decisions regarding environmental policy.

Thus, valuation models limited to economic concerns pose ethical difficulties for environmental advocates. Kelman suggests that "there are good reasons to oppose efforts to put dollar values on nonmarketed benefits and costs" (1990, p. 129). The validity of such models depends on several assumptions. First, one must assume that economists can control for all dimensions of quality other than the presence or absence of the nonmarketed entity. Second, one must assume that the nonmarketed entity affects all people equally and that all people have the same constraints. For example, when assessing the value of "peace and quiet" by comparing the selling price of equivalent homes near an airport and in a more distant neighborhood, the monetary value imputed to "peace and quiet" will be inappropriately low if some people have different perceptions or needs than others. Those who hear less noise, or who cannot drive, will take the house by the airport at less of a discount than the "average" person. Third, we must assume that there is no difference between the price a person is willing to pay to get something and the price the same person is willing to pay to avoid giving up something.

Fourth, and of fundamental significance, basing the value of publicly owned resources, such as wildlife, on capitalist axioms requires the assumption that citizens do not differentiate between values expressed in private transactions and those expressed in public policy decisions. If this assumption is correct, then people who drive their cars to work would oppose public transit and those who golf on courses made affordable by taxpayer-subsidized irrigation intended for food production would oppose reform of water laws. Fisher Linda Herrington would have either accepted Exxon's demand to desist from picking up oily debris or its request that she resign her post as head of the bird rescue operation. Empirical support for these assumptions is lacking.

Most important, Kelman points out that "the very statement that something is not for sale affirms, enhances, and protects a thing's value in a number of ways" (1990, p. 134). Pricelessness says that the thing is valued for its own sake, whereas something on the market (whether hypothetical or real) is valued instrumentally, as a means for achieving a more important end. Being not for sale does more than reflect the quantity of the thing's valuation. Rather, it signals a thing's distinctive quality and "expresses our resolution to safeguard that distinctive value" (p. 135). In contrast, the very act of pricing a nonmarketed entity may reduce its perceived value. The contemporary Western aversion toward buying and selling humans (including the purchase of babies by presumably loving adoptive parents) is based on the judgment that this act diminishes human worth. For many people, part of wilderness's value comes from its position as a repository of qualitative values found only in noneconomic sectors. If wilderness is a resource held in public trust by the government, then private economic motives provide not only inadequate but also politically inappropriate means for determining policy. Mill's familiar declaration reminds us that public policy should be tied to the needs of the least powerful: "The State, while it respects the liberty of each man in what specially regards himself, is bound to maintain a vigilant control over his exercise of any power which it allows him to possess over others" ([1859] 1947, p. 106). Economic valuation models for wilderness, however, postulate the value of a public resource as defined by those who exert the greatest control over the cycle of payment and nonpayment capacity.

In suggesting that economic models provide an inappropriate basis for determining publicly owned natural resource management policy, we are not advocating that analysis of ecological problems should begin from other causes within society, then proceed to assign blame for damages. Luhmann (1989) argues that rather than assigning moral responsibility, such analyses only provide exculpation by determining innocence. He claims that they create a "rhetoric of anxiety," which can always be used for moral justification but simply achieves more anxiety. The aftermath of the *Exxon Valdez* disaster illustrates such anxiety. Rather, we are suggesting that attempts to respond to ecological problems must recognize multiple causality in environmental conflict and that this understanding can be incorporated only when decision rules explicitly integrate values instantiated in the programs of all function systems.

The discourse used to define or evaluate anything has the potential to invent that thing in fundamental ways. To begin with, it creates knowledge about that thing and guides appropriate responses to it. For example, if wilderness is defined primarily in economic terms, it becomes an economic resource. As an economic resource, its value is determined completely by its relationship to the cycle of capacity and incapacity to make payments. Thus, its concrete naming has determined its abstract nature, while insidiously naturalizing existing patterns of domination as they relate to resources that belong to the public. As Bowker and Stoll (1988) discovered, wealthier respondents value the endangered whooping crane more than do those with lower incomes. Findings such as these have led to the general conclusion that persons with lower incomes care less about environmental policy than do more wealthy persons and may, therefore, be safely ignored when management policy is debated. The significance of Bowker and Stoll's claim is somewhat diminished, however, when one realizes that respondents' valuations of whooping cranes were determined by the number of dollars they were willing to pay to ensure that bird's continued existence. Perhaps poorer respondents offered to pay fewer dollars to "buy" the whooping crane's existence than did wealthier respondents simply because poor people have less discretionary income with which to "buy" anything. If management policy for publicly owned natural resources is to reflect more than the relative abilities of various segments of society to control economic cycles, then programs from function systems in addition to the economy must be called upon to guide management decisions. Valuation of wilderness, as well as other environmental issues, must be based on analyses of the relationships to law, science, politics, religion, and education, as well as to the economy. Cultural motifs discovered in the relationships between these systems could induce the formulation of value terms that are more consonant with the concept of public trust and less likely to legitimate both local and global patterns of inequity that characterize relations within human societies, and relations between human society and its environment.

The potential danger of exclusively relying on economic valuation models for establishing wildlife conservation policy parallels problems experienced by agricultural conservationists. Tarla Rai Peterson (1986) has suggested that the impact of U.S. farm conservation rhetoric has been constrained by philosophical determinism. Because conservation's position within American agriculture's hierarchy of values depends on its connection with short-term profits, its application has been erratic at best, for farmers can reject conservation practices without rejecting the ecological principles upon which they are based. Because agricultural conservationists have relied on an economic connection between "conservation" and "agriculture," as soon as nonconservational production appears to be more efficient or immediately profitable, farmers replace conservation with less responsible farming practices—until another environmental crisis emerges.

Wilderness managers may actually reinforce the primal importance of short-term capital gain by marketing wilderness conservation and preservation on the basis of economic valuation. By depending so completely on association with economic motives, conservationists limit the possibilities of their rhetoric to a utilitarian perspective. When wilderness conservation and preservation are reduced to only an instrument for achieving an economic goal, policies that endanger fragile or rare habitats should not be attributed to rejection of the principles upon which conservation is based, but rather to limitation of wilderness's relative significance in an economically determined hierarchy of values. Thus, reliance on total valuation models for justifying natural resource conservation and preservation may ultimately do more harm than good. Although public outcry promoted awareness of short-term economic restitution's inadequacy as an appropriate response to the *Exxon Valdez* disaster, the term *disaster* was required to justify the additional obligation. We hope this chapter will encourage the reader to question the primacy of economic valuation techniques even in less disastrous environmental advocacy settings.

Notes

1. A consortium of oil companies operating the North Slope oil fields in Alaska formed the Alyeska Pipeline Service Company in 1970 to build and operate the Trans-Alaska Pipeline. At the time of the *Exxon Valdez* spill, the following oil company subsidiaries owned shares: BP [British Petroleum] Pipelines (Alaska), Inc., 50 percent; Arco Pipe Line Company, 21 percent; Exxon Pipeline Company, 20 percent; Mobil Alaska Pipeline Company, Amerada Hess Pipelines Corporation, Unocal Pipeline Company, and Phillips Alaska Pipeline Corporation, 9 percent. Alyeska was responsible for North Slope oil from the first pipeline pump station to the tankers in Port Valdez. Although shipowners typically were responsible for the oil once it was in the tankers, Alyeska was charged with the initial response to a sea spill (Laycock et al., 1989, p. 87).

2. Alyeska, whose contingency plan was designed specifically for this location, gradually slipped out of the picture. Exxon brought in its own equipment, took command of Alyeska's equipment, and attempted to implement Alyeska's contingency plan. Transferring the oil remaining in the damaged tanker to another, and skimming oil from the sea, had been the foundation of this plan. Before Exxon personnel could transfer the oil, however, they had to locate appropriate pumps. They brought skimming attachments from San Francisco and from England. When the skimming attachments arrived, it was discovered that no one had developed an effective method of transferring the skimmed oil to a collection barge. Exxon determined that dispersants were the only practical means of controlling the oil, but the *Exxon Valdez* had crashed in a zone where dispersants were not recommended. Additionally, local fishers feared that dispersants would destroy the fishery (Davidson, 1990).

References

Behar, R. (1990, March 26). Exxon strikes back. *Time,* pp. 62-63.

Berger, P.L., & Luckmann, T. (1966). *The social construction of reality: A treatise in the sociology of knowledge.* Garden City, N.Y.: Doubleday.

Bergstrom, J.C., Stoll, J.R., Titre, J.P., & Wright, V.L. (1990). Economic value of wetlands-based recreation. *Ecological Economics, 2,* pp. 129-47.

Binkley, C.S., & Miller, R.S. (1980). Survivorship of the whooping crane, *Grus americana. Ecology, 61,* pp. 434-37.

Bitzer, L. (1968). The rhetorical situation. *Philosophy and Rhetoric, 1,* pp. 1-14.

Bowker, J.M., & Stoll, J.R. (1988). Use of dichotomous choice nonmarket methods to value the whooping crane resource. *American Journal of Agricultural Economics, 70,* pp. 372-81.

Brookshire, D.S., Eubanks, L.S., & Randall, A. (1983). Estimating option prices and existence values for wildlife resources. *Land Economics, 59*, pp. 1-15.

Burke, K. (1966). *Language as symbolic action: Essays on life, literature, and method.* Berkeley: University of California Press.

————. (1984). *Permanence and change.* 3rd ed. Berkeley: University of California Press.

Came, B., Quinn, H., & Lowther, W. (1989, April 10). Tragedy on a reef. *Maclean's*, pp. 76-77.

Crutchfield, J.P., Farmer, J.D., Packard, N.H., & Shaw, R.S. (1986). Chaos. *Scientific American, 255*, 9, pp. 46-57.

Davidson, A. (1990). *In the wake of the Exxon Valdez.* San Francisco: Sierra Club Books.

Dayton, L. (1990, March 24). Alaska's silent spring. *New Scientist*, pp. 25-26.

Environmental Assessment Panel. (1990). *Northern diseased bison: Report of the Environmental Assessment Panel.* Panel Report no. 35. Hull, Que.: Federal Environmental Assessment Review Office.

Giddens, A. (1979). *Central problems in social theory.* London: Macmillan.

————. (1984). *The constitution of society: Outline of the theory of structuration.* Berkeley: University of California Press.

Hackett, G., Hager, M., Drew, L., & Wright, L. (1989, April 17). Environmental politics. *Newsweek*, pp. 18-19.

Hvenegaard, G.T., Butler, J.R., & Krystofiak, D.K. (1989). Economic values of bird watching at Point Pelee National Park, Canada. *Wildlife Society Bulletin, 17*, pp. 526-31.

Kelman, S. (1990). Cost-benefit analysis: An ethical critique. In T.S. Glickman & M. Gough (Eds.), *Readings in risk* (pp. 129-36). Washington, D.C.: Resources for the Future.

Laycock, G. (1990, September). The disaster that won't go away. *Audubon*, pp. 106-13.

Laycock, G., Dold, C.A., Soucie, G., Luoma, J., Gilliland, J.R., & Dawson, T. (1989, September). The baptism of Prince William Sound. *Audubon*, pp. 74-91.

Loomis, J.B., Peterson, G., & Sorg, C. (1984). A field guide to wildlife economic analyses. *Transactions of the North American Wildlife and Natural Resource Conference, 49*, pp. 315-24.

Luhmann, N. (1989). *Ecological communication.* J. Bednarz Jr. (Trans.) Chicago: University of Chicago Press.

Lyon, K.S., & Keith, J.E. (1984). Analyzing values of fish and wildlife populations. *Transactions of the North American Wildlife and Natural Resource Conference, 49*, pp. 356-65.

Mill, J.S. ([1859] 1947). *On liberty.* Arlington Heights, Ill.: Harlan Davidson.

Miller, J.R., & Hay, M.J. (1984). Estimating substate values of fishing and hunting. *Transactions of the North American Wildlife and Natural Resource Conference, 49*, pp. 345-55.

Peterson, T.R. (1986). The will to conservation: A Burkeian analysis of Dust Bowl rhetoric and American farming motives. *Southern Speech Communication Journal, 52*, pp. 1-21.

Randall, A., & Stoll, J.R. (1983). Existence value in a total valuation framework. In R.D. Rowe & L.G. Chestnut (Eds.), *Managing air quality and scenic resources at national parks and wilderness areas* (pp. 265-74). Boulder, Colo.: Westview Press.

Scott, E. (1991). Each year since the cleanup, the oil comes back. *Earth Island Journal, 6*, 3, pp. 30-31.

Stoll, J.R. (1983). Recreational activities and nonmarket valuation: The conceptualization issue. *Southern Journal of Agricultural Economics, 15*, pp. 119-25.

————. (1986). Methods for measuring the net contribution of recreation to national economic development. In B.L. Driver & G. Peterson (Eds.), *A literature review: Values and benefits* (pp. 19-33). Washington, D.C.: President's Commission on Americans Outdoors.

Stoll, J.R., & Johnson, L.A. (1984). Concepts of value, nonmarket valuation, and the case of the whooping crane. *Transactions of the North American Wildlife and Natural Resource Conference, 49*, pp. 382-93.

Ulrich, H., & Probst, G.J.B. (Eds.). (1984). *Self-organization and management of social systems.* New York: Springer-Verlag.

U.S. Fish and Wildlife Service. (1986). *Whooping crane recovery plan.* Albuquerque, N.Mex.: U.S. Fish and Wildlife Service.

U.S. General Accounting Office. (1989). *Adequacy of preparation and response to Exxon Valdez oil spill.* Report to Congressional Requesters no. B-236137. Gaithersburg, Md.: U.S. General Accounting Office.

Walsh, R.G., Loomis, J.B., & Gillman, R.A. (1984). Valuing option, existence, and bequest demands for wilderness. *Land Economics, 60,* pp. 14-29.

10
Liberal and Pragmatic Trends in the Discourse of Green Consumerism
M. Jimmie Killingsworth & Jacqueline S. Palmer

Frequently viewed as a countercultural movement born of political discontent in the 1960s and 1970s, environmentalism has, in the last decade, won wide support in the United States as a collective search for a clean human habitat and a lifestyle that brings prosperity without threatening the continued existence of other life-forms and ways of life (Hays, 1987; Dunlap, 1989). As Walter Truett Anderson observed, "Practically everybody today is some kind of environmentalist" (1990, p. 52). Indeed, the American people show signs of accepting environmental awareness as a core element of a national ethos.

As it grows in strength, however, environmentalism is also diversifying into a new set of possible political positions and discourses, creating the conditions for discord or dilution of purpose among environmentalists (Norton, 1991). Where there was once environmentalism now appears a multiplicity of environmental-isms. Old-fashioned conservationists must compete for public attention with the greens (German, British, and American varieties), the deep ecologists, the social ecologists, the ecofeminists, and the advocates of sustainable development, to name only a few of the more prominent categories of contemporary environmentalism.

For the rhetorical analyst, this diversification means that, instead of accounting for a single environmentalist ethos or political character, one must allow for a number of possible "subject positions," "perspectives," or "discourse communities," each with its own characteristic discourse practices and political agendas (Killingsworth & Palmer, 1992a). As we demonstrated in an earlier essay (Killingsworth & Palmer, 1992b), one means of clarifying a particular political perspective is to analyze its use of instrumental discourse as a means of winning adherents and orchestrating social action. Instrumental discourse—the kernel of which is the simple imperative sentence "Do this"—gives directions, makes policy, and aspires to influence readers at the level of habitual behavior. The rhetoric of instrumentality not only informs technical manuals and school textbooks but also appears in a culture's wisdom literature, proverbs, maxims, and religious litanies (Killingsworth, 1992). Despite broad differences in tone and style, both technical literature and ritual instruction aim to set up routines of action that embody a system of values. A computer manual, for example, builds upon the performative values of efficiency and personal growth, while a guide to yogic meditation enacts and extends the values of spirituality and mindfulness. While the ends may differ, the rhetorical means are similar, as is the desired psychological outcome of habituation.

In our earlier work, we described the instrumental genre of green how-to books—guides to living the good life according to the values of environmentalism, books like the best-selling *50 Simple Things You Can Do to Save the Earth* (EarthWorks Group, 1989)—as politically unaligned. We argued that, while various political perspectives may co-opt such instrumental discourses, developing them for their own particular purposes, the genre itself appears neutral—or, at best, "protopolitical"—a characterless cipher waiting to have life breathed into it by an informing political ethos, or a half-formed offspring of environmental consciousness awaiting instruction by a more mature environmentalist ideology with clear political aims.

In this essay, however, we want to consider the possibility of a closer link between the genre and a unique political identity. The green how-to books, we suggest, may well represent an independent and fully coherent political ethos—specifically a perspective that accepts environmental concern as a value but treats environmental protection as an issue, albeit a perpetual issue (or, in rhetorical terms, a topic), but not as an informing principle of a total ideology. Thus, though we follow the terminology employed by a number of the books themselves in labeling this perspective "green consumerism," we argue that much of the appeal of the green consumer books actually rests upon their refusal to espouse any particular ecological-*ism*. While both environmentalists and political conservatives may view green consumerism as a field ripe for harvest, one group hoping to capitalize upon and expand the "greenness" of the movement, the other group seeking to make the most of its "consumerism," neither can entirely capture its distinct political character, for the movement has grown up in the midst of radical and conservative appeals and has chosen a different approach.

Our aim is to explore green consumerism through an analysis of its discourse paths. In the first part of this chapter we offer a theoretical basis for a more specific analysis, which appears in the second part. First, we consider political theory, charting the alternative perspectives of environmental politics in order to demonstrate the possibility that green consumerism does not adhere to any widely recognized eco-political ideology, but rather perpetuates an issue-oriented approach to environmental protection most compatible with the traditional norms of liberal democracy. Second, we consider a pragmatic theory of discourse to show how the instrumental writing of green consumerism replicates and shapes cultural norms in a dialectical fashion, offering a discourse that, by being sustainable over the long term, differs strongly from the institutionalized means of dealing with issues in the political forum and mass media. Following this theoretical treatment, we examine some of the most popular examples of the green how-to books, focusing on their special functionality within a political and cultural context. We conclude by speculating on the cultural power of green consumerism as compared with the more traditional discourses on environmental politics.

Our overall contention is that the green consumer books may well serve a vital function. Instead of attempting to impose sweeping changes upon government, industry, or culture, consumer-based environmental concern quietly transforms civil society from the inside out. Lacking the impressiveness of what might be called macropolitical discourses—whether the consciousness-raising essays

of John Muir, Aldo Leopold, and Rachel Carson or the newer political theory of
the deep ecologists, eco-anarchists, and ecofeminists—the green how-to manuals
hammer away at the micropolitical level, reflecting the hope that the books can
change individuals' habits of action as a first step toward reforming public policy.
Though seemingly superficial, the effects of such discourse may go deep, creating
meaning at the ritual level of habitual performance, affecting daily practices
like shopping, housecleaning, and neighborly conversation. While macropolitical
discourse tends to be favored by traditional standards of eloquence and by the
tendency to divide environmental politics into abstract categories of contentious-
ness—environmentalism versus developmentalism, for example (Killingsworth &
Palmer, 1992a)—micropolitical discourse should not be overlooked by students
of discourse and human ecology. Unlike impassioned polemics and sensational
news reports, both of which may be powerful but ephemeral in their appeal, the
discourse of habit formation is less likely to be vulnerable to the fluctuations of
interest that characterize the usual public response to political issues as presented
in political campaigns and media coverage. Insofar as it becomes rooted in daily
experience, green consumerism creates a clear basis for the further production
and refinement of environmental values, a reminder of the primacy of environmental issues in ongoing political discussions.

In this light, the "salvation" promised by titles like *50 Simple Things You Can
Do to Save the Earth* signifies not so much the salvation of the physical Earth but
rather the salvation of earthly consciousness, a state of mind, or state of being,
that depends upon a continuing connection of human identity with the material
Earth.

Theoretical Considerations

The first part of our argument is that the green consumer books suggest a coherent political perspective on environmental politics. To demonstrate that green
consumerism is both politically significant yet different in outlook from other environmentalist ideologies, we draw upon two related conceptual bases—first political theory and history, then a pragmatic theory of discourse.

Green Consumerism on the Map of Environmental Politics

In the kind of taxonomic exercise that becomes ever more frequent in a society
where "everyone is an environmentalist of some kind," Anderson (1990,
pp. 52–53) identifies "four distinct wings" of the environmental movement. In
our earlier analysis of the green how-to books (Killingsworth & Palmer, 1992b),
we showed how each of these groups has met special needs by modifying the
instrumental genre. In various ways, they all use instrumental discourse to encourage and direct their adherents toward a stepwise accomplishment of primary social and political objectives. The "wings" of the environmental movement
comprise

1. *Politicos,* Washington lobbyists and special-interest groups who continue
the work begun in the early years of reform environmentalism with appeals

to big government to control big business. Publications include *Ecotactics: The Sierra Club Handbook for Environmental Activities* (Mitchell & Stallings, 1970) and *Blueprint for the Environment* (Comp, 1989).

2. *Greens,* also known as deep ecologists (Devall, 1988; Manes, 1990) and social ecologists (Bookchin, 1990), radicals who argue for fundamental changes in sociopolitical structure and individual lifestyle. Publications include *Ecodefense: A Field Guide to Monkeywrenching* (Foreman & Haywood, 1993) and the original *Monkey Wrench Gang* (Abbey, 1975).

3. *Grassroots activists,* community members associated with local projects for environmental improvement and "ecodefense" against specific dangers resulting from industrial pollution or land development. Publications include *Not in Our Backyards!: Community Action for Health and the Environment* (Freudenberg, 1984) and *Fighting Toxics: A Manual for Protecting Your Family, Community, and Workplace* (Cohen & O'Connor, 1990).

4. *Globals,* groups like the Worldwatch Institute (see Brown, 1991) who, unlike other groups that react against overdevelopment, instead support proactive efforts for "sustainable development," with planning extending from local communities to the ecology of the Earth as a whole life system. Publications include the monthly magazine *Worldwatch.*

The range of instrumental writings covered by these established political perspectives is indeed broad, and the number of books produced is extensive. But the instrumental genre is not exhausted by the combined efforts of the politicos, the greens, the grassroots groups, and the globals. Some of the green how-to books—*50 Simple Things You Can Do to Save the Earth* (EarthWorks Group, 1989) and *The Green Consumer* (Elkington, Hailes, & Makower, 1988), for example—remain free from a clear identification with any of the political action groups in Anderson's classification.

The greatest difference between the instrumental discourses of the various environmentalisms and the discourse of green consumerism lies in the implied relation of author to audience. The environmentalist publications tend to "write down" to the audience, from a position of greater knowledge (or experience) to those of lesser knowledge or experience. Even the grassroots books, while recognizing that local conditions vary widely, take the perspective of experienced trainers sharing their expertise with neophyte activists. By contrast, the authors of books like *50 Simple Things You Can Do to Save the Earth* make an effort to come across not as experts or experienced veterans facing new recruits but as equals, as friendly neighbors offering "tips" for readers instead of systematic instruction (Killingsworth & Palmer, 1992b). The difference in tone, though subtle at times, is crucial. While the various environmentalists work toward positioning consumers (readers) on a map of allegiances in environmental politics, the green consumer books imply a pluralistic aim, leaving open the possibility that an environmentally concerned public does not necessarily have to subscribe to any of the environmentalisms listed by Anderson. Nor does such a public have to choose a consistent stand from among the ideologies formulated by any number of other political analysts in environmental studies (see, e.g., Dryzek & Lester, 1989). Indeed, any of the ideological types could pursue the actions recommended in the green consumer books without abandoning their preferred values and solutions. All would likely see some benefit in shopping wisely and fixing a leaky toilet to save water, for instance.

This very agreeableness may seem to throw us back upon the conclusion that green consumerism is politically neutral, or that it is merely an innocent collection of actions untouched by critical philosophy. Then again, it may lead us to wonder about the clarity of the distinctions among the environmental ideologies, much in the manner of the pragmatist Bryan Norton (1991), who has argued that environmentalists—considered in light of their actions rather than the explanations they give for their actions—agree on more than they seem willing to admit and should therefore seek unity rather than differentiation in the political forum. In our view, however, the ideological differentiation that may not work at the micropolitical level continues to make sense at the macropolitical level, the level of inquiry attacked by political scientists like John Dryzek and James Lester. The different environmental ideologues may agree about household actions, that is, but not about collective actions, such as the degree of governmental intervention or centralization necessary in environmental control. Thus the distinctions between the ideologies would remain clear even according to the pragmatic rule of clarity developed by C.S. Peirce, who asserts that "what is tangible and practical" lies at "the root of every real distinction of thought, no matter how subtle it may be," that "there is no distinction of meaning so fine as to consist in anything but a possible difference of practice" (1991, p. 168).

We would argue, furthermore, that, like the other political perspectives, green consumerism also remains distinct at the macropolitical level. The tendency to focus exclusively on actions that would attract a broad base of practical support suggests that, instead of being politically neutral and philosophically ungrounded, the green consumers are (like Norton himself) philosophically pragmatic and politically liberal. Nothing in the green consumer books requires that the audience abandon the primary values of Western liberal democracy. Indeed, the books appeal to readers who have developed an environmental conscience and are seeking the means of implementing their newfound values without betraying a fundamental commitment to the established political norms of private property and individual freedom. By contrast, the more fully developed environmentalist ideologies have historically called such norms into question. "Limits to growth," as conceived by the politicos, the greens, the globalists, and even most grassroots activists, almost always entail fairly strong corresponding restrictions upon personal autonomy. Early in the history of modern environmentalism, the liberal values of individualism and the institution of private property were called into question by such influential environmentalists as Paul Ehrlich, Barry Commoner, and Garrett Hardin. Deep ecology, social ecology, and ecofeminism have followed suit (Merchant, 1992).

The popularity, indeed the very appearance, of the green consumer books suggests that, while environmental concern has swept the country, there has not been a concomitant subscription to environmentalist ideology (of any kind). The public opinion polls tend to support this interpretation. Three decades of data suggest a widespread acceptance of the need for environmental protection but offer no evidence that environmentalist ideology has ever enjoyed a wide margin of public favor. On the contrary, the discrepancy between survey respondents' stated commitment to environmental values and their voting behavior hints that the great majority of the American people continue to see environmental protection as an issue, not as the foundation of a "total ideology" as defined by Daniel Bell, "an all-inclusive system of comprehensive reality . . . a set of beliefs in-fused with passion" that "seeks to transform the whole of a way of life" (1988, pp. 399–400).

Perhaps the most famous case supporting this view involves the public response to the Reagan Administration's attempt to weaken the federal system of environmental control. As Riley Dunlap (1989) has shown in an extensive meta-analysis of the literature on surveys and polls, after government established regulatory laws and agencies in the late 1960s and early 1970s, public interest in environmental issues peaked around Earth Day 1970, then appeared to be dying back (see also Downs, 1972). As Dunlap argues compellingly, however, it is only natural that public interest should have declined if the public felt that the problem of environmental degradation was being effectively addressed by the federal government. The proof of this assertion lies in the extraordinary shift of public opinion against the Reagan Administration on the issue of environmental concern. The data from polls and surveys conducted during the 1980s showed an undeniable increase in public support for governmental regulation of environmental interactions, a trend that ran directly against the policies of Secretary of the Interior James Watt. Memberships in and contributions to environmental interest groups—the great enemy of American progress, in Watt's view—increased. The trend of public support for environmental protection continues to climb upward (see Hays, 1987; Norton, 1991; Killingsworth & Palmer, 1992a).

After rejecting the ill-conceived antienvironmentalism of the Reagan Administration, however, the American people continued to vote Republican, electing Ronald Reagan to a second term in 1984 and George Bush to a first term in 1988. The data on the impact of environmental issues on voting behavior—though "limited" and "controversial" (Dunlap, 1989, p. 129)—do suggest the possibility that, if Reagan had not fired Watt and other offending officials and softened his stand against environmentalism, he might have lost votes. Thus, Dunlap argues, "The pro-environmental consensus may be 'permissive,' but it might not tolerate . . . a blatant rejection of its goal of environmental protection" (1989, p. 132).

But the very "permissiveness" of the "environmental consensus" also implies that most voters supported the overall program of the Reagan Administration, a program that aggressively defended the rights of individual freedom and private property and did not view the ideological slant of this program as incompatible with environmental values. Thus, environmental protection has, as Dunlap argues, "achieved the status of an important value among Americans" (1989, p. 133), but it is not a value that, once achieved, brings the prevailing ideology of liberal democracy into question, as most environmentalists suggest it must.

A political value is nonetheless something more than an issue, it would seem, largely on the basis of its persistence. Issues come and go. They are essentially problems that, once solved, can be safely set aside. But values persist as cultural norms and ethical standards long after issues depart.

The appearance of the green how-to books in American bookstores at the end of the 1980s, as Reagan was leaving office, may have represented a (conscious or unconscious) effort to establish environmental values solidly within a public whose interest in environmental issues was cresting at the highest level in twenty years. The self-help discourse of the how-to books, we argue, is uniquely suited to meeting such a goal in the context of a liberal democracy. By influencing the daily habits of a large segment of the general population, the actions recommended in the green consumer books produce a local significance in the form of ritual performance, influencing the identity formation of individuals and ultimately becoming a driving force in the marketplace.

Instrumental Discourse and Ritual Action

Following C.R. Miller (1984) in identifying genre with social action, then, we see the proliferation of the green how-to books in the late 1980s as representing a shift toward a broader acceptance of environmental values and a search for proactive solutions to environmental problems that can be undertaken by individuals at the local level. In our earlier analysis (Killingsworth & Palmer, 1992b), we argued that realizing the political significance of the how-to genre requires us to abandon the traditional analysis of genres that distinguishes strongly between the aims of rhetorical (or persuasive) discourse and the aims of instrumental (or directive) discourse (Kinneavy, 1980; Beale, 1987). According to the traditional scheme, persuasive discourse serves political ends by presenting arguments that win converts, shore up weaknesses in communal spirit, or attack enemies and outsiders, while instrumental discourse presents techniques for solving problems within a social framework assumed to be more or less stable.

Recent discourse theory, as well as the hermeneutic strand of rhetorical analysis connected with Kenneth Burke (1969a, 1969b), inclines away from such binding distinctions. As Walter Beale (1987) suggests, no single genre or discourse "aim" can be considered absolutely stable over time and distance. Genres shift grounds, invade one another's territory, and transform themselves constantly according to the demands of social change. On such grounds, Beale presents a pragmatic model of the motives of discourse that allows for broad and flexible generic fields rather than insisting upon rigid categories. On this scheme, persuasive discourse drifts toward instrumental discourse insofar as it stipulates steps toward the technical solution of a problem or gives instructions on how to carry out procedures. And insofar as instrumental discourse advocates sociopolitical ends, it drifts toward persuasive discourse. The instrumental tends to be means-oriented, while the persuasive tends to be ends-oriented. To concentrate entirely upon ends without allowing for means, however, tends to result in rhetorical stridency; to focus exclusively on means without considering ends makes a discourse seem reductive, mechanical, and uncritical (Killingsworth & Palmer, 1992a).

From the perspective of rhetorical criticism, the instrumental discourse of the how-to book—familiar in computer manuals and best-selling books on how to lose weight or do your taxes or improve your psychological well-being—may seem politically irresponsible because it has few mechanisms for criticizing that which it represents. This critique has been amply developed in a century-long attack of critical theory (from Weber to Habermas) on "instrumental rationality," the modus operandi of modern bureaucracy. Though not necessarily committed to the social agenda of instrumentalism—scientific management, "Taylorism," and the vulgar pragmatism of "whatever works is right"—instrumental discourse at least raises the specter of social control and manipulative politics (Killingsworth, 1992).

The right mix of persuasion and instrumentality depends upon the ability of the producers of discourse to line up their goals and activities with those of the discourse users. If a writer and a reader share the same ends, then the problem of persuasion is minimal. The writer needs only to instruct the reader on the means of achieving mutual goals, perhaps adding ends-oriented rhetoric here and there for the purpose of encouragement and community reinforcement. Such is the social context of the how-to manuals produced by established environmentalist organizations—the politico's *Ecotactics* (Mitchell & Stallings, 1970), for example, or the grassroots group's *Not in Our Backyards!* (Freudenberg, 1984).

If the writer and the reader have different goals, however, then the writer either must attempt to lead the reader to accept new goals before elaborating the means of achieving the goals—thus deferring instrumental discourse until the work of traditional rhetoric is completed—or must discover means that satisfy divergent goals. If the writer can show mutual interest or compatibility of ends, then the work of recruitment or conversion may be undertaken within a primarily instrumental discourse. If, for example, the writer's goal is to reduce carbon emissions and the reader's goal is primarily to save money, the writer could produce a set of instructions on energy savings for the home that could well meet both goals. This strategy—the effort to establish an implicit argument for mutuality of ends within an instrumental discourse that focuses on means—hints at the kind of questionable compromise and possible manipulation that raises serious questions about the motives of the writers of the green how-to books (Killingsworth & Palmer, 1992b; Muir 1992).

In focusing on sociopolitical motives of authors in our previous analysis, however, we failed to address the indirect or symbolic effect of the actions recommended by green consumerism. This effect becomes clear when we consider the very special character of the books' implied audience. These are readers who have accepted the need to act on behalf of environmental protection but are lost about what to do. They can send money to environmental protection groups, or they can join local groups and pursue local projects. But occasions for direct political action tend to be limited and ephemeral, much like the issues that thematize such actions and the press coverage that reports on them. Moreover, the apocalyptic turn in the treatment of global environmental problems may also breed frustration at the local level. The problems come to seem so large as to make local or small-scale action seem utterly inadequate.

The reader who turns to a book that promises "simple things to do" to "save the Earth" is thus a reader with a hunger for action who is frustrated or confused about how to satisfy that hunger. Such a reader is in a state of transition, having grown dissatisfied with old habits of action (with being "part of the problem"). This person is ready to change but must locate strategies or rules of action. As Dennis Pirages has suggested, "rules for action" form an essential part of "a transition strategy for reaching preferred futures" (1977, p. 8). While the various environmentalisms set their sights on distant futures ("Utopias," as Pirages says), green consumerism looks to the transition, offering ways of acting that reinforce the mentality of the convert and, to use the language of pragmatism, "fix belief" by making it habitual, bringing consciousness into contact with daily practice and with the body (Peirce, 1991, 144–59).

It was Aristotle who first suggested that, while the intellect grows with teaching, "moral virtue comes about as a result of habit," that "states of character arise out of like activities" (1980, p. 28). Building upon the implicit pragmatism in Aristotle, on the way to developing a monism that connects consciousness to bodily sensation, the father of philosophical pragmatism, C.S. Peirce, anticipated the call for strategic "rules for action," arguing that belief itself "involves the establishment in our nature of a rule of action, or . . . *habit.*" In this view, belief appears as a "stadium for mental action, an effect upon our nature due to thought, which will influence future thinking." Peirce goes on to say that "different beliefs are distinguished by different modes of action." Hence habits are connected with periods of transition in modes of action, which, for the pragmatist, are identical with modes of being and identity. Peirce argues that, "since belief is a rule for action, the application of which involves further doubt and further thought, at the same time it is a stopping-place, it is also a new starting-place for thought. That is why I have permitted myself to call it thought at rest, although thought is essentially an action" (1991, p. 166).

The consequences of this way of thinking for our evaluation of green consumerism are considerable. Peircean pragmatism suggests that, while ideological environmentalism may have created a reservoir of commitment within the American public, it may have not given sufficient attention to providing a mechanism of transition whereby old habits are transformed into new habits of action. Green consumerism fills this gap by offering a means of fixing belief on a mass scale, so that the ordinary householder and not just the environmental activist and outdoor enthusiast has a habitual way of reinforcing awareness and commitment. It takes the old habitual "stopping places" of shopping and housecleaning and treats them as "starting places" for new associations and flash points of emerging identity.

As starting places for thought, habits are, in Peirce's general theory of signs, symbolic. A symbol is defined as a sign that, through a habitual association, connects one thing with another. The word *dog*, for example, is connected with the furry mammal not by any existential or necessary relation but by a linguistic habit of association. The association may have come about, however, by means of another kind of sign, which Peirce calls an index. An indexical sign forms an actual existential link with its object. The classic example is the pointing finger (the "index finger"). It locates its object by directing attention to an actual physical presence. A child usually learns a symbolic system like language by starting with indexical references. The voice of the parent, for example, forming the word *apple* at the same moment that a sweet red object is handed over causes a habit of association to form in the child's mind. With repetition, the symbolic relation takes on a life of its own, leaving the original indexical relation behind or retaining only a vestige of it, thereby creating the possibility for conceptual and abstract thought. The index-symbol relation thus represents what Peirce believed was an essential continuity between the physical and the mental. One outcome of this thinking is the idea that a new set of experiences, complete with new physical associations or indexes, may give rise to a completely new mentality (Peirce, 1991, pp. 251–52; Killingsworth & Gilbertson, 1992).

Green consumerism sets up the possibility for the development of just such a chain of semiotic connections. The householder begins to associate daily actions with the prospect of "saving the Earth." This key concept of environmentalism thus becomes rooted in habitual physical experience. Instead of a distant phenomenon—associated with the faraway and disappearing rain forests or with the equally distant processes of Washington government—the idea of saving the Earth becomes something the environmental enthusiast can experience locally and feel with the body, the starting place of all consciousness and all human action.

The theoretical understanding of the relation between thought and habitual action established in pragmatic semiotics is strongly reinforced by the study of ritual in cultural anthropology. Particularly instructive for our purposes is the work of Mary Douglas, whose ideas on the cultural significance of pollution and taboo have already been applied fruitfully by Neil Evernden (1992) to problems of definition and classification in environmentalist thinking. Douglas's concept of ritual (1966) as a system of social symbolism applies equally well to the study of the instrumental discourse of green consumerism.

"As a social animal," Douglas writes, the human being "is a ritual animal," so "it is impossible to have social relations without symbolic acts" (1966, p. 62). Moreover, as Douglas sees it (following Durkheim), there is not a significant difference between religious and secular rites insofar as each serves the ends of sociability. Regardless of whether it is the acceptance of a wafer at mass or the washing of dishes after dinner, actions that are regularly undertaken beginning in childhood, with the instructions of one's parents and teachers, tend to become "symbolic enactments" that provide "a focusing mechanism, a method of mnemonics and a control for experience" (Douglas, 1966, pp. 58–63). The wafer at mass reminds one of the Savior's body—as well as one's own body—of suffering and sacrifice turned to nourishment, physical and spiritual. The ritual of washing dishes brings to mind a string of associations connected with family practices and prohibitions, the need for cleanliness and the avoidance of dirt and disorder. "Dirt is the by-product of a systematic ordering and classification of matter," says Douglas (p. 35). "This idea of dirt takes us straight into the field of symbolism and promises a link-up with more obviously symbolic systems of purity." What Douglas calls "pollution behavior" is "the reaction which condemns any object or idea likely to confuse or contradict cherished classifications" (p. 36). Bugs belong out-doors, not in the house. Germs must remain clear of the body if we are to remain well. Foaming chemicals should not appear in the streams along the hiking trail.

The list of things to do, substances to avoid, and habits to cultivate in the green consumer books brings to mind the ritual practices and taboos given at such great length in the books of Leviticus and Deuteronomy. As Douglas's fasci-nating analysis of these Hebrew Scriptures shows, no interpretation based on utility or medical materialism can sufficiently account for the full range of the many rules and prohibitions relating to diet and ritual worship. The rules only make sense as a symbolic system that reinforces the cultural identity of the Hebrew people, a chosen people with whom God has formed a covenant. With each ritual slaying of an ox, with every avoidance of pork, with each ritual cleans-ing of a woman after childbirth or a priest before he enters the temple, the cultural identity asserts itself, like a prayer of affirmation. Transmitted from gen-eration to generation, the practices seal the identity over time. They are simple things that can be easily performed and even taught to children, who in following the rules and taboos come to identify with their parents and their people. Thus writes the author of Deuteronomy: "You shall put these words of mine in your heart and soul, and you shall bind them as a sign on your hand, and fix them as an emblem on your forehead. Teach them to your children, talking about them when you are at home and when you are away, when you lie down and when you rise" (11.18–19, NRSV).

In Douglas's interpretation, ritual aligns attitude with action to produce wholeness (or holiness). Therein lies the power of ritual action, a power not always apparent to those seeking radical social change through traditionally political means. From the perspective of radical politics, references to "healing the Earth" or "saving the Earth" in the green consumer books may seem unnecessarily meta-physical, pretentious, and overblown. If everyone on Earth did all fifty things, would the Earth be "saved" in any material sense? But perhaps this question misses the point. The real issue has to do with participation at many levels and over long periods of time. As we noted in *Ecospeak* (Killingsworth & Palmer, 1992a), a rheto-ric of sustainability is not well served either by radical polemics that demand changes now (and must therefore seem to have failed when compromises occur) or by a popular press that thrives on reporting "big news"—disasters and scientific breakthroughs—rather than the day-to-day "small news" of a culture suffused with environmental awareness. Where the polemics and the news fall short, in this sense, the green consumer books may have made a real contribution.

Examples of the Discourse of Green Consumerism

We turn now to some specific examples of the green how-to books in an attempt to show that manuals like *The Green Consumer* (Elkington, Hailes, & Makower, 1988) and *50 Simple Things You Can Do to Save the Earth* (EarthWorks Group, 1989) both reflect and extend public participation in a form of action that, while limited to the realm of the householder, remains politically significant. Our approach is to attempt an answer to a pragmatic question: What do these books do for the readers who take them up? In other words, what social, political, and cultural purposes do they serve?

Green Consumerism Inspires Readiness for Action

The authors who originated the concept of green consumerism in their book *The Green Consumer* (Elkington, Hailes, & Makower, 1988) urge their readers to "join the green consumer movement" in a front-cover blurb on their book for children (Elkington et al., 1990). Their use of the word *movement* suggests a forthright acknowledgment of the political significance of their work. But they have in mind a new kind of "movement." The term usually connotes a direct involvement in political affairs, but the approach of green consumerism is indirect, in that it does not move citizens to take to the streets and demonstrate, or even to vote. It concentrates instead upon doing what one can within the role of "consumer," a role formerly associated with political passivity. (Hence the notion arose of the "consumer advocate," a person who speaks for those who are unwilling or unable to speak on their own behalf.)

The key premise of *The Green Consumer* is that, by making slight adjustments in daily practices, ordinary citizens can produce powerful effects—if not in government, then in the political economy of the marketplace. In this arena, the consumer, though often unconscious of power, holds a considerable position of influence, the authors remind us, a position whose power, if tapped, may in fact exceed the power of the vote: "You may be surprised at how easy it is to make your voice heard in the marketplace. The marketplace is not a democracy; you don't need a majority opinion to make a change. Indeed, it takes only a fairly small portion of shoppers—as few as one person in ten—changing buying habits for companies to stand up and take notice" (Elkington, Hailes, & Makower, 1988, pp. 9–10).

The message is clearly crafted to combat feelings of powerlessness and to move people out of their anxious despair over the state of the environment. References to the ease of social action predominate in the early pages of *The Green Consumer,* standing in open defiance against the prevailing notion that individuals and small groups can make but an insignificant difference in the great global mess. "By choosing carefully," the authors insist, "you can have a positive impact on the environment without significantly compromising your way of life. That's what being a Green Consumer is all about" (Elkington, Hailes, & Makower, 1988, p. 5).

Terms of degree abound in the definition of a *green product,* as one that

is not dangerous to the health of people or animals

does not cause damage to the environment during manufacture, use, or disposal

does not consume a disproportionate amount of energy and other resources during manufacture, use, or disposal does not cause unnecessary waste, due either to excessive packaging or to a short useful life

does not involve the unnecessary use of or cruelty to animals

does not use materials derived from threatened species or environments [Elkington, Hailes, & Makower, 1988, p. 6]

Concentrating on the elimination of the "disproportionate," the "unnecessary," and the "excessive," the green consumer movement appeals to the common sense of moderation and compromise, accepting "one step better" when no "perfect solution" appears on the horizon (Elkington, Hailes, & Makower, 1988, p. 7).

The implicit pragmatism of the movement refuses to allow the absence of the ideal to stand in the way of the workable. The point is to begin to do one's part in the overall scheme of things. The introduction to *50 Simple Things You Can Do to Save the Earth* takes up the same theme: "*50 Simple Things* empowers the individual to get up and do something about global environmental problems. No point in letting the news reports and magazine coverage drive you to despair; even the most intractable environmental problems march toward a solution when everyday people get involved" (EarthWorks Group, 1989, p. 6).

The references to moderation and simple actions in both books further suggest that, while resisting the temptation of conservatism to go back to the days of pre-environmentalist innocence, green consumerism nevertheless allows the emerging political consciousness to take one step at a time. The reader does not have to live in caves, or even go camping, but can maintain the role of consumer in a liberal technological society.

Green Consumerism Creates a Ritual Foundation for Identity

The reader is nevertheless led to make significant changes. One rhetorical aim of green consumerism is to get inside the ritual practices of daily secular life and redirect important symbolic associations. In *50 Simple Things You Can Do to Save the Earth,* forty-four of the fifty recommended actions have to do with cleaning, buying and preparing food, and maintaining the home—fields of practice traditionally associated with ritual behavior as defined in cultural anthropology. Taboos in the green consumer books also include the usual objects of ritual prohibition—foods (such as beef grown in pastures reclaimed from the tropical rain forests) and vessels for food (such as styrofoam and plastic packaging). The books thus provide for a sustainable course of action, both in the form of ritual or tabooed behaviors (in which the individual constructs a mental dialogue of attitude and action) and in conversation between household partners, parents and children, and teachers and pupils. In addition to the forty-four actions devoted to diet and household maintenance, five other "things to do" involve communication (stop junk mail, make a telephone call, spread the word) or transportation (carpool to work, drive less), two fields that represent symbolic extensions of the individual into the discursive social field.

Actions connected with daily home practice and discursive relations are the classic province of ritual performance for several reasons. First, such practices are undertaken on a regular basis so that they have strong mnemonic potential; they strike the mind regularly, creating waves of association and identification. Second, because they are easy to do, many can participate, not just lobbyists, experts, and idealists (the high priests and prophets of modern culture), but the layperson and the child as well. Indeed, the green consumer movement has great appeal for children and their mentors in the schools; dealing with everyday experiences, the recommended activities have the kind of relevance for which elementary school teachers search in preparing their lessons. Third, practices involving diet and cleanliness have traditionally been associated with salvation and healing. The clean is holy, that which counteracts the diseased and the dirty. As the old saying goes, "Cleanliness is next to godliness." This set of associations is built right into the English language. The Old English word *hal* is the root for a surprising range of deeply associated words: *holy, whole, well, heal,* and *healthy.* The holy is the embodiment of wellness and health (or wholeness) in spiritual physiology. Thus, when the woman touches Jesus' garment in the New Testament story, seeking a cure for an unceasing menstrual flow (which has made her ritually unclean, a social pariah), Jesus proclaims that her faith has made her "well" or "whole" and therefore holy, at one with God and the community (Mark 5.25–34). The simple act of touching the hem of the Master's garment reveals a state of mind, a longing for atonement, and the result is not only a physical healing but also a restoration of the woman to her social place, a creation of the possibility of cultural healing and identification. Wholeness—or holiness in this biblical sense—involves the means by which the body is reconnected to the network of tradition, culture, and significance.

In a culture that puts a premium on having a place of one's own, the daily affairs of household management become yet more meaningful. Little acts around the house are flash points of both personal and social identity. One person's use of electronic gadgets points toward an identification with the culture of technology. Another's commitment to organic gardening suggests a longing for some connection with the Earth obscured by technological culture. Each flip of a switch or each spading of compost can be rich with symbolic associations for the individual. Every little action becomes a sign, like a religious token or a ritual act, that points to a larger truth, a more distant connection.

In *50 Simple Things,* the authors have found an ingenious way of reinforcing such connections. Following the conventions of magazine journalism and the design of technical manuals, they have reserved a single, thematically unified page (or two-page spread) for each "thing to do." Designed to accommodate a browsing reader, each page defines a problem in a "Background" section, then gives facts relevant to solving the problem ("Did You Know") and finishes with a series of practical tips. Chapter 10, "Home on the Range," for example, says that large home appliances like stoves, refrigerators, and air conditioners "consume 7% of the nation's total electricity—the equivalent of more than 50% of the power generated by all of our nuclear power plants." The text goes on to give more specific facts for other appliances and finally (only a half page later) provides a list of tips on how to buy and maintain different kinds of appliances to achieve surprisingly large energy savings (EarthWorks Group, 1989, pp. 30–31). Further information can be obtained, we are told, from the American Council for an Energy-Efficient Economy. The address, major publications, and cost of publications from this organization are also given. The rhetorical efficiency, brevity, and wit of the treatment maximize the opportunity for associations to develop. Now our large appliances, which we use every day, become less "transparent" to us. They are revealed to be part of a huge energy complex, a system that connects our daily actions to nuclear power plants, for example. Their demand on energy is large enough to render credible the proposition that every householder can make a difference in the overall system simply by cleaning the dryer's lint trap, buying an electronic ignition system for a gas stove, or checking the settings on temperature controls.

More important yet, from a political or cultural perspective, is the idea that each time we take one of these measures we appear to ourselves a bit more environmentally conscious than we used to be. Our new habits, as they develop, become the basis for a new moral character.

Green Consumerism Encourages Critical Awareness

We suggested in our survey of environmentalist political theory above that green consumerism appears at first glance to subscribe to a program of "free-market conservatism," much in the manner of the early conservation movement. The green consumer books do indeed reiterate many of the favored themes of Wise Use conservationism. They predictably advocate programs for reducing home energy consumption, for example. But they go beyond the argument that, in saving energy, one saves money while simultaneously preserving resources for future use. Energy conservation becomes part of a larger campaign to curtail effects like global warming and acid rain, the dimensions of which suggest the ultimate inadequacy of Wise Use policies.

Representing a later stage of environmental awareness than conservationism, with its ties to nineteenth-century progressive ideology (Hays, 1959), the green consumer movement is concerned with industrial and technological output as well as the use of resources, with the prevention of pollution and the preservation of the Earth "as we know it," as well as overall economic efficiency. "We're all familiar with the sensible home economics and good foreign policy of saving energy," write the authors of *50 Simple Things*, for example. "The argument goes: If we insulate our homes, we'll help keep OPEC at bay. Plus, we'll save so much money on utility bills that we'll recoup our investments in a year or two. . . . Not a bad argument. . . . But it doesn't take the environment into account. As a result, many Americans have no idea whether saving energy—or water—makes an ecological difference. Will a dab of caulk around your drafty windows really have an effect on our shattered environment? The answer is a resounding yes" (EarthWorks Group, 1989, p. 17). The implication is that, while many conservation practices associated with Wise Use are still worth cultivating in the householder's practice, a new critical attitude is needed. Now, when applying the dab of caulk, we should not merely congratulate ourselves on saving money but should also bring to mind the fate of the Earth.

Green consumerism thus participates in the new critical consciousness shaped by environmentalism. In this vein, *The Green Consumer* includes a two-page sidebar entitled "How the American Way of Life Is Destroying the Earth." This section relates how consumer culture, in satisfying a seemingly endless appetite for new products, has assented, often unwittingly, to the accumulation of great piles of harmful by-products—an astounding amount of garbage, measured in per-person quantities; air pollution, exacerbated by the individual buying and driving of gas-guzzling cars; and water and soil pollution, arising largely from the pesticide use and landfilling connected with large-scale and irresponsible mass consumption (Elkington, Hailes, & Makower, 1988, pp. 8–9).

It is this drive to foster critical awareness that distinguishes books like *The Green Consumer* and *50 Simple Things* from green how-to books that take a more conservative stance. In the interest of preserving the status quo or undermining environmentalist gains in public favor, conservative interests have sought to capitalize on the popularity of the new genre while dulling the critical edge of green consumerism. One such author is the newspaper columnist Heloise, who suggests that there is nothing really new in adherence to environmentally friendly practices. In her book *Hints for a Healthy Planet,* Heloise says that her household advice columns "have always emphasized the 'Three R's'—Recycle, Reuse, and be Resourceful" (Heloise, 1990, p. 11). Her "Three R's" resonate strongly with advice in *The Green Consumer*—almost to the point of plagiarism—but differ significantly from the three R's as they appear in *The Green Consumer:* "*Refuse,* Reuse, and Recycle" (Elkington, Hailes, & Makower, 1988, p. 40). To "refuse" to buy plastic products or those that are overpackaged, as *The Green Consumer* recommends, hints at boycotting, a practice that is evidently too close to the radical fringe for Heloise's traditional readership, or so she appears to believe.

A motive similar to that of Heloise appears to inform the advertising campaigns of several large corporations, especially the fast-food and packaging industries. An indication of corporate nervousness about green consumerism appears in efforts to co-opt the green how-to genre. Not satisfied with taking the critical edge off of green consumerism, in the manner of Heloise, some companies have sought to turn the apparent power of consumer awareness to their own ends. In one case we noted in our earlier essay (Killingsworth & Palmer, 1992b), a television advertisement for plastic garbage bags, the kind of product that *The Green Consumer* urges people to refuse, offers a free guide to "living better in the environment." The video for the ad shows Boy Scouts filling plastic garbage bags with aluminum cans to be recycled, as the benign ghosts of Indian warriors smile in the background. The voice-over says, "Don't get mad, get moving!" The sentence plays upon the company's well-known slogan of many years—"Don't get mad, get Glad!"—in a clear attempt to derail the environmentalist critique and head off the progress of the green consumer toward radical anger. Local small-scale actions, like recycling and reusing plastics and metals, become not a path to fuller awareness and political action, but an impediment to it. The implicit message is "Get moving, but not too far!"

By contrast, the critical attitude of green consumerism admits the possibility of radical passion. If it does not openly advocate a particular environmentalist ideology, it certainly leaves the way open for a fuller involvement in radical environmentalism.

Green Consumerism Directs Individuals toward Communal Action

Closely related to the inclination to critical awareness—and another point on which the original green consumer books differ from their ideological competitors—is green consumerism's willingness to provide pathways that lead the individual to consider not just home economics but also the big picture of the political economy. While maintaining the conventional division between home economics and political economy obviously serves the interest of household advisors like Heloise, it works against the argument of green consumerism that the householder's actions can really make a difference on the global scale and thus undermines the subtle semiotic connections the books develop between the actions of the consumer and the fate of the planet. Another connection that the green consumer books seek to support is that between the individual and the community. In this interest, the books point the way toward a broader political participation, usually without trying too hard to control the direction of the reader along ideological paths.

Beginning with its simple directions for green living at home, *50 Simple Things* widens the circle of participation with each new section, concluding with "things to do" in the public realm: number 49 is "Stay involved"; number 50, "Spread the word" (EarthWorks Group, 1989, pp. 94–96). "Some activists," the authors admit, "worry that books like this one will lull people into believing that doing a few positive things for the environment is enough." They insist that "it isn't" and encourage their readers to go beyond their individual households in their work to save the Earth (p. 94). To smooth the way, *50 Simple Things,* like the other green consumer books, provides an extensive list of organizations and information sources. By "networking" with other environmentalists, the novice moves outside the comfortable sphere (and attendant quietism) of home and garden.

In the overall context of environmentalism, the gradualist approach shared by the green consumer books—the effort to transform the politics of the consumer one step at a time—complements the shock tactics and sensationalism of groups like Earth First! and Greenpeace, thus creating a point of entry for citizens suspicious or fearful of rapid societal change. This gradualism shows in the books' organization. Each begins with a short section that educates the audience on the extent and severity of environmental problems. Refusing to dwell on the negative, however, the authors move swiftly into a program of action, which they present as positive things an individual can do toward curing environmental ills. These recommended actions are partial, no doubt, as the authors freely admit. "While perfect solutions are lacking," says *The Green Consumer,* "it is better to do something than nothing" (Elkington, Hailes, & Makower, 1988, p. 8). The books conclude with recommendations for action in the wider political arena.

At least one of the green how-to books, *The Green Lifestyle Handbook: 1001 Ways You Can Heal the Earth,* edited by Jeremy Rifkin (1990), the controversial opponent of recombinant DNA experiments, not only urges critical awareness but also advances a more ideologically committed deep ecological program. The reader is led from instructions on how to shop and keep house to chapters on how to expand one's ecological consciousness, such as "Spiritual Dimensions: The Role of Religion" by the cultural historian and green guru Thomas Berry. Moreover, in the tradition of the politicos and grassroots activists, the book includes a set of five chapters on community organizing, complete with guides for understanding, influencing, and using environmental law. The book presents these chapters last, thereby challenging readers who have gradually expanded their awareness of ends to expand the means by which they pursue their program of personal environmentalism. Working through home economics, then, the book progresses to questions of lifestyle—practicing ecological sensitivity in diet, gardening, personal investments, recreation, health, and spirituality—emerging at last into community organization with full political involvement. From this approach, an image of environmentalist character development takes shape. The initiate moves through three stages: first, home practice with enhanced awareness, then extensive revision of lifestyle involving both ideology and practice, and finally, participation in political activity at the community level.

But Rifkin's *Green Lifestyle Handbook* belongs to the second round of green consumer books (it was published in 1990 in the wake of the success enjoyed by *The Green Consumer* and *50 Simple Things* in 1988 and 1989) and goes considerably farther in directing the reader toward clear ideological choices. Whereas the earlier books tend to leave open the degree and type of political participation, thus respecting the individual reader's possible wariness about ideological environmentalism, *The Green Lifestyle Handbook* tends to push a bit and thereby risks narrowing its audience to those readers who begin the book with a fairly strong commitment to sweeping political change.

In any case, the very attempt on the part of both the greenest environmentalists and the most consumer-oriented conservatives to make effective use of the green how-to genre contradicts the common notion that consumers represent a relatively passive or unimportant political interest—a crowd of victims in need of protection. With a full-scale campaign in the self-help market for trade paperbacks, the green consumer movement openly acknowledges the power of the consumer and seeks to tap this reserve of political subjectivity. While willingly pointing the unconverted or the recently converted to a fuller involvement in environmentalist politics, however, green consumerism tends to set a limited goal for itself, serving a transition function in spreading the values of environmentalism but stopping short of ideological indoctrination.

Closing the Gap between Consumerism and Environmentalism

A fundamental point of environmentalist criticism is that those of us who live in advanced technological cultures have lost our connection with the land and thereby sow the seeds of our own destruction, creating effects in the environment that we do not see or understand but for which we will eventually pay dearly. The critique assumes that consumers have only the vaguest ideas about how food gets to the store or water to the faucet. Restoring the missing associations that connect us to the land is one of the primary aims of environmentalism, and in this sense, it is a cultural movement as well as a political ideology.

Though protective of the liberal ethos and thus inclined to be politically cautious, if not conservative, when compared with the more radical standards of environmentalist politics, green consumerism nevertheless makes a strong contribution to cultural environmentalism. The green consumer movement follows feminist theory in refusing to overstate the division between the personal and the political (see Diamond & Orenstein, 1990). In making its mark upon the everyday actions of the householder, green consumerism may be transforming environmental politics at a deep level, rather than merely a low level.

Above all, it creates an opening for broader participation by decreasing the physical distance between the ordinary householder and the key symbols of environmental protection. In the past, environmentalism was identified primarily with the so-called wilderness ethic, a perspective with limited opportunities for public participation, essentially a culture of hikers, mountaineers, and outdoor types—the typical Sierra Clubber (Nash, 1982; Killingsworth & Palmer, 1992a, pp. 31–48). Taking the wilderness enthusiast as typical of the environmentalist character, conservative critics like James Watt have been able to claim with seeming justice that environmentalism represented not a broad social movement but a "special-interest group"—the wilderness lobby. But if the environmentalist experience encompasses such everyday activities as shopping and home repair, activities that involve greater numbers of people and yet do not exclude those who have had a rich and varied life in the outdoors (all of whom, like it or not, are also consumers), then the charge of exclusiveness falls aside as opportunities for broader participation increase.

One might object that, while the wilderness experience and the ethics that grow from it are deep, the experience and subsequent ethos of the householder are shallow. But such a critique is likely to be the product not of a careful critical analysis but of a Romantic metaphysic that sees nature as sublime and society as a diminished state of being and that favors philosophical idealism over philosophical pragmatism. This view may well represent an attitude unfit for democratic politics, an attitude based upon what John Dewey called "a spectator theory of knowledge," which denigrates participatory interpretations of action because they embrace the world of common life and work. "The depreciation of action, of doing and making, has ever been cultivated by philosophers," writes Dewey in a famous passage. "But while philosophers have perpetuated the derogation [of practical activity] by formulating and justifying [that derogation], they did not originate it." Their harsh judgment of practical activity, Dewey insists, originated in class struggle, which allotted the hardest work, that which involved the body most heavily, to the lowest classes. Drawing upon the subsequent association of work and the body with the vilified classes, philosophers "glorified their own office . . . by placing theory so much above practice" (Dewey, 1929, pp. 4–5). The association of environmentalism with leisure time (hiking and camping) and with natural mysticism (communion with the sublime) could be said to constitute a similar elitism, a flight from the world of practice and daily experience.

Because mysticism and the aesthetic sublime—like prophecy, another familiar mode of environmentalist discourse—are associated with distance from the common experience of life, their political potential is minimal in a democracy that seeks a broad base of support. Like the reporting of big news, moreover, the experience of mysticism and aesthetic epiphany are difficult to sustain over time. And without a sustainable discourse, environmentalism has little hope of surviving, except as an occasional object of interest, the stuff of crisis management and sensationalist journalism.

The green consumer movement has hit upon a provocative alternative in the attempt to create ritual signs out of household items and actions, to redirect the symbolic energy as well as the practical activity of householders by relating ordinary work to environmental values. With great effectiveness and with some irony, it uses the technological context of everyday life—the very things that seem to alienate us from the Earth—to embed saving the Earth as a theme in human consciousness and political life.

References

Abbey, E. (1975). *The monkey wrench gang.* New York: Avon.

Anderson, W.T. (1990, July-August). Green politics now come in four distinct shades. *Utne Reader,* pp. 52-53.

Aristotle. (1980). *The Nicomachean ethics.* D. Ross, J.L. Ackrill, & J.O. Urmson (Trans.). New York: Oxford University Press.

Beale, W.H. (1987). *A pragmatic theory of rhetoric.* Carbondale: Southern Illinois University Press.

Bell, D. (1988). *The end of ideology.* Cambridge, Mass.: Harvard University Press.

Berry, T. (1990). Spiritual dimensions: The role of religion. In Rifkin, 1990, pp. 87-91.

Bookchin, M. (1990). *Remaking society: Pathways to a green future.* Boston: South End Press.

Brown, L.R. (Ed.). (1991). *The world watch reader: On global environmental issues.* New York: W.W. Norton.

Burke, K. (1969a). *A grammar of motives.* Berkeley: University of California Press.

———. (1969b). *A rhetoric of motives.* Berkeley: University of California Press.

Cohen, G., & O'Connor, J. (Eds.). (1990). *Fighting toxics: A manual for protecting your family, community, and workplace.* Washington, D.C.: Island Press.

Comp, T.A. (Ed.). (1989). *Blueprint for the environment: A plan for federal action.* Salt Lake City: Howe Bros.

Devall, B. (1988). *Simple in means, rich in ends: Practicing deep ecology.* Salt Lake City: Peregrine Smith.

Dewey, J. (1929). *The quest for certainty.* New York: G.P. Putnam's Sons.

Diamond, I., & Orenstein, G.F. (Eds.). (1990). *Reweaving the world: The emergence of eco-feminism.* San Francisco: Sierra Club Books.

Douglas, M. (1966). *Purity and danger: An analysis of the concepts of pollution and taboo.* London: Ark.

Downs, A. (1972). Up and down with ecology: The "issue-attention cycle." *Public Interest, 28,* pp. 38-50.

Dryzek, J.S., & Lester, J.P. (1989). Alternative views of the environmental problematic. In J.P. Lester (Ed.), *Environmental politics and policy: Theories and evidence* (pp. 314-30). Durham, N.C.: Duke University Press.

Dunlap, R.E. (1989). Public opinion and environmental policy. In J.P. Lester (Ed.), *Environmental politics and policy: Theories and evidence* (pp. 87-134). Durham, N.C.: Duke University Press.

EarthWorks Group. (1989). *50 simple things you can do to save the Earth.* Berkeley, Calif.: EarthWorks Press.

Elkington, J., Hailes, J., Hill, D., & Makower, J. (1990). *Going green: A kid's handbook to saving the planet.* New York: Puffin Books.

Elkington, J., Hailes, J., & Makower, J. (1988). *The green consumer.* New York: Penguin.

Evernden, N. (1992). *The social creation of nature.* Baltimore: Johns Hopkins University Press.

Foreman, D., & Haywood, B. (Eds.). (1993). *Ecodefense: A field guide to monkeywrenching.* 3rd ed. Tucson, Ariz.: Earth First! Books.

Freudenberg, N. (1984). *Not in our backyards!: Community action for health and the environment.* New York: Monthly Review Press.

Hays, S.P. (1959). *Conservation and the gospel of efficiency: The progressive conservation movement, 1890-1920.* Cambridge, Mass.: Harvard University Press.

———. (1987). *Beauty, health, and permanence: Environmental politics in the United States, 1955-1985.* Cambridge: Cambridge University Press.

Heloise. (1990). *Hints for a healthy planet.* New York: Perigee.

Killingsworth, M.J. (1992). Realism, human action, and instrumental discourse. *Journal of Advanced Composition, 12,* pp. 171-200.

Killingsworth, M.J., & Gilbertson, M.K. (1992). *Signs, genres, and communities in technical communication.* Amityville, N.Y.: Baywood.

Killingsworth, M.J., & Palmer, J.S. (1992a). *Ecospeak: Rhetoric and environmental politics in America.* Carbondale: Southern Illinois University Press.

———. (1992b). How to save the Earth: The greening of instrumental discourse. *Written Communication, 9,* pp. 385-403.

Kinneavy, J.L. (1980). *A theory of discourse: The aims of discourse.* New York: W.W. Norton.

Manes, C. (1990). *Green rage: Radical environmentalism and the unmaking of civilization.* Boston: Little, Brown.

Merchant, C. (1992). *Radical ecology: The search for a livable world.* New York: Routledge.

Miller, C.R. (1984). Genre as social action. *Quarterly Journal of Speech, 70,* pp. 151-67.

Mitchell, J.G., & Stallings, C.L. (Eds.). (1970). *Ecotactics: The Sierra Club handbook for environmental activists.* New York: Pocket Books.

Muir, S. (1992). Shaping patterns of environmental interaction: Mystification and commodification in the greening of America. In C.L. Oravec & J.G. Cantrill (Eds.), *The conference on the discourse of environmental advocacy* (pp. 232-46). Salt Lake City: University of Utah Humanities Center.

Nash, R. (1982). *Wilderness and the American mind.* 3rd ed. New Haven, Conn.: Yale University Press.

Norton, B.G. (1991). *Toward unity among environmentalists.* New York: Oxford University Press.

Peirce, C.S. (1991). *Peirce on signs.* Chapel Hill: University of North Carolina Press.

Pirages, D.C. (1977). Introduction: A social design for sustainable growth. In D.C. Pirages (Ed.), *The sustainable society: Implications for limited growth* (pp. 1-13). New York: Praeger.

Rifkin, J. (Ed.). (1990). *The green lifestyle handbook: 1001 ways you can heal the Earth.* New York: Holt.

11

The Mass Media "Discover" the Environment: Influences on Environmental Reporting in the First Twenty Years

David B. Sachsman

The American mass media—and the media around the world, for that matter—did not think in terms of the "environment" until the end of the 1960s. Before that time newspapers and television stations would cover a week-long smog alert or, better yet, a river that was on fire, but the story produced had no real ecological connotation. It was an event story, unlinked to any concept of the global environment (Sachsman, 1973, p. 2).

There were books about the environment, most notably Rachel Carson's *Silent Spring,* published in 1962, and there were magazine articles. But for the most part there were no environmental reporters. The environment was not a journalistic beat. Instead, there was science writing and some health reporting. In the late 1960s some of these journalists would add the environment to their science and health beats, while others would go so far as to put the environment first and begin calling themselves environmental reporters (Sachsman, 1973, p. 2).

It is hard today for people to imagine that there was once a time when there was no "there" there, when the "environment" basically did not exist in American consciousness. Although Philip Shabecoff, in *A Fierce Green Fire* (1993), says the first wave of environmentalism occurred around 1900, with the creation of national parks and forests, that environmental movement was about conservation, not pollution (Perrin, 1993). It was about "nature," not environmental health risk. It took a long time for a global, risk-oriented concept to capture the American imagination.

The mass media and the public until the late 1960s accepted pollution as part and parcel of industrial society. This was no accident. Professional public relations specialists working for corporate America actively promoted this viewpoint. They supplied the press with plenty of good-news press releases concerning the successful efforts of corporations that were taking care of pollution problems (Sachsman, 1973, p. 3).

A case in point concerned the new plant that International Harvester built in Memphis after World War II. The Memphis works burned coal, and its big smokestacks spewed smoke, soot, and cinders. International Harvester built the plant in open fields, but not long afterward new homes were constructed adjacent to the works. The air pollution around these houses was so bad that laundry hung out to dry turned black, and windows had to be permanently closed. Before the homeowners had time to react to the dilemma, they received a letter from the plant manager saying that the corporation was looking for a solution. Company representatives actually went door to door assuring the citizens that something would be done (Marston, 1963).

For three long years, no pollution controls were installed, no homeowners went to the press, and no newspapers covered the situation. The locals were apparently satisfied by the corporation's contention that it had spent $68,000 on improvements. Much of this money was used to purchase seventeen acres around the plant to serve as a green belt, to catch low-level debris, a pollution control that is generally considered a very good investment for the company involved. Finally International Harvester spent $71,900 installing a device to trap most of the residue coming from the plant powerhouse. To mark the installation, the company held a community meeting and press conference glorifying its efforts to solve air pollution. The first newspaper story ever carried on the issue was headlined "IH Spends $71,900 to Be a Good Neighbor" (Marston, 1963, pp. 207–9).

Throughout the 1940s and 1950s, in those isolated cases in which a few citizens resisted companies on such issues as land use, they faced professional public relations practitioners and local media that generally accepted industry arguments. For example, when U.S. Rubber decided to construct a ninety-acre research center in a residential neighborhood in Wayne Township, New Jersey, it found that it would first have to bring about a change in the community's zoning ordinance. The corporation began its public relations campaign with a press release, stressing that the buildings would be set back from the property lines, the tract would be landscaped, and there would be no offensive odors, traffic problems, or water pollution. The corporation sent personal letters to local opinion leaders and community and state officials. Company representatives met with various civic groups, distributed booklets explaining rubber research, and invited residents to visit other rubber laboratories. Ten property owners filed a suit to prevent the change in zoning, but the corporation won big, with the press, the local government, and most of the people in its pocket (Wilks, 1957).

In the International Harvester and U.S. Rubber cases, the companies were the only ones producing press releases. The environment was a business story in the period following World War II, the public relations–supplied good-news story of corporate America. By the late 1960s this picture had changed. The mass media were receiving environmental releases not only from industry and industry-related institutions but also from government agencies and officials, citizen-action pressure groups, and other institutions such as universities. The rise of environmental awareness in the 1960s was probably caused partly by what Richard W. Darrow, president of the Hill and Knowlton public relations company, in 1971 called the Great Ecological Communications War, the war between conflicting public relations sources. As Darrow explained to the Economic Council of the Forest Products Industry, corporate America was on the defensive: "The hour is later, Communications Time than it is Mountain Standard Time, for you and me and our colleagues at the control points of industry. We will do those things that earn us attention and gain us understanding, or we will live out the remainder of our professional lives in the creeping, frustrating, stultifying, stifling grasp of un-realistic legislative restraints and crippling administrative restriction. A public that ought to understand us—and thank us for what we are and what we do—will instead clamor for our scalps" (1971, p. 18).

The rise of environmental awareness and the subsequent environmental in-formation explosion was caused partly by what Darrow would call "unrealistic legislative restraints and crippling administrative restriction," the realization by politicians that the environment was a comparatively safe issue. President Lyndon B. Johnson was one of the first national political figures to understand that being against pollution made for good public relations. Johnson said in his message to Congress on February 8, 1965:

> In the last few decades entire new categories of waste have come to plague and menace the American scene. These are the technological wastes—the by-products of growth, agriculture and science. . . .
> Almost all these wastes and pollution are the results of activities carried on for the benefit of man. A prime national goal must be an environment that is pleasing to the senses and healthy to live in.
> Our government is already doing much in this field. We have made significant progress. But more must be done. [Burton, 1966, pp. 207–8]

As public officials began to talk about the environment, the mass media began to treat it as a serious government story, and the general public became in-creasingly aware that important issues were involved. Since the 1960s government officials and agencies have been directly involved in environmental decision making, and they have churned out an enormous number of environmental press releases, which have been received and used by the mass media (Sachsman, 1973, p. 7). More and more, government officials realized that the environment was more than a fad and that they would have to add actions to their words. As Walter J. Hickel explained: "When I took office in 1969 as Secretary of the Interior, pollu-tion was no longer a joke; this fact was made clear by the nature of my confirma-tion hearings. The subject was aggravating millions of Americans; frustration and hostility were growing. The nation was desperately looking for leadership, and I decided that we should take the lead" (1971, p. 65).

While government was finding the environment to be a comparatively safe issue, the established environmental activist groups were learning that good public relations made for solid press coverage, including some investigative reporting. Furthermore, new activist groups were developing, with a separate citizen-action organization for virtually every issue, and all were trying to reach the public through the press. Shabecoff (1993) says that 1970 was the time of the second wave of environmentalism, a white, middle-class, Earth Day environmen-tal movement. (He says that by the 1980s a third wave of environmentalism was launched by thousands of local environmental groups, which included a mix of social classes, women, African Americans, Native Americans, and others [Perrin, 1993].) Other institutions were also involved. Universities, programs, and insti-tutes would become centers of discussion and study concerning environmental matters, and these institutions would produce their own share of press releases (Sachsman, 1973, p. 8).

The environment exploded onto the front pages and the airwaves in 1969. The dramatic Santa Barbara Channel–Union Oil leak and the flood of conflicting environmental press releases caused print and broadcast editors to take seriously their own local problems of air and water pollution. Night after night network viewers watched oil-soaked birds dying in the arms of California college students, and local journalists everywhere went looking for their own local sidebar, their own local human-interest ecological crisis. The story that had begun simmering with Carson's *Silent Spring* in 1962, that government had discovered in 1965, and that had begun to boil in the late 1960s because of the outbreak of the environmental communications war of conflicting press releases now in 1969 finally found its way onto the front-page agenda of the mass media (Sachsman, 1973, p. 2).

In 1969 the *New York Times* followed the lead of others in creating an environment beat, a practice that would be followed by major newspapers across the country. In the same year *Time* and *Saturday Review* began regular sections on the environment, *Look* devoted almost an entire issue to the environmental crisis, *Life* greatly increased its coverage of the topic, and *National Geographic* offered a nine-thousand-word article on environmental problems. At the start of the new decade, Paul Ehrlich's book *Population Bomb* was a best-seller, and the *CBS Evening News with Walter Cronkite* was presenting an irregular feature called "Can the World Be Saved?" (Rubin & Sachs, 1971, chap. 2, p. 1).

The PR Influence and the Role of Environmental Reporters

The mass media in the early 1970s were faced with the overwhelming task of sorting through a barrage of environmental information and deciding what news to carry about ecological issues. Much of this information came in the form of press releases. Scott M. Cutlip, a leading public relations scholar, had estimated in 1962 that some 35 percent of the content of newspapers came from PR practitioners: "Today's public news media do not have the manpower, in terms of numbers or in terms of mature specialists, to cope with the broadening spectrum of news and the deepening complexity of news subject matter. More and more the news gathering and reporting job is abandoned to the public relations man who supplies the information in neat, easy-to-use packages." Cutlip was particularly concerned about complex "new areas of news" such as science, health, education, and social welfare: "Study will show that these areas are covered in a large degree by the PR man, not the aggressive, investigative reporter" (1962, p. 68).

The American news media in the 1970s and 1980s employed fewer reporters than most readers, viewers, and listeners might have imagined. It was not unusual for a newspaper to have fewer than one hundred journalists, for a television station to have fewer than forty news people, or for a radio station to depend on a handful of newscasters. In contrast, there were hundreds of thousands of public relations people in the United States, many times the total number of journalists. The owners and managers of the news media recognized that they needed the information supplied by public relations practitioners. They were part of a symbiotic relationship, similar to that which exists between lobbyists and members of Congress.

An assumption commonly made in the 1970s and 1980s was that a specialized reporter could be of significant value in achieving solid news coverage of a particular topic. The idea was that a specialist would either come to the position with knowledge about the subject area or would develop such expertise while being assigned to specialized duties. Another assumption sometimes made was that specialized reporters were aggressive, investigative reporters. While some specialists did investigative reporting, the two terms were not synonymous.

Research conducted in the San Francisco Bay Area in the early 1970s had much to say about the influence of public relations on environmental news coverage and the roles played by specialists in that coverage. At that time the Bay Area had twenty-eight daily newspapers, six television stations, and eight radio stations with independent news operations. Fifteen of these media employed a total of sixteen environmental reporters. Ten were reporters who spent at least 25 percent of their time on environmental reporting and had an environment beat. Two were science reporters actively covering the environment, and four were general reporters covering the environment less than 25 percent of the time, who could nevertheless be considered to have an environment beat by virtue of their attitude, and the attitudes of their editors, toward their work. Half of these news people were assigned to environmental reporting after 1969, presumably in response to the increased public interest in the environment and the increased number of meetings and other activities involving environmental issues. Most of them were hired from within their news organizations, and in almost half the cases, regular local environmental coverage grew out of a political beat that included meetings of agencies with responsibilities concerning environmental affairs. At the time, most of the representatives for media that did not have an environmental specialist said that they could not afford one. But many also said that they did not need one. They said that the number of local environmental stories did not justify a specialist or that the subject could be handled just as well by a general assignment reporter (Goodell, 1971).

The research conducted in the Bay Area in the early 1970s found that public relations efforts significantly influenced environmental coverage. The sheer number of press releases was impressive—eleven reporters received 1,347 releases in eight weeks—although what looked like a flood in 1971 appeared quite normal in the 1980s. Back in 1971 something like 42 percent of the environmental press releases came from government agencies and officials, 23 percent from corporations and industry-related institutions, 17 percent from institutions such as universities, and 17 percent from activist groups. In a twelve-day content analysis of environmental coverage in twenty-five of the Bay Area media, government sources were the bases of information most often identified within the stories. Sources from institutions like universities, corporations, and activist groups were also regularly identified, while industry-related institutions were rarely named (Sachsman, 1976).

The tendency of journalists to treat government as the "official source" probably explained to some degree the reliance by reporters on government sources in environmental stories. But the fact that more than 40 percent of the press releases in the 1971 study came from government showed that the environment was a government story partly because government-supplied materials dominated the environmental information explosion. The twelve-day content analysis in 1971 counted 1,002 items about the environment in twenty-five daily media. The content analysis coding sheets were then carried back to these media, where forty-one different journalists were interviewed about them. These reporters and editors were able to provide information about 887 of the items. More than 40 percent of these stories came from the wire services, the syndicates, and the networks, all important providers of environmental news. A total of 474 items (53 percent) were not press service or network stories, or letters to the editor. The forty-one journalists were able to reconstruct the origins of 200 of the 474 non-wire, nonnetwork stories. They found that 105 stories (53 percent) had been influenced by public relations, including 51 on which no further research had been done by journalists (46 rewritten press releases, 3 stories from a PR service, and 2 films supplied by sources). Reporters did additional research for 54 of the 105 items influenced by public relations; 28 of these came from press releases, while 26 came from telephone calls or personal contacts (Sachsman, 1976). Cutlip's estimate that about 35 percent of media content comes from public relations practitioners was supported by this research, and there is no indication that the influence of public relations lessened—in any area, let alone environmental affairs—in the 1970s or 1980s.

The environmental reporters, business editors, and other editors responsible for environmental news who were interviewed in the 1971 study all said they received environmental press releases. The environmental specialists knew and used many public relations environmental sources. Reporters and editors at media with no environment beat said they did not have enough environmental contacts. According to the 1971 study, specialized reporters were valuable not only because of their expertise but also because they had many public relations contacts; they were magnets that attracted environmental information (Sachsman, 1976).

The research found that in most cases journalists of every description saved and/or used large numbers of environmental press releases, while environmental specialists received more releases than other journalists. In addition to having more information than other reporters or editors, specialists were allowed more time to originate environmental items. More information plus more time equaled more environmental coverage. While more of their stories were influenced by public relations, they also wrote more enterprise stories (Sachsman, 1973, pp. 278, 283–86).

Throughout the 1971 study there was little indication of any stigma being attached to public relations. While then as now public relations in some areas was considered press agentry, this negative connotation is not generally applied to environmental affairs. This was after all a scientific news topic, with scientists representing all sides of the equation. Perhaps because journalists tended to respect scientists—industry scientists as well as government and university scientists—environmental press releases were often viewed as statements by "official" representatives. In the area of environmental affairs, the term *official source* applies not only to government but to others as well (Sachsman, 1976).

Since specialized reporters either started with some expertise or developed it along the way, they tended to sort the environmental mail wisely. Now, with more than twenty years of hindsight, it seems clear that specialized environmental journalists in the 1970s and 1980s contributed quality as well as quantity and were of real value in providing solid reporting.

The environment established itself as a subject of continuing importance in the 1970s and 1980s. Given the recognition that existed in the 1990s of the seriousness of many environmental problems, one might have expected that the environment beat at the typical newspaper or television station would have grown into a whole department of people, like business or sports. Environmental affairs had become a journalistic staple, and environmental stories were written for the general public (Cantrill, 1993, p. 14, citing McGeachy, 1989). The environment in the early 1990s was very much a part of the day-to-day coverage provided by local-beat reporters, who watched over everything that occurred within their geographic areas, and general assignment reporters, who were sent out to a traffic accident one night and a speech the next. Though many large daily newspapers had science and environmental specialists, smaller media generally did not. And even where specialists were employed, they were not normally sent out on first-day breaking stories concerning tanker truck accidents or fires with noxious fumes. Thus the everyday environmental accident, like the zoning board meeting (which may also be an environmental story), was the province of the local-beat and general assignment reporters, who were often the youngest and least-experienced journalists. They knew how to cover their beats and the accidents and fires that come with the territory, but they generally did not know much about science (Sachsman, 1991, p. 7; Sachsman, Sandman, Greenberg, & Salomone, 1988, p. 287). Government officials remained the most important sources of environmental information (Friedman, 1990). Corporate sources supported technical answers to environmental problems, and some audience members found environmental coverage to be more institutional and conservative than activist (Cantrill, 1993, p. 15, citing Allen & Weber, 1983, and Schlechtweg, 1992).

While there were many more environmental specialists in the 1990s than in the 1970s, they still averaged fewer than one per news organization, and the job of environmental reporting remained the work of general reporters. The difference was that in the early 1990s covering the environment was a basic (sometimes even central) function of general reporting. While general reporters knew little about science, many of them attended school in the 1970s and 1980s, decades of environmental awareness very different from earlier generations. David McIntosh quotes Wayne R. Agner, managing editor of the *Santa Maria (Calif.) Times:*

"Invariably, environmental issues tie themselves to other issues. In the past, we'd cover planning or traffic issues in a purely governmental way. We're now more concerned with the environmental impact" (McIntosh, 1992, p. 5).

Environmental Risk Communication

In the early 1970s "environmental communication" was about environmental news sources and their links to environmental reporters. It was about activists and government officials, corporate press releases, and scientific studies, and how environmental stories were presented to the mass media and through the mass media to the general public. Most people did not think of environmental communication as health communication, and certainly not as risk communication, a concept that did not become important until the 1980s (Sachsman, 1991, pp. 3–4).

From the 1970s on, the list of specialized science communication fields became longer and more detailed: environmental communication, health communication, risk communication, press coverage of science and technology, environmental risk reporting, and more. From a health perspective, environmental communication became a subset of health communication. From an environmental perspective, many forms of health communication fit within environmental communication. The development of these closely related science communication fields was clearly on the right track, since it was important to learn the most effective ways of explaining radon to homeowners, the problems of asbestos to school board members, and any number of health and environmental issues to various publics (Sachsman, 1991, p. 1).

These science communication problems were society-driven issues. They were important to governments, industries, and citizens' groups. Various institutions and organizations were willing to support research about them, not only experimental and quantitative research but also descriptive and qualitative studies. These were action areas, meaning that there was interest and even money in them. In this regard such communication research tended to be more like medical research than studies about communication theory and process. This research often started with a real-life problem: explaining the risk of cigarette smoking, showing workers the dangers of benzene, reporting environmental affairs. And it often began with descriptive research: this is what government officials know about leaking underground storage tanks; this is how suburban newspapers cover environmental accidents. These studies also tried to find real-life solutions: this is how to train reporters to cover spills; this is how to teach workers about avoiding occupational exposure to hazardous substances. The basic theory sometimes came later, but this work was not by any definition antitheoretical. It was about problem solving and about looking at and describing the big picture. Sometimes it involved experiments, controls, and theoretical models, but often it was about trying things out in the real world until they actually worked and the specific problem was solved (Sachsman, 1991, p. 2).

Risk communication was the newest of these specialized fields. It deserves to be considered its own area of study, because it really is an environmental and health communication and perception field and not a subset of risk assessment.

Risk assessment is that scientific, statistical field that tells us the probability of getting sick from drinking tap water, or of getting cancer because of a toxic waste dump. Risk assessment talks about one-in-a-million possibilities and the scientific planning that should result from statistically known risks (Sachsman, 1991, p. 2). Risk communication, though, is about explaining risk and people's perceptions of risk. Risk communication research tells us that one in ten million can seem like a big number and that we want absolute protection from an explosion at a nuclear power facility and from getting cancer from toxic waste dumps. But it also says that many of us are willing to take the chance of exposing ourselves to pesticides in order to have a perfect lawn. Risk communication explores perceptions, and even choices (we choose to spray our bushes), not the statistical science of risk assessment (Sandman, Sachsman, & Greenberg, 1988).

By the mid-1980s, from any kind of academic perspective, it no longer made sense to discuss environmental communication or environmental reporting in a vacuum. In fact, increasingly the terms in academic use were *environmental risk communication* and *environmental risk reporting,* and *risk* generally referred to the health risk involved in the environmental issue. But while academics integrated the new field of risk communication into environmental communication and environmental reporting, only a handful of the most specialized environmental reporters thought about the scientific degrees of environmental risk. The local-beat and general assignment reporters who did the everyday environmental reporting as a basic part of their jobs appeared to use the terms *environmental reporting* and *environmental risk reporting* interchangeably. Did they see the big picture? Did they understand the interrelationship between environmental journalism and health reporting? Did they see that the degree of risk was a central question in most environmental issues?

In 1985, the Hazardous Substance Management Research Center in New Jersey began funding a joint project of Rutgers University and the University of Medicine and Dentistry of New Jersey–Robert Wood Johnson Medical School for the design and implementation of a program of continuing education for print and broadcast journalists on risk assessment. Soon afterward, a second project was added to teach scientists serving as news sources how to provide environmental risk information to the media. The five-year efforts of "the Rutgers group," so called by Victor Cohn in *Reporting on Risk* in 1990, provided much information about current environmental reporting practices in the 1980s.[1]

News Values Influence Coverage

The Rutgers group learned just how little journalists knew or cared to know about "scientific degree of risk" through almost every aspect of the research—including an initial study of environmental news stories in New Jersey newspapers (Sandman et al., 1987, pp. 3–57). Later the group studied two years' worth of network television evening news coverage of environmental risk, again learning that environmental coverage was determined by traditional journalistic news values (timeliness, proximity, prominence, consequence, and human interest) rather than by scientific risk. The group further found that the television networks were also guided by the availability of dramatic visual images and by geographic factors, such as cost and convenience. Not only was "geography" more of a factor than scientific risk, but apparently it sometimes had more effect than did the broadcasters' own news values (MacDougall, 1977; Greenberg et al., 1989a, 1989b).

One of the first things learned by the Environmental Risk Reporting Project was that local-beat reporters and general assignment reporters and their editors, rather than the environmental specialists, were the ones who had to be reached with educational programs and information about environmental risk in order to improve environmental risk reporting. The local reporters were the ones on the front line of environmental journalism, and they needed and wanted help (Sachsman, Sandman, Greenberg, & Salomone, 1988).

The Rutgers group's first study in 1985 analyzed the strengths and weaknesses of the "best" environmental risk stories published in New Jersey newspapers the previous year. The group examined all of these articles—submitted as the best by their papers' editors—rather than a random sample, because they did not want to take a cheap shot. The group did not wish to evaluate "typical" environmental risk stories, but rather to find the problems that existed within even the best environmental coverage. How good is the best? This is precisely the question the project wanted to answer (Sandman et al., 1987, pp. xiii, 52).

The study found that these articles did not provide very much information about degree of risk. When risk was reported, the result was usually more alarming than reassuring. The information about risk in the articles was provided by government, industry, and unattributed sources, rather than by uninvolved scientific experts. The reporting about environmental risk tended to speak in extremes rather than degrees. A noxious substance was present or not present. Situations were risky or not risky. Middle-of-the-road or tentative positions expressed by sources were rarely cited. The articles were generally accurate, with few errors of fact. Problems involved the omission of information. No intentional bias was found; flagrant distortions did not characterize the articles (Sandman et al., 1987, pp. 99–100).

Perceived bias in environmental reporting—when no intentional bias exists—is apparently a feeling by experts from government, industry, and other institutions and groups that is a reaction to journalism's tendency toward extremes, its reliance on particular extreme sources rather than those that express intermediate positions, and its translation of technical jargon into volatile common language. In short, what scientists and representatives from government, industry, and other groups view as bias may just be the normal tendencies of journalism. Journalists tend to cover environmental affairs when problems and risks are present. For many media, good news or even intermediate degrees of risk may not be newsworthy at all (Sandman et al., 1987).

The Rutgers group's study of network evening news coverage of environmental risk was a more conventional content analysis. The aim of the project was to try to evaluate the state of environmental coverage on network television, for a twenty-six-month period from 1984 into 1986. The study found a total of 564 environmental risk stories, an average of five per week across all networks, or only about one in every four network news broadcasts. In a time period that included the Bhopal tragedy (sixty-one stories were done on the deaths resulting from the Union Carbide chemical release in India), news about environmental risk took up only 1.7 percent (13.8 hours) of the news time of the three network newscasts (Greenberg et al., 1989a, pp. 120–21).

Visual Impact and Geography Influence Network News

Scientific risk had little to do with the environmental coverage presented on the nightly news. Rather, the networks appeared to be using traditional journalistic news values and the television criterion of visual impact to determine their coverage of environmental matters. These journalistic standards pointed reporters to events rather than issues and to the dramatic rather than the chronic. Even when journalists covered chronic risk issues, they often required "acute" new and timely information—a news peg or hook on which to hang the story. The researchers concluded: "Network evening news coverage surely tends to reinforce the public's overestimation of the health impact of acute risk events and underestimation of most chronic risk issues. The public's conception of risk is almost certainly distorted by television's focus on catastrophes and its dependence on films" (Greenberg et al., 1989a, pp. 123, 125).

Next the Rutgers group compared the networks' coverage of hazardous waste (eighteen stories) and oil or gas spills and leaks (eighty-six stories) with scientific evidence regarding waste sites and oil and gas releases. They knew that the three networks had major broadcast centers or news bureaus in a handful of cities in only eight states plus the District of Columbia, and they wondered: "How much does geography guide news selection at the expense of risk?" They found that thirteen of the eighteen hazardous waste stories (72 percent) were in news-center states, compared with 29 percent of the sites, and that the average airline distance from the nearest news bureau to the twelve televised locations was 133 miles, compared with an average distance of 192 miles to the Environmental Protection Agency's 112 priority sites. Sixty-three of the eighty-six oil or gas spill stories (73 percent) came from news-center states, while only about one-third of the accidents and the potential for accidents occurred in these states. The group concluded that the networks were not only much more likely to cover hazardous waste problems and oil or gas leaks in news-center states, but they were about twice as likely to cover an environmental risk issue that is near a news bureau than one that is farther away. Geographic considerations such as cost, time, and convenience clearly affect television network news coverage of environmental risk (Greenberg et al., 1989b, pp. 269, 272–73).

Improving Environmental Risk Reporting

The group's research about environmental news coverage provided the foundation for work on improving environmental risk reporting. In the early 1970s the Bay Area study had found that specialized environmental reporters could help provide quality and quantity coverage. The study had argued for enterprise and interpretive reporting for experienced, expert environmental specialists with the time to investigate, interpret, and report in depth (Sachsman, 1976). In the 1980s the need for specialized environmental reporters remained clear, but it no longer appeared that the problems in environmental risk coverage would be solved if the number of specialists was doubled or even quadrupled. The environment had become a story for virtually every local beat and general assignment reporter, and nothing short of training all of them to be environmental specialists would do the trick. Furthermore, many editors also needed environmental training: witness the networks' paltry 1.7 percent interest in environmental risk and Liz McGeachy's study of three general-orientation magazines, which found that only 0.9 percent of their pages addressed the environment in 1986 (McGeachy, 1989, p. 12). Fortunately, general assignment, local-beat, and student reporters have consistently demonstrated their willingness to learn about environmental risk. They will attend classes and workshops on everything from interviewing, writing, reporting, and editing—the process of journalism—to specific environmental issues and the concept of scientific risk assessment (Sachsman, 1990; 1976, p. 60). The Rutgers group spent several years conducting workshops and seminars and designing associated teaching materials, including its videotape, *Covering an Environmental Accident* (Sachsman, 1985), and its Environmental Reporter's Handbook (Sachsman, Greenberg, & Sandman, 1988), which grew out of its "environmental risk press kit." The handbook idea would continue to be of particular value.[2]

The researchers spent the same years—in their second project on risk communication for environmental news sources—working with the technical, scientific, corporate, regulatory, and community news sources whose responsibility it is to inform the media, and thus the public, about environmental risk. The group concluded: "The more productive approach in the long run may be to teach news sources to better understand and deal more effectively with the media. . . . Environmental news sources who empathize with journalists and are willing to teach reporters about their specific fields can help make mass media coverage of environmental risk as accurate and professional as the American public deserves" (Sachsman, Sandman, Greenberg, & Salomone, 1988, pp. 295–96).

Influences on the Environmental Agenda

Who set the agenda in the United States in the 1970s and 1980s? Who determined what the American people were thinking about? Since the 1960s the argument has developed that the media have the ability to decide what the public will think about and discuss (Lowery & De Fleur, 1988).

Television's role during the 1969 Santa Barbara Channel–Union Oil leak, putting on the air night after night heartbreaking pictures of suffering, dying oil-soaked birds, influenced newsrooms across the country to find their own local environmental crises and put them on display. This was a clear demonstration of the agenda-setting function of the media, with the press and public throughout the nation immediately responding to television's image of the situation. Further-more, in this case television was so powerful that it not only told the public what to think about, it also told many Americans how to feel and what to think. Doro-thy Nelkin says:

> The actual influence of the press . . . will vary with the selective interest and experi-ence of readers. In esoteric areas of science and technology where readers have little direct information on preexisting knowledge to guide an independent evaluation (e.g., the effect of fluorocarbons on the ozone in the atmosphere), the press, as the major source of information, in effect defines the reality of the situation for them. During the period of maximum press coverage of the ozone controversy, for ex-ample, 73.5 percent of the general public had heard about this highly technical issue, previously remote from their experience, for the first time in the press. [1987, p. 77]

When there is an environmental dilemma, and the mass media define it as a crisis, then the act of treating it as a crisis and putting it on the air or the front page sets the agenda in those terms. If the people representing an involved institu-tion or organization do not seem to be responding appropriately to a situation, then the perception of risk may actually be amplified. For example, as Robin Gregory writes, "The appearance of incompetence or callousness on the part of risk managers is likely to foster a split between the perpetrators and victims of an accident (e.g., witness public anger at the delayed response of the Exxon company to the March 1989 Alaska oil spill)" (1991, p. 5).

The question of media agenda-setting is more complex when we consider chronic environmental problems for which the networks rarely pull out all the stops. Do the mass media set this agenda, providing detailed information about such issues? (Atwater, Salwen, & Anderson, 1985, p. 397). The environment by the early 1990s was an intensely local story, covered by local reporters and influenced by the public relations efforts of many local, regional, and national groups, or-ganizations, companies, and government entities. For these myriad local environ-mental stories it was possible to argue that the conflicting elements in the community—the conflicting public relations news sources—supplied the infor-mation and the pressure that caused the media to set the agenda as they did, and that it was the news sources that really set the public agenda through their influ-ence on the press. Cutlip's assertion in 1962 that public relations practitioners are responsible for a significant number of the stories carried by the media appeared to be true during the 1990s as well, in particular for complex subjects such as the environment (just as he thought).

If public relations people were providing environmental stories to the media, then the public relations practitioners—or the conflicts among them—were influencing the priorities that set the media agenda that decided what the public thought about. Government sources, then, may really have been the prin-cipal agenda-setters, since often they were the suppliers of the most public in-formation about the environment. And the companies and all those activist groups also influenced the media agenda, as well as trying to reach the public and affect government action through other, more direct forms of communication.

C.N. Olien, P.J. Tichenor, and G.A. Donohue make the point much more broadly and emphatically: "Appealing as the Fourth Estate view of media has been for centuries in Western political thought, it is largely a myth in terms of social structure. Newspapers, television, radio and magazines are organized agencies within an interactive and interdependent matrix, responding to centers of power in various other institutional areas. In modern society, government and business are the dominating institutions and the media are as influenced by those power centers as are other agencies" (1989, p. 22).

Sometimes one source prevails and has the most influence over a particular story. Often journalists benefit from the wide range of information supplied by conflicting news sources. While public relations efforts have had as great an influence on environmental stories as on any area of the news, the conflicting nature of the environmental sources tends to give journalists enough information so that they can take charge of their own agenda. Also, the dominance of journalistic news values such as prominence, proximity, timeliness, and human interest over scientific values such as "degree of risk" helps to keep the agenda-setting power in the hands of journalists, who set the agenda not just because news sources told them to but also because Rock Hudson was dying (prominence) or because a chemical truck was leaking on a local street (proximity).

Through more than twenty years of environmental coverage, journalists have stuck to their own news values (plus dramatic visual images and geographic factors such as cost and convenience) rather than moving toward or emphasizing "importance," the one value they share with science. By maintaining their own standards, they have kept control of their own agenda, and it has been this media agenda of prominence, proximity, timeliness, and human interest as well as consequence that has influenced what the public has thought about, if not what people thought.

Notes

An early draft of this manuscript, which was called "The Mass Media and Environmental Risk Communication: Then and Now," was presented at the Conference on Communication and Our Environment, Big Sky Conference Center, Big Sky, Montana, July 23–25, 1993.

1. "Dr. Michael R. Greenberg, professor of urban planning and public health; Dr. Peter M. Sandman, professor of environmental journalism and director, Environmental Communication Research Program, Rutgers University; and Dr. David B. Sachsman, dean, School of Communications, California State University at Fullerton. For convenience I have referred to them in several instances as 'the Rutgers group' or 'the Rutgers professors,' since Sachsman is a former New Jerseyite. They are authors of two highly recommended manuals" (Cohn, 1990, p. 61).

2. Witness the rise of other handbooks, such as Robert Logan's *Environmental Issues for the '90s: A Handbook for Journalists* (1992).

References

Allen, C.T., & Weber, J.D. (1983). Public environmental knowledge: A statewide survey. *Journal of Environmental Education, 18,* pp. 31-39.

Atwater, T., Salwen, M.B., & Anderson, R.B. (1985). Media agenda-setting with environmental issues. *Journalism Quarterly, 62,* pp. 393-97.

Burton, P. (1966). *Corporate public relations.* New York: Reinhold.

Cantrill, J.G. (1993). Communication and our environment: Categorizing research in environmental advocacy. *Journal of Applied Communication Research, 21,* pp. 14, 15, 66-95.

Cohn, V. (1990). *Reporting on risk.* Washington, D.C.: Media Institute.

Cutlip, S.M. (1962, May 26). Third of newspapers' content PR-inspired. *Editor and Publisher,* p. 68.

Darrow, R.W. (1971). *Communication in an environmental age.* New York: Hill & Knowlton.

Friedman, S.M. (1990). Two decades of the environmental beat. *Gannett Center Journal, 4,* pp. 17.

Goodell, R. (1971). Environment: The last fad? The metaethic of Bay Area media. Unpublished paper, Stanford University.

Greenberg, M.R., Sachsman, D.B., Sandman, P.M., & Salomone, K.L. (1989a). Network evening news coverage of environmental risk. *Risk Analysis, 9,* pp. 119-26.

———. (1989b). Risk, drama, and geography in coverage of environmental risk by network TV. *Journalism Quarterly, 66,* pp. 267-76.

Gregory, R. (1991). Risk perceptions as substance and symbol. In L. Wilkins & P. Patterson (Eds.), *Risky business: Communicating issues of science, risk, and public policy* (pp. 1-10). New York: Greenwood Press.

Hickel, W.J. (1971, October 2). The making of a conservationist. *Saturday Review,* pp. 65-67.

Logan, R.A. (1992). *Environmental issues for the '90s: A handbook for journalists.* Washington, D.C.: The Media Institute & The Radio Television News Directors Foundation.

Lowery, S.A., & De Fleur, M.L. (1988). *Milestones in mass communication research.* 2nd ed. New York: Longman.

MacDougall, C. (1977). *Interpretative Reporting.* 7th ed. New York: Macmillan.

Marston, J.E. (1963). *The nature of public relations.* New York: McGraw-Hill.

McGeachy, L. (1989). Trends in magazine coverage of environmental issues, 1961-1986. *Journal of Environmental Education, 20,* pp. 6-13.

McIntosh, D. (1992, September). Stalking the green beat. *Presstime: The Journal of the Newspaper Association of America,* p. 5.

Nelkin, D. (1987). *Selling science.* New York: W.H. Freeman & Co.

Olien, C.N., Tichenor, P.J., & Donohue, G.A. (1989). Media and protest. In L. Grunig (Ed.), *Environmental activism revisited: The changing nature of communication through organizational public relations, special interest groups, and the mass media* (pp. 22-39). Troy, Ohio: North American Association for Environmental Education.

Perrin, N. (1993, January 24). Think globally, act locally. Review of Philip Shabecoff, *A fierce green fire. New York Times Book Review,* p. 8.

Rubin, D.M., & Sachs, D.P. (1971, September). *Mass media and the environment.* (Vol. 2). Report of project supported by National Science Foundation Grant GZ-1777. Stanford, Calif.: Stanford University.

Sachsman, D.B. (1973). Public relations influence on environmental coverage (in the San Francisco Bay Area). Unpublished doctoral dissertation, Stanford University.

———. (1976). Public relations influence on coverage of environment in San Francisco area. *Journalism Quarterly, 53,* pp. 54-60.

————, producer. (1985). *Covering an environmental accident.* Television program.

————. (1990, February 15). The five W's of environmental risk coverage. Paper presented at the American Association for the Advancement of Science Pre-Meeting Workshop for Journalists, New Orleans.

————. (1991, May 25). Environmental risk and the mass media. Paper presented at the annual meeting of the International Communication Association, Chicago.

Sachsman, D.B., Greenberg, M.R., & Sandman, P.M. With L. Fuerst, A. Cooke, & L. Van-Leer. (1988). *Environmental reporter's handbook.* Newark, N.J.: Environmental Risk Reporting Project, Hazardous Substance Management Research Center, New Jersey Institute of Technology.

Sachsman, D.B., Sandman, P.M., Greenberg, M.R., & Salomone, K.L. (1988). Improving press coverage of environmental risk. *Industrial Crisis Quarterly, 2,* pp. 287, 295-96.

Sandman, P.M., Sachsman, D.B., & Greenberg, M.R. (1988). *The environmental news source: Providing environmental risk information to the media.* Newark, N.J.: Risk Communication Project, Hazardous Substance Management Research Center, New Jersey Institute of Technology.

Sandman, P.M., Sachsman, D.B., Greenberg, M.R., & Gochfeld, M. (1987). *Environmental risk and the press: An exploratory assessment.* New Brunswick, N.J.: Transaction.

Schlechtweg, H.P. (1992). Framing Earth First!: The MacNeil-Lehrer Newshour and Redwood Summer. In C.L. Oravec & J.G. Cantrill (Eds.), *The conference on the discourse of environmental advocacy* (pp. 262-87). Salt Lake City: University of Utah Humanities Center.

Shabecoff, P. (1993). *A fierce green fire.* New York: Hill & Wang.

Wilks, R. (1957). A rubber research lab moves into a small town. In A.H. Center (Ed.), *Public relations ideas in action* (pp. 63-66). New York: McGraw-Hill.

12
Media Frames and Environmental Discourse: The Case of "Focus: Logjam"

Harold P. Schlechtweg

The upsurge in mass media interest in environmental issues over the past several years has not stilled critics, who charge that media organizations systematically shield government and corporations from responsibility for actions harmful to the environment (Lee & Solomon, 1991). For their part, journalists argue that few environmental advocates understand the conventions of daily journalism, which, they claim, limit what even the fairest and most aware correspondent can communicate about environmental controversies (Stocking & Leonard, 1990).

While media critics often look at the informational content of newscasts and newspaper articles, or evaluate the "balanced" reporting of controversies, a continuing thread of research in media studies highlights the significance of frames, or the frameworks of interpretation and evaluation that guide journalists in reporting the news (e.g., Hall, 1974, 1975, 1982; Tuchman, 1978; Gitlin, 1980; Davis & Walton, 1983; Bird & Dardenne, 1988; Griffin, 1992). The term *frame* or *framework* is borrowed from Erving Goffman, who used it to designate what he called "schemata of interpretation," or the cognitive maps people use to organize their reality (1974, p. 21). Frames direct our attention to experiential phenomena and then sort these phenomena into meaningful categories: identifying, labeling, and defining events and occurrences. Frames help people to cope with new or problematic experiences by relating them to taken-for-granted, preexisting categories of interpretation.

Media frames simplify the process of reporting complicated stories because they supply journalists with ready-made conceptual schemes for organizing the flood of information that washes over the newsroom each workday. Thus, Todd Gitlin has defined media frames as "persistent patterns of cognition, interpretation, and presentation, of selection, emphasis, and exclusion, by which symbol-handlers routinely organize discourse, whether verbal or visual" (1980, p. 7). Media frames may also help audiences understand news items, by relating new information to familiar interpretations learned from previous encounters with news programs (Morley, 1980, p. 8; Lewis, 1991, p. 143). But as critics point out, frames often obscure much that is unique about activities and events while silencing explanations, arguments, and perspectives that do not fit within the media's framework of interpretation.

It is this last possibility that makes the news media's interpretive frameworks a continuing concern of critics and relevant to the discussion of environmental discourse. Pointing out that the same event can be signified, or represented, in different ways, Stuart Hall (1982) argues that media discourses instantiate choices based on unstated assumptions about the world and what counts as true. These assumptions, which guide journalists' choices about newsworthiness and how newsworthy events will be reported, are not haphazard. Rather, they follow an associative logic implying a particular notion of "how the world works" and, on this basis, constructing relationships among events and actors.

The associative logic informing a news frame can be a powerful ideological factor in the reporting of public controversies (Lewis, 1991, pp. 143–48). Media frames define the issue in contention and constrain opposing arguments to the established terms of debate. Those arguments that attempt to change the terms of reference by defining the issue otherwise can easily be dismissed as "straying from the point" (Hall, 1982, p. 81). Aberrant arguments, those that fall "outside" the logic of the news item, can be edited out of the newscast or article. If reported, they will lack intelligibility within the discursive logic governing the story.

By closely attending to the verbal and visual elements of the televisual text, it is possible to tease out the assumptions, and the unifying logic, that shape its content. This type of analysis is not directed at the level of informational content but attempts to discern the system or framework of interpretive categories that govern the production of the news item as a meaningful message. Mapping the media's interpretive frameworks is a useful exercise because it can help explain why some aspects of a news story receive special attention while others are neglected. But close textual analysis is only an important first step. To understand how a newscast or newspaper item intersects with public discourse, we have to go "beyond the text" to consider the discursive—and hence social, cultural, and historical—context in which it is embedded. It is at this point that the limitations imposed by a media frame, and its implications and significance for environmental discourse, become apparent.

The importance of media frames to environmental discourse is illustrated in the following analysis of "Focus: Logjam," a news feature about the radical environmental group Earth First! and the 1990 "Redwood Summer" protests against the timber industry on California's North Coast. "Focus: Logjam" was broadcast on the evening of July 20, 1990, on public television's *MacNeil-Lehrer Newshour*. The *MacNeil-Lehrer Newshour* attempts to provide a more issue-oriented alternative to early evening network news programs. The one-hour program includes a brief "news summary" of the day's major events, followed by a more detailed look at usually three "focus" stories. The focus on a limited number of major stories each night can allow correspondents as much as ten minutes for a taped report, almost half the time of an entire network news broadcast, which comprises only twenty-two minutes (without the commercials). This format permits *MacNeil-Lehrer* to probe issues in more depth and be less constrained by the event-orientation of the nightly network news (Hoynes, 1994).

One might think that the issue-centered approach of *MacNeil-Lehrer*, coupled with public television's mission to provide a forum for debate and controversy (Carnegie Commission, 1967), would result in a greater diversity of perspectives in a story about environmental conflict.[1] Yet with respect to "Focus: Logjam," such was not the case. While the news item is formally balanced, quoting spokespeople from opposing groups, the controversy is framed in a way that excludes Earth First!'s definition of the issues in dispute. Moreover, the news frame constructs Earth First! as a violent threat to the physical safety and livelihood of timber workers, enhancing the potential for violence in an already volatile situation. Significantly, the broadcast occurred the evening before the largest demonstration of Redwood Summer, an event many feared might erupt into violence between environmentalists and industry supporters.

The essay that follows is in two parts. The first section recounts the background of Redwood Summer and the discursive context in which adversaries sought to influence public opinion about the controversy. The second section is a detailed textual analysis of "Focus: Logjam" that maps the interpretive frame organizing the news item. In analyzing the program, I am primarily interested in the use of images, language, and sound (and their combination) as evidenced in the televisual text, and not in making judgments about the intentions of the reporter or producers of "Focus: Logjam." The purpose is not to register another indictment of media bias but rather to illustrate the consequences of a particular media frame for a specific instance of environmental discourse.

Redwood Summer

The May 1, 1990, issue of the *Earth First! Journal* trumpeted the call for eco-activists to flood California's redwood forests, putting their bodies on the line in defense "of the most famous eco-system in the world." From May into September, activists would nonviolently blockade logging roads, climb giant trees to prevent their being logged, and peacefully picket timber corporations on California's North Coast. The article promised that Redwood Summer would be the biggest national mobilization yet of Earth First! activists. The call specified that all participants were to take part in nonviolence training. Property destruction, threats, and verbal and physical violence against loggers or police were prohibited. Any would-be participant not in agreement with nonviolence as "*the*" principal concern during the actions would be barred from Redwood Summer (Cherney, 1990, p. 1).

Redwood Summer actions occurred in logging communities deeply polarized between environmental activists and forest workers. The *Earth First! Journal* reported more than one hundred direct action protests in the North Coast's Humboldt, Mendocino, and Sonoma counties during the preceding four years (Cherney, 1990, p. 1). Shortly before the Redwood Summer actions were announced, Eel River Sawmills and Louisiana Pacific announced mill closings and layoffs affecting several hundred workers. Company officials blamed environmentalists for the layoffs, citing cutbacks in Forest Service timber sales, efforts to protect the forest habitat of the northern spotted owl, the potential impact of statewide ballot initiatives to restrict logging, and litigation by North Coast environmentalists (Forster, 1990). The mill closings, coupled with the success of environmental lawsuits blocking the logging of old-growth redwoods, touched off an emotional backlash among timber workers who believed their jobs were threatened by efforts to save the forests (Stein, 1990c, p. A39).

The polarization between local Earth First! activists and many timber workers is further exacerbated by cultural differences (Anders, 1990). According to Brian Jaudon (1990), who spent time with both groups, North Coast Earth First! members tend to be "back to the landers," who fled the cities in the late 1960s and early 1970s. Many in this group continue to follow countercultural lifestyles. The region's older residents, including most timber workers, have roots that go back generations and still regard the "back to the land" group as outsiders. Jaudon believes that the most significant difference dividing the two groups, however, is the Earth First! members' biocentric approach to nature, which views humankind as only one species among many. Most loggers take an anthropocentric perspective, viewing human concerns as paramount.[2]

Divisions have also been sharpened by timber industry campaigns aimed at sparking a countermovement (Anderson, 1989). The countercampaign emphasizes that the economic future of timber towns, and the "way of life" of timber workers, require the continued logging of old-growth forests and the continued "wise use" of new-growth "timber resources" (Tomascheski-Adams, 1989). During 1989 and 1990 the Wise Use countermovement mobilized thousands of industry supporters in California and the Pacific Northwest in demonstrations blaming environmentalists for layoffs and mill closings.

Environmental groups charged that timber corporations were engaged in a logging "frenzy," culminating decades of mismanagement in which private and public forestlands were leveled with little regard for the consequences. In opposition to the arguments of the Wise Use movement, environmental organizations pointed to automation and the log exports as the most immediate threat to mill workers' jobs. Moreover, they charged that the record timber "harvests" were unsustainable, with disastrous consequences for future employment in the industry and the survival of species dependent on the remaining ancient forests. Citing provisions of the Endangered Species Act, environmental groups asked the U.S. Forest Service to declare large tracts of public old-growth forest off-limits to logging, in order to protect the habitat of threatened wildlife such as the northern spotted owl and the marbled murrelet (Gup, 1990).

In northern California, Georgia Pacific, Louisiana Pacific, Simpson Timber, and the Pacific Lumber Company (PALCO) accelerated the pace of logging on their privately owned industrial timberlands. During the summer of 1990, newspaper reports cited figures from the California Department of Forestry that showed the timber giants cutting down trees more than twice as fast as they could be replaced by new growth (Gilliam, 1990; Pelline, 1992, p. C7). Angered by a state report that timber companies were overcutting their Mendocino County forestlands by more than 250 percent, the county board of supervisors formally requested the State Board of Forestry to restrict logging in the county to "sustainable yield" (O'Rourke, 1990).

PALCO's decision to clear-cut three thousand acres of virgin redwoods provided the initial impetus for Redwood Summer. PALCO was saddled with a $550 million debt following a junk bond–financed Wall Street takeover in 1986 ("Redwoods Will Go," 1986). The Humboldt County company, formerly praised by environmentalists for selective cutting and sustained yield logging practices, doubled its timber harvests to increase short-term profits and to pay off debts associated with the buyout (Campbell & Malarkey, 1989). When PALCO announced plans to log its Headwaters forest, the largest ancient redwood grove not protected by national or state parks, Earth First! decided to act.

Local activists Judi Bari and Darryl Cherney hoped to win over timber workers by committing Redwood Summer to the goal of sustained yield in areas of new growth, rather than blanket opposition to logging. In his call to Earth First! members to join Redwood Summer, Cherney advised those who wanted to participate in the protests that organizers disapproved of bigotry toward timber workers (Cherney, 1990, p. 1). Speaking to members of the local press, Bari argued that the battle was not between timber workers and environmentalists but between giant logging corporations and North Coast communities. The same logging practices that were destroying forest ecosystems were also destroying workers' jobs ("Earth First Seeks," 1990).

These attempts by Earth First! to bridge the divide between forest workers and environmentalists were hampered, however, by the group's past associations with tree spiking, its continued incitement to "monkey wrenching," and strategic interventions by adversaries at pivotal moments in the controversy. The only reported tree-spiking injury occurred on May 8, 1987, at a Louisiana Pacific sawmill in Cloverdale, in Sonoma County. A mill worker was nearly killed when his saw struck a nail in the log he was cutting and disintegrated in a hail of flying shrapnel. California Earth First! disclaimed responsibility, and no members of Earth First! were implicated in a subsequent police investigation. Nevertheless, the *Los Angeles Times* called the event "environmental terrorism" and denounced Earth First! ("Environmental Terrorism," 1987). Similarly, the *Eureka Times-Standard,* published in Humboldt County's largest city, ran the headline "Earth First Blamed for Worker's Injury" (Gravelle, 1987). But a possible alliance between Earth First! and timber workers was also damaged by national Earth First! cofounder Dave Foreman's widely publicized response to the tragedy: "I think it's unfortunate that somebody was hurt, but, you know, I quite honestly am more concerned about old-growth forests, spotted owls and wolverines and salmon— and nobody is forcing people to cut trees" (Stammer, 1987). Following the Cloverdale incident, the *Eureka Times-Standard* continued to associate local Earth First! activists with new alleged or threatened incidents of tree spiking, despite an acknowledged lack of evidence ("Coward's Way," 1989).

Bari knew that no alliance with timber workers was possible while Earth First! appeared to endorse tree spiking (Bari, 1992). While the national organization was unlikely to change its position, grassroots Earth First! chapters pride themselves on their independence and insist on making their own policies. Bari persuaded Earth First! activists in northern California and southern Oregon to call publicly for an end to tree spiking in order to "stop fighting the victims and concentrate on the corporations" (Earth First! 1990). Bari and other Earth First-ers, however, continued to endorse "monkey wrenching," or equipment sabotage (Bari and Cherney, 1990).

Given the level of polarization in North Coast communities, it is not surprising that environmental activists were victims of violence (Martel, 1990). In the most serious incident, on May 24, 1990, a pipe bomb exploded in the Subaru station wagon carrying Cherney and Bari to a speaking engagement in Oakland. Cherney received minor injuries, but Bari's pelvis was shattered. She was permanently disabled (Bari, 1992). In an action that seemed to many activists to punish the victim for the crime, Cherney and Bari were arrested the next day and charged with illegal possession and transportation of explosives. Following almost two months of speculation in the press about the pair's guilt or innocence, charges were dropped on July 18 for lack of evidence.

According to newspaper accounts, the bomb blast intensified worries about violence and counterviolence in North Coast logging communities connected with Redwood Summer (Ronningen & Grabowicz, 1990; Harris & Grabowicz, 1990; Cockburn, 1990; Stein, 1990a). Such fears were at their height on the evening of July 20, when *The MacNeil-Lehrer Newshour* broadcast Spencer Michaels's "Focus: Logjam" report. Earth First! had called for a rally in Fort Bragg in Mendocino County the next day, and thousands of supporters were expected to attend. The pro-industry Community Solidarity Coalition had called for a counter-demonstration to be held on the other side of town. Like Earth First! they expected several thousand participants. Fearing the worst, and hoping to prevent a riot, law enforcement agencies mobilized 425 police to keep the rival demonstrations separated (Stein, 1990b).

On the day of the actions, several hundred angry pro-industry counter-demonstrators pushed around police lines and confronted the Redwood Summer protesters. When violence appeared likely, Redwood Summer organizers invited representatives of the group of loggers to address their rally. As four workers mounted the stage to address the crowd, the counterdemonstrators muffled their catcalls. Quick thinking by protest organizers had defused the situation, allowing the moment to pass without incident (Haddit, 1990).

Method

The analysis of "Focus: Logjam" that follows uses strategies adapted from Kenneth Burke's cluster-agon methodology (Burke, 1964; Burke, 1973, pp. 1–137; Rueckert, 1982, pp. 83–111), combined with techniques of close analysis of visual texts developed in film and television studies.[3] According to Burke, a cluster-agon analysis discloses the "motive" of the author or agent of discourse. Motive, Burke wrote in *The Philosophy of Literary Form*, is identical with the agent's perception of the situation or context of action (1973, p. 20). By this Burke means that every perception of a situation always incorporates taken-for-granted notions of how the world works (and thus the anticipated consequences of action) and of the forms of behavior appropriate to different circumstances. A cluster-agon analysis, by directing attention to the agent's subjective situation, discloses the taken-for-granted categories or assumptions that underlie and structure discourse.

"Focus: Logjam," of course, has a corporate, not an individual "author," but this is not a problem if we "textualize" Burke's treatment. An approach based on discourse analysis requires us to rethink the relationship between cognition and discourse, between author and text. According to new theories of cognition, we do not think first and then speak or write about what is "on our minds"; rather, speaking and writing are themselves cognitive activities. Discourse does not "represent" a preexisting mental activity; it is a manifest, public exhibition of thought in action. The aim of discourse analysis is not to get at the hidden mental processes "behind" a text but to lay bare the interpretive assumptions that are incorporated *in* the text as structuring elements.[4] Burke's methodology provides a way to dig these assumptions out of the text, whether that text is produced by a single author, as in the case of most literary works, or by a news organization, as in the case of a televisual text such as "Focus: Logjam."

In analyzing the transcript and videotape of "Focus: Logjam," I first made an index of key terms and images. In a cluster analysis, selection of key terms is guided by frequency of use and intensity (evocation of a strong emotional response and/or moral judgment). *Terrorist* is an example of a key term that carries a strong "charge." Next I made a concordance, an exhaustive list of the verbal and visual contexts in which the listed key terms are used. The concordance highlights the "indexicality" of language, or the variation in meaning according to the context of usage. Where the key "term" is an image, I noted the verbal context and adjoining images. Where the key term is linguistic, I noted the visuals contextualizing the term and the verbal context.

The video portion of "Focus: Logjam" contains ninety-eight editing cuts and includes archival footage and other clips, probably acquired from news video services.[5] I decided to take each camera shot, or the visuals between editing cuts, as my visual unit of analysis. Thus, I took apart the video portion of the newscast along the "seams" made by the item's producers, who constructed the program by "sewing" together a large number of visual fragments, or camera shots. Because the camera shots making up "Focus: Logjam" were selected by the program's producers, and not by me, attention to them underscores the constructed nature of the televisual text. Television is not a window on objective reality. As Michael Griffin points out, news production is "a process of piecing together narratives using visual fragments and sound bites from a variety of sources" (1992, p. 124). In other words, news images are carefully selected and always imply a perspective on the event being covered.

In evaluating the visual portion of the broadcast, I thus paid special attention to shot selection (e.g., long shot, medium shot, close-up), camera angles, background-foreground relationships, visual juxtapositions, and verbal interpretations of the image by the voice-over. Editing, camera techniques, and voice-overs can amplify visual images. "Focus: Logjam" often combines several of these techniques, powerfully "loading" particular images. By the same token, a previously "charged" image can, if repeated, amplify words associated with it.

The concordance of contexts does more than contextualize the use of key terms and images. It also allows the analyst to discern "equations" that are useful in interpreting the meanings of terms. An equation occurs when words with very different dictionary meanings function interchangeably within an overall textual economy of meaning. This is illustrated when very different key terms substitute for each other within the same or a similar context. The concordance, by highlighting the repetition of contexts, signals the equations among key terms, visual and verbal, situated within the same or similar contexts. With the help of equations, we can build up clusters of terms that form what Burke calls operational synonyms.

The final step is to chart the relationships between the different clusters. Mapping the relationships between clusters of key terms discloses the interpretive frame that governs the selection of words, sounds, and images and organizes their presentation in the news item. The analysis below illustrates how the televisual text clusters key terms by constructing equations and how it constructs relationships between clusters. While key terms are positioned in associational clusters, they are also structured in relations of causation, dependency, and opposition. The structural description of terms and their relationships highlights the frame that "Focus: Logjam" imposed on the events of Redwood Summer.

"Focus: Logjam"

No media broadcast occurs in a vacuum. Any North Coast audience for *The MacNeil-Lehrer Newshour* was already informed by discourses that counterposed environmentalism to jobs and environmentalists to loggers. More ominously, it was an audience polarized by fear of the other. In this situation, however naively, Earth First! organizers had attempted to open up a dialogue with timber workers by arguing a common interest in sustained-yield logging in new-growth forests. But the frame that "Focus: Logjam" imposed on the events of Redwood Summer undercut this attempt at dialogue. Within the oppositional logic governing the news item, the claim that radical environmentalists and loggers share common interests is incomprehensible. What may be worse, especially as the item was broadcast the evening before the Fort Bragg rallies, is that "Focus: Logjam" constructs violence as a reasonable response to Redwood Summer protests.

The news item took up nine minutes and forty seconds of the *MacNeil-Lehrer Newshour* on July 20, 1990. Anchor Robert MacNeil's introduction lasted twenty-two seconds, framing the story as a report on "new tactics" used by "some" environmentalists in their battle with the logging industry. MacNeil's introduction anticipates "Focus: Logjam"'s differentiation of Earth First! from mainstream environmental groups such as the Sierra Club, who agree with Earth First! on the issues, "while not endorsing any illegal actions." Such differentiation distinguishes acceptable environmental activism from that which is unacceptable. MacNeil's introduction also anticipates and underscores the succeeding report's central focus on the tactics of Earth First! rather than the issues in dispute.

Excluding MacNeil's opening introduction, the newscast is divided into five structurally distinct segments. The segments are the reporter's introduction, a section focusing on Earth First! tactics, a cursory account of the issues in dispute, a summary section refocusing attention on Earth First! tactics, and a concluding section that contextualizes the conflict with regard to three environmental initiatives on the California ballot.

Reporter's Introduction

The report, filed by correspondent Spencer Michaels of public station KQED–San Francisco, begins with three clusters of images: logging trucks moving on the highway and in industrial settings, pristine redwood forests, and environmentalist protests. These establishing shots image key terms, as the reporter's voice-over identifies the antagonists as Earth First! and the logging industry. The reporter also tells us that the stakes in the battle are, on the one hand, "small-town economies" and, on the other, "the tall beautiful redwood which grows nowhere else in the world." The viewer learns that these "small-town economies" depend on "lumber." The viewer also learns that while "business is humming," the "cut" is threatened by Earth First! a group *also* active this summer "demonstrating in the woods and at lumber mills."

Naming Earth First!

Having established initial oppositions and made reference to what is at stake, the report moves quickly to a discussion of the tactics of Earth First! Michaels's voice-over acknowledges that there is nothing new about conflicts between environmentalists and loggers. But what makes Earth First! and Redwood Summer "unusual," he claims, are the "methods" used and the "apocalyptic attitude" of Earth First! toward cutting redwoods. According to the reporter, Earth First! is "radical," "apocalyptic," and distinguished from other organizations by its "methods." What are these methods? "Earth First! thinks civil disobedience is essential." Significantly, Michaels's voice-over omits the modifier *nonviolent,* a term usually coupled with *civil disobedience.*

The term *civil disobedience* is a key expression in the newscast's lexicon. A dictionary defines the term as "the refusal to obey the government in order to influence legislation or policy, characterized by nonviolent techniques such as boycotting, picketing, etc." (*Random House* 1980). Civil disobedience is a tactic of organizations representing social movements, including environmental organizations, which hope thereby to change governmental or institutional practices. The use of civil disobedience by Earth First! does not stamp the group as "unusual." As Burke notes, however, people rarely use a word in its dictionary sense. Rather, a word's overtones, or its connotative meanings, are revealed in the company it keeps (Burke, 1973, p. 35). As we will see, "Focus: Logjam" transforms the dictionary meaning of *civil disobedience,* infusing the term with a new set of connotations.

The broadcast cuts to a medium close-up of Earth First! organizer Rick Clininger. Clininger insists that environmentalists must engage in confrontations to win a hearing. He stresses, however, that confrontations must be nonviolent.

While Clininger addresses the issue of tactics, he does not use the term *civil disobedience*. The televisual text has now twice refused the customary association of civil disobedience with nonviolence. The double omission weakens the usual connotative linkage. Furthermore, the images that accompany Clininger's voice-over disrupt the association he attempts to establish between nonviolence and confrontation. The effect is to undermine his address.

The viewer hears Clininger's voice-over as the camera cuts to a scene of heated verbal dispute. Clininger's last words, "We're proving here in this county that we can have confrontations in a nonviolent manner," must compete with the image's angry sound track. A new sound track of ambient crowd noises, high-pitched voices, and screams overpowers the first and segues into a scene of conflict. A group of men appear to be forcing open a gate against the resistance of others, as crowd noise crescendos in the background.

The combination of images and sound track questions the commitment of Earth First! to nonviolence and weakens connotative links between nonviolence and civil disobedience. The broadcast now moves to take advantage of this effect by constructing a new association. The correspondent's voice-over cuts through the crowd noise: "Earth First! has a record of civil disobedience, injuring private timberlands, sabotaging logging machinery, and at least talking about and writing about putting metal spikes in trees so they can't be logged." Repeating the term *civil disobedience* in this context transforms the word's overtones, associating it with the cluster "injuring" equals "sabotage" equals "tree spiking."

There is no question that tree spiking endangers forest workers. While some Earth First! chapters continue to advocate the practice, Bari and other North Coast Earth First! members denounced tree spiking and banned the practice at Redwood Summer actions. Nevertheless, "Focus: Logjam" uses tree spiking to define Earth First! metonymically, reducing the group to one aspect of its practice symbolized by the newscast. The sequence of voice-over and visuals are reproduced here:

VOICE	*IMAGE*
Michaels: Earth First has a record of civil disobedience, injuring private timberlands, sabotaging logging machinery,	(Cut 22) Medium shot. Crowd noise. Excited voices. Protesters are trying to open a gate.
and at least talking and writing about	(Cut 23) Extreme close-up of the blunt edge of an ax or hatchet pounding a spike into bark.
putting metal spikes in trees so they can't be logged.	(Cut 24) Close-up of a man in a yellow slicker who seems to be trying to get a spike out of a tree. Camera zooms in to extreme close-up of his hand.

Earth First! organizer Judi Bari on the eve of Redwood Summer	(Cut 25) Medium shot of tree with sign spray-painted on it. The day-glow orange message reads "Nai . . . " [Nail?]. Camera zooms in to extreme close-up on spike head. Simultaneously, the voice-over says, "Judi Bari."
denounced tree spiking as dangerous	(Cut 26) Close-up of two hands holding a paper with the printed words (center screen): "Warning the timber in this area has been spiked."
but she didn't rule out other contro . . .	(Cut 27) Long shot. Circle of people holding hands in road, blocking a logging truck.
. . . versial actions.	(Cut 28) Medium shot. Police arresting protester. Protester is center screen.

The sequence of words and images is powerfully enhanced by the camera work. The repeated use of close-ups intensifies the images and amplifies connotative associations. The visual image of cut 25 almost nails Judi Bari to the tree, as the camera zooms in from a medium shot to an extreme close-up on the spike head while the voice-over speaks her name. A powerful associational linkage, "Earth First!" equals "tree spiking" equals "Judi Bari," is established, overriding the tardy disclaimer. If we understand the images as scene and the voice-over as action, then the scene functions as a rhetorical amplification, "almost a kind of 'natural chorus'" (Burke, 1964, p. 299). Furthermore, the juxtaposition of images of tree spiking with images of civil disobedience reinforces the reporter's voice-over association of civil disobedience with "injuring," "sabotage," and "tree spiking."

The news feature follows this sequence with a verbal and visual quotation from Judi Bari. Bari uses three key terms, *terrorist, equipment sabotage,* and *tree spiking*. She denies that Earth First! is a terrorist organization, she opposes tree spiking, and she supports equipment sabotage. She also predicts that Earth First! members will be remembered as heroes in some future time. Structurally, the quotation is important because it introduces the word *terrorist* into the program's vocabulary. The quotation associates Bari and Earth First! with terrorism, even though Bari attributes the accusation to others and denies that Earth First! is a terrorist group.

The Bari quotation bridges the previous account of tree spiking and an account of the Oakland bombing that injured Bari and a companion in May. Bari's name is the key term that appears in both contexts. As with the sequence about tree spiking, camera work amplifies the bombing, constructing a discourse about Earth First! in which "the bomb" figures as another defining metonymy. The camera amplifies by means of repetition and intensity of focus, alternating shots of Bari's bombed-out car with close-ups of metal fragments. Without the voice-over, the bombed vehicle could be any wrecked car, the injured Earth Firsters the victims of any automobile accident. With accompanying narration, however, Bari and her companion are constructed not as victims but as suspects:

VOICE	IMAGE
Michaels: Late in May a bomb exploded in a car in which Bari	(Cut 30) Medium long shot of crumpled foreign car. Car is badly smashed up.
and another Earth Firster were	(Cut 31) Close-up. Piece of bent metal in road (center screen).
riding.	(Cut 32) Close-up. Another part in the road.
Bari was severely injured. Police first accused them	(Cut 33) Medium shot. Two emergency medical technicians wheel injured person on a stretcher. They move from bottom screen right to screen left. Camera follows stretcher.
of transporting the bomb, but in mid-July, prosecutors announced they would not press charges. Earth First!	(Cut 34) Medium shot. Medics put patient on stretcher. Camera moves back to take in more of the action.
insists the bomb was an attack by those who oppose them. Whi . . .	(Cut 35) Medium shot. Protesters holding hands. Camera pulls back to long shot of crowd.
. . . le the case has not been solved,	(Cut 36) Medium long shot of the front of the bombed-out car.
timber people are not taking chances.	(Cut 37) Long shot. Truck stacked with plywood on road moves screen left to screen right. Lumber in background. Truck goes off screen right.
Fearing sabotage and vandalism,	(Cut 38) Close-up shows a sign for a Georgia Pacific Wood Products facility.
Georgia Pacific canceled its mill tours	(Cut 39) Medium close-up. A sign says: "No scheduled mill tours available at this time."
and closed its forestland to the public,	(Cut 40) Close-up. Notice reads: "No Trespassing."

a move lobbied for by logging contractor Mike Anderson.

Anderson: A lot of those contractors have a lot of money invested in this equipment, ·

and we could lose our whole source of livelihood, you know, overnight, if the wrong people are around to tamper with it.

(Cut 41) Medium long shot. Logging truck. Workers loading logs on truck at center screen. Anderson's voice-over begins in this cut.

(Cut 42) Close-up of Mike Anderson looking screen left. Anderson is in the middle of a work area. Truck is being loaded behind him. "Mike Anderson, Logging Contractor" is superimposed on the screen.

This sequence encourages the suspicion that Bari may have bombed herself, a suggestion that may deprive her of viewer sympathy. The association of Earth First! with "sabotage" and "tree spiking," forms of terrorism according to the industry and news media, lends plausibility to Bari's arrest for transporting explosives. The decision of the prosecutors not to press charges is left an enigma. Michaels's voice-over implies that the timber industry remains convinced of Bari's guilt: "While the case has not been solved, timber people are not taking chances." The broadcast constructs a causal relationship between the bombing and Georgia Pacific's decision to cancel mill tours and close forestlands to the public.[6] These actions, apparently caused by the bomb, are then associated with fears of equipment sabotage. Tree spiking and sabotage are "signs" of Earth First! This whole sequence occurs "under the sign" of Earth First!'s "methods." As we will see, the equation, or set of linked and equivalent terms, "bomb" equals "tree spiking" equals "sabotage," provides legitimating premises for counterviolence and invites the viewer to adopt a "law and order" perspective on the conflict.

The sequence names the adversaries of Earth First! as the "timber people." A corporation, Georgia Pacific, is personified as the people who work in the industry. These workers are an undifferentiated mass, a "people." The logic that animates "Focus: Logjam" does not distinguish between owners and employees or workers and bosses. It is the "timber people" who are "not taking chances." Within the framework of oppositions that organize the newscast's depiction of reality, Earth First's argument of a community of interest between local environmentalists and timber workers against the industry is unintelligible.

The broadcast frames the contractor's quotation against a background of men working. His concern that the "wrong people" might tamper with logging equipment follows the account of the bombing and its aftermath. Anderson's words remind listeners that "timber people" depend on the industry for their "livelihood."

The report continues with a verbal description of "base camps" in the woods run by Earth First! and images of chow lines and tents. The language here is militaristic, reinforcing the terrorist imagery. The newscast differentiates the "veterans" from the "new recruits." The "recruits" equal "newcomers" equal "volunteers," and these people are "young and idealistic." The "veterans" instruct "newcomers" in "survival techniques."

The ultimate equation invokes the devil term *violence.* Do these "young and idealistic" "recruits" know what they are getting themselves into? Michaels raises the issue of violence with a "volunteer": "There are some people afraid of Earth First!—that it might turn violent." Honna Metzger testifies to the nonviolence of Earth First! but her defense is undercut by the word superimposed on the screen under her name: "Volunteer." Metzger is "young and idealistic," a "volunteer"; the "veterans" might respond differently. The reporter's use of the word *violent* is charged with significance: he has named Earth First!

Cathedral or Timber Crop?
The first section of the newscast focuses on the tactics of Earth First! The second addresses the issues. Michaels's voice-over asserts, "One of the main aims of these activities is to reduce the redwood cut, which threatens the economies of the lumber towns." The reporter's assertion is an argument, not an undisputed fact, and like all arguments it responds to alternative possibilities. The alternative, however, is unrepresented. Michaels introduces Tom Loop, a man "who sells firewood and has lived on the North Coast all his life." Thus credentialed, Loop tells us that North Coast inhabitants enjoy a "way of life" that depends on timber "companies" that provide a lot of "jobs." But Earth First! argues that "overcutting," not the activities of environmentalists, threatens forest worker "jobs," "town economies," "livelihoods," and the North Coast "way of life." The point here is not that the argument of Earth First! is right and that of "Focus: Logjam" is wrong, but that the argument of Earth First! is unrepresented.

Instead, "Focus: Logjam" presents the issue as a conflict between environmentalists who view the forest as a "cathedral" (in the words of Anderson) and "timber people" who regard trees as a "timber crop" to be "harvested." Anderson, previously identified as a logging contractor, is further credentialed as a member of the State Board of Forestry. His comments are illustrated with images of forest workers "harvesting" the "timber crop."

Anderson and Tom Deer, Georgia Pacific's resource manager in Fort Bragg, provide reasoned rebuttals to the broadcast's straw-man version of environmentalist arguments. We are never told why "sustained yield" is important to environmentalists. Rather, the audience learns that Georgia Pacific's cut is "market driven" and that overcutting when "consumer demand" is high is balanced by undercutting "when the market is low." The combined imagery of "harvest" and "market" constructs the industry as agribusiness and the "market" as a natural force, the invisible hand that balances every interest. This naturalization of the market absolves Georgia Pacific and its human overseers from responsibility for their actions.

"Regular People Get Pretty Nervous"

The camera cuts to a long shot of logs, a crane, and industrial smoke, signaling an abrupt transition to what seems, at first, the broadcast's conclusion. The reporter's voice speaks over long shots of industrial images similar to those with which the report began. The voice-over returns attention to the "record" of Earth First!: "Even though demonstrations have been peaceful so far, logger Jerry Philbrick looks at the past actions of Earth First! and worries." This sequence follows:

VOICE	IMAGE
Philbrick: There's been severe	(Cut 85) Medium shot of logging truck moving screen right to screen left.
equipment damage over the years. There's been tree spiking where a man almost lost his life in Cloverdale. Now we've got the bomb going on.	(Cut 86) Medium close-up of Philbrick leaning on the cab of a logging truck as he speaks.
I mean when people start fooling around with bombs	(Cut 87) Insert. Close-up of yellow flyer taped inside a store window: "Save our timber jobs, families, communities, and our sustainable forests. Solidarity."
regular people get pretty nervous, and then they start beefing up their own defenses and they get, you know, a little bit irate, so we're expecting what could be the worse half the time.	(Cut 88) Medium close-up of Philbrick.

The reporter's introduction to the sequence references the past actions of Earth First! It was foreshadowed by his earlier description of Earth First!'s "record of civil disobedience, injuring private timberlands, sabotaging logging machinery, and . . . putting metal spikes in trees so they can't be logged." These earlier references intensify the negative connotations of "past actions." This charge projects forward, imparting a hyperintensity to Philbrick's already loaded statement. The words "peaceful so far" are heavy with the portent of an *unpeaceful* future. The yellow flyer (inserted during editing), reminds us that "jobs," "town economies," "a way of life," and "livelihoods" are threatened by "terrorists." The charged trinity of "bomb," "tree spiking," and "sabotage" positions Earth First! outside the law, justifying defensive counterviolence. Philbrick's observation that when "regular people get . . . nervous . . . they start beefing up their own defenses" is stated as a matter of fact. The "beefing up" of "defenses" is a commonsense response of "regular people" confronted by outlaws whom the police seem powerless to control.

Contextualizing the Conflict

One expects "Focus: Logjam" to end on this note of confrontation, since the words and images summarize the story. The camera shots and Michaels's words reference the beginning of the report, providing a unifying element. Instead, the image cuts from Philbrick to correspondent Michaels. Michaels contextualizes the newscast with new information: "Tensions run high, but soon the entire state of California will have to choose sides." Michaels tells us that there are three timber-related initiatives on the California ballot. One is sponsored by the attorney general, the second initiative is industry-sponsored, "but the most controversial measure, sponsored by environmentalists, feared by industry and workers, is called Forests Forever."

The ballot initiatives are an ambiguous resolution at best, because "Focus: Logjam" has emphasized the criminal "tactics" of Earth First! rather than the issues in dispute. The final image, however, offers a less ambiguous resolution while powerfully condensing earlier themes. Complementing the broadcast's focus on Earth First! tactics, the image further depoliticizes the issue by re-emphasizing the organization's criminality. But the visual text goes beyond this to suggest that an alternative to violent self-defense remains possible. If the police will respond more forcefully, the justified vigilante action of "regular people" may still be avoided:

> (Cut 98) Medium shot of arrest scene. Police handcuff long-haired demonstrator in brown leather jacket. A police dog momentarily wanders into the camera frame. The camera zooms in on the handcuffs. Close-up and freeze-frame on handcuffed wrists. Fade to black.

The above analysis shows how key terms, both words and images, selected on the basis of frequency and intensity, cluster and form operational synonyms within the televisual text. While terms are linked in associational clusters, such clusters are also structured in relations of opposition, dependency, and causation. The discussion that follows, a structural description of clusters and their relationships, discloses the thematic frame that "Focus: Logjam" imposed on the events of Redwood Summer.

There are seven principal clusters associated with the two adversaries in this struggle. The four clusters linked to "regular people" provide a more complete description of this group than the three clusters associated with Earth First! Another key difference is that the identity terms associated with Earth First! are heavily charged with negative connotations. For these reasons, I call "regular people" the protagonist of the newscast, and Earth First! the antagonist.

I have labeled the protagonist's four clusters identity terms, state-of-mind terms, value terms, and terms that describe activities and institutions. Each category comprises a cluster of key terms drawn from the text. I distinguish between value terms and activity terms even though, conceptually, some terms would seem to fit within either category. For example, the terms *livelihood* and *timber harvest* seem to be synonymous. But as used in the news item, the terms belong to differ-

TABLE 12.1: The Key Terms of "Focus: Logjam"

Protagonist's Key Terms		
Identity Terms		**Value Terms**
Loggers		Small-town economies
Workers		Jobs
Timber people		Livelihood
Regular people		Way of life
State-of-Mind Terms		**Activities and Institutions**
Nervous		Not taking chances
Scared		Lumber
Irate		Companies
Timber harvest		Industry

Antagonist's Key Terms		
Identity Terms	**Activities**	**Value Terms**
Apocalyptic	Civil disobedience	Tall, beautiful
Radical	Confrontation	Redwoods
Wrong people	Sabotage	Cathedral
	Terrorist	Tree spiking

ent conceptual clusters, with the value term *livelihood* depending for its realization on the activity term *timber harvest.* I have not listed key images because they are essentially visual analogues of the language.

"Focus: Logjam" establishes a clear relation of dependency between the protagonist's value terms and the terms referring to activities and institutions. Thus, "small-town economies" depend on "lumber." "Jobs" are provided by "companies." A logging contractor's "livelihood" depends on the "timber harvest." The "way of life" of "regular people" depends on the timber "industry."

There is no cluster of key terms describing the antagonist's state of mind. Michaels uses the word *angered* to describe Earth First! members in the broadcast's introduction, but this usage occurs only once. In contrast, "Focus: Logjam" supplies us with a variety of negative and pejorative identity terms for Earth First! A second principal cluster contains key value terms. The antagonist's activities are listed in the third category.

The identity clusters of protagonist and antagonist are in clear opposition. "Focus: Logjam" positions "violent" Earth Firsters in confrontation with "regular people." This opposition, for audiences who identify with the timber workers as depicted in the news item, forestalls identification with Earth First! The protagonist's activity term "timber harvest" is opposed by the antagonist's value term "cathedral" or "redwoods." But recall that the protagonist's "way of life" depends on the "timber harvest." Thus the value clusters of the adversaries are also, although less directly, in opposition.

The broadcast establishes causal relationships between clusters. This is interesting because the force of causality moves in only one direction—from antagonist to protagonist. In their relations, the antagonist acts on the protagonist. The antagonist causes the state of mind of the protagonist through the antagonist's identity and actions. For example, "timber people" are "not taking chances," and "regular people" are "pretty nervous" because the antagonist is "playing with bombs." According to Michaels, "people" are "scared" that Earth First! might "turn violent." As the active agent, Earth First! is repeatedly positioned as the party responsible for violence and disruption. One wonders why the timber industry, whose accelerated logging operations precipitated Redwood Summer, is not at least co-responsible for the ensuing protests.

The Consequences of Framing

As I have demonstrated here, media frames can have important consequences for environmental discourse, consequences that go beyond the amount and quality of information dispensed or the balanced presentation of opposing speakers. The frame that "Focus: Logjam" imposes on the events of Redwood Summer constructs a reality in which two adversaries, radical environmentalists and the ordinary people who live in northern California's small towns and rural areas, are locked in a desperate struggle. At stake in this battle are, on the one hand, the way of life of generations of forest and mill workers and, on the other, ancient, cathedrallike forests. Thus the binary, oppositional logic that informs the frame constructs a cleavage between radicals and regular people, between ecosystem and livelihood. This may be a valid perspective on the conflict, but when a perspective on reality is unreflectively represented as "the way things really are," instead of one point of view, then alternative perspectives are silenced. The arguments of Earth First! fall outside the newscast's frame of reference and are not represented in the verbal text. Had they been reported, they would not have made sense within the frame of reference provided by the news item.

This is not to say that newscasters can do without frames or that this would be desirable. As Hall notes, every account of reality is necessarily partial and reflects social influences. The decision to represent one aspect rather than another is always subject to criteria other than that embedded in the material itself (Hall, 1974, p. 23). Burke argues a similar case for naming, which singles out an object or event as "such-and-such" rather than "something other" (1973, p. 4). We cannot eliminate naming without eliminating vocabulary itself as a way of encompassing reality. But as Burke suggests, we can test the accuracy of names or, as in this case, the *inclusiveness* of interpretive frameworks by subjecting them to the collective practices of analysis and discussion. Critical attention to media frames, in turn, may serve the public good by opening up the mediated space of public discussion to a greater diversity of thought about environmental issues.[7]

But this analysis also raises issues that go beyond the constraints and enablements that specific media frames afford particular environmental discourses. Once the initial oppositions are delineated and the adversaries named, "Focus: Logjam" begins constructing the identities of the opposing forces. According to "Focus: Logjam," Earth First! is a violent, terrorist organization, willing to use sabotage, bombs, and tree spiking to save redwood forests. Earth First! so the program says is a threat to the physical safety of timber workers and to the industry on which they, their families, and their communities depend for economic survival.

These depictions take for granted what is contested and reinforce the original defining opposition. But they also provide a defensive rationale for violence. As "Focus: Logjam" points out, given the apparent failure of the police to control the (reputed) lawlessness of Earth First! regular people will understandably "start beefing up their own defenses." The newscast aired on July 20, 1990, the evening before the largest demonstrations of Redwood Summer. In this context, it provided justification in advance, should there be violence against Earth First! members at the demonstration—a protest, it should be recalled, whose organizing principle was nonviolence.

Notes

The author thanks Christine Oravec for her valuable feedback and Ramona Liera-Schwichtenberg for suggestions about the analysis of televisual texts.

1. William Hoynes, who conducted a six-month study of the program, warns that it is problematic to argue that *MacNeil-Lehrer* will provide more diverse news solely because it is sheltered from commercial pressures. He attributes a lack of diversity to pressures arising from the need to attract corporate sponsors because of inadequate public funding, the failure to insulate public television from political pressures, and, importantly, journalistic norms that are shared with commercial broadcasting (Hoynes, 1994, pp. 84-87).

2. For an ethnographic account of Earth First! in Oregon, see Lange, 1990. Jentri Anders (1990) studied "back to the land" communities in Humboldt County for more than ten years. Her book also discusses the response of longer-term residents to the counter-cultural lifestyles of many "back to the landers."

3. I believe this is the first time a cluster-agon analysis has been used to address both the verbal and the visual aspects of a televisual text. For an additional explanation of the method as applied to language, see Berthold, 1986. For an application to a purely visual medium, see Reid-Nash, 1984.

4. Summarizing new approaches to cognition and discourse analysis is beyond the scope of this chapter; the reader is referred to Harre, 1992.

5. At least half a dozen companies maintain archives and sell video news footage. Some of the same clips used by "Focus: Logjam" were used in an NBC *Today* news report (also broadcast on July 20). NBC also sells clips from its news video archives. See *Motion Picture, TV, and Theater Directory,* 1990.

6. Actually, Georgia Pacific canceled its mill tours on May 10, two weeks before the bombing (Mencher, 1990).

7. The mass media exhibit a distinct tendency to frame protests such as Redwood Summer as disruptions of civil order rather than as statements about social problems

(Hall, 1974; Gitlin, 1980). But there is also persuasive evidence that, given the appropriate circumstances, social movements have achieved some success in expanding media frames to encompass their views (Gamson & Wolfsfeld, 1993).

References

Anders, J. (1990). *Beyond counterculture: The community of Mateel.* Pullman: Washington State University Press.

Anderson, R. (1989). Get involved. . . . Your survival depends on you. Remarks to the fortieth annual Sierra Cascade Logging Conference, Anderson, Calif.

Bari, J. (1992, August 27). Interview by author, Willits, Calif.

Bari, J., & Cherney, D. (1990, May 1). Ukiah burning. *Earth First! Journal,* p. 6.

Berthold, C. (1986, Winter). Kenneth Burke's cluster-agon method: Its development and an application. *Central States Speech Journal, 27,* pp. 302-9.

Bird, E.S., & Dardenne, R.W. (1988). Myth, chronicle, and story: Exploring the narrative qualities of the news. In J.W. Carey (Ed.), *Media, myths, and narratives: Television and the press* (pp. 67-86). Newbury Park, Calif.: Sage.

Burke, K. (1964). Fact, inference, and proof in the analysis of literary symbolism. In L. Bryson, L. Finkelstein, R.M. MacIver, & R. McKeon (Eds.), *Symbols and values: an initial study,* Thirteenth Symposium of the Conference on Science, Philosophy, and Religion (pp. 283-306). New York: Cooper Square.

————. (1973). *The philosophy of literary form.* 3rd ed. Berkeley: University of California Press.

Campbell, J., & Malarkey, T. (1989, November 17). Letter from the president and the vice chairman of Pacific Lumber Company to "customers, suppliers and friends."

Carnegie Commission on Educational Television. (1967). *Public television: A program for action.* New York: Bantam Books.

Cherney, D. (1990, May 1). Freedom riders needed to save the forest: Mississippi summer in the California redwoods. *Earth First! Journal,* pp. 1, 6.

Cockburn, A. (1990, May 31). A chill falls on Redwood Summer. *Los Angeles Times,* p. B7.

Coward's way to solve problems. (1989, October 29). *Eureka (Calif.) Times-Standard,* p. 4. Editorial.

Davis, H., & Walton, P. (1983). Death of a premier: Consensus and closure in international news. In H. Davis & P. Walton (Eds.), *Language, image, media* (pp. 8-49). Oxford: Basil Blackwell.

Earth First! (1990, April 11). Northern California Earth First! renounces tree spiking. Press release.

Earth First seeks peaceful protest, logging blockade. (1990, April 22). *Eureka (Calif.) Times-Standard,* p. 3.

Environmental terrorism. (1987, July 23). *Los Angeles Times,* sec. 2, p. 4.

Focus: Logjam. (1990, July 20). *The MacNeil-Lehrer Newshour.* Videotape, MLNH780.

Forster, D. (1990, March 29). Sawmill layoffs announced. *Eureka (Calif.) Times-Standard,* pp. 1, 8.

Gamson, W.A., & Wolfsfeld, G. (1993, July). Movements and media as interacting systems. *Annals of the American Academy of Political and Social Sciences, 528,* pp. 114-25.

Gilliam, H. (1990, September 2). Logging: Not a clear-cut case. *This World,* pp. 7-8, 15-16.

Gitlin, T. (1980). *The whole world is watching: Mass media in the making and unmaking of the New Left.* Berkeley: University of California Press.

Goffman, E. (1974). *Frame analysis.* Cambridge, Mass.: Harvard University Press.

Gravelle, M. (1987, May 16). Earth First blamed for worker's injury. *Eureka (Calif.) Times-Standard,* p. 1.

Griffin, M. (1992). Looking at TV news: Strategies for research. *Communication, 13,* pp. 121-41.

Gup, T. (1990, June 25). Owl vs. man: In the Northwest's battle over logging, jobs are at stake but so are irreplaceable forests. *Time,* pp. 56-65.

Haddit, M. (1990, July 30). Dialogue in Fort Bragg. *Santa Rosa (Calif.) Press Democrat,* p. B5.

Hall, S. (1974). Media power: The double bind. *Journal of Communication, 24,* 4, pp. 19-26.

———. (1975). The "structured communication" of events. *Getting the message across.* Paris: UNESCO.

———. (1982). The rediscovery of "ideology": Return of the repressed in media studies. In M. Gurevitch, T. Bennett, J. Curran, & S. Woolacott (Eds.), *Culture, society and the media* (pp. 56-90). London: Methuen.

Harre, R. (Ed.). (1992). New methodologies: The turn to discourse. *American Behavioral Scientist, 36,* 1. Special issue.

Harris, H., & Grabowicz, P. (1990, May 26). Two leaders of Earth First! were aware of the device, police say. *Oakland Tribune,* pp. A1, 14.

Hoynes, W. (1994). *Public television for sale.* Boulder, Colo.: Westview Press.

Jaudon, B. (1990, November). When a tree falls in the forest. *Sojourners,* pp. 10-17.

Lange, J.I. (1990). Refusal to compromise: The case of Earth First! *Western Journal of Speech Communication, 54,* pp. 473-94.

Lee, M.A., & Solomon, N. (1991, January-February). And that's the way it is: Media coverage of the environment has surely increased, but is it doing justice to the issues or doing the bidding of media owners? *E Magazine,* pp. 38-43, 65-67.

Lewis, J. (1991). *The ideological octopus: An exploration of television and its audience.* New York: Routledge.

Martel, B. (1990, October). We have different fears. *Country Activist,* p. 2.

Mencher, B. (1990, May 10). G-P ends mill tours, fears sabotage. *Mendocino (Calif.) Beacon,* p. 1.

Morley, D. (1980). *The nationwide audience.* London: British Film Institute.

Motion picture, TV, and theater directory. (1990, Spring). Tarrytown, N.Y.: Motion Picture Enterprises Publications.

O'Rourke, L. (1990, June 6). County supervisors vote for sustainable yield timber harvesting. *Mendocino (Calif.) Beacon,* p. 3.

Pelline, J. (1992, July 13). Timber shortage chops industry. *San Francisco Chronicle,* pp. C1, 7.

The Random House Dictionary. (1980). New York: Ballantine Books.

Redwoods will go to cut MAXXAM debt. (1986, August). *Forest Industries,* p. 9.

Reid-Nash, N.K. (1984). Rhetorical analysis of the paintings of Hieronymus Bosch. Unpublished doctoral dissertation, University of Denver.

Ronningen, J., & Grabowicz, P. (1990, May 25). Injured activists are organizers of summer long protests. *Oakland Tribune,* pp. A1, 14.

Rueckert, W.H. (1982). *Kenneth Burke and the drama of human relations.* 2nd ed. Berkeley: University of California Press.

Stammer, L.B. (1987, May 15). Environmental radicals target of a probe into lumber mill accident. *Los Angeles Times,* sec. 1, p. 3.

Stein, M.A. (1990a, June 4). Activists put safety first after blast. *Los Angeles Times,* p. A3.

———. (1990b, July 22). Earth First! and loggers in face off. *Los Angeles Times,* p. A1.

———. (1990c, May 26). Police hold Earth First! pair in blast. *Los Angeles Times,* pp. A1, 39.

Stocking, H., & Leonard, J.P. (1990, November-December). The greening of the press. *Columbia Journalism Review,* pp. 37-43.

Tomascheski-Adams, L. (1989). Letter to President Bush by the executive coordinator of the West Coast Alliance for Resources and the Environment. In A. Gottlieb (Ed.), *The Wise Use Agenda* (pp. 60-68). Bellevue, Wash.: Free Enterprise Press.

Tuchman, G. (1978). *Making news.* New York: Free Press.

Contributors

Connie Bullis is an associate professor and chair of the Department of Communication at the University of Utah. She received her Ph.D. from Purdue University. Her interests are in the processes and outcomes of human/nature relationships as they are enacted through communication. She has published in several professional books and journals and has worked with environmental issues such as wilderness management, personal lifestyle choices, wildlife legislation, and corporate environmental auditing.

James G. Cantrill is an associate professor in the Department of Communication and Performance Studies at Northern Michigan University. He received his Ph.D. from the University of Illinois and is known for his social scientific studies of environmental communication and the cognitive processing of environmental advocacy. He has published in several professional journals and books and serves as an environmental communication consultant for U.S. federal and state as well as Canadian provincial governments. Long associated with coordinating national conferences on environmental discourse, he currently focuses his efforts on assisting agencies and communities to establish environmentally sustainable policies in the Great Lakes region.

Donal Carbaugh is a professor of communication at the University of Massachusetts. He received his Ph.D. from the University of Washington. His main interest is the exploration of natural and cultural worlds from an ethnographic and communication perspective. He has held academic appointments at Linacre College, Oxford, at the Universities of Jyvaskyla and Tampere in Finland, and at the University of Pittsburgh. His most recent book is *Situating Selves: The Communication of Social Identity in American Scenes.*

Norbert Elliot is a professor of English in the Department of Humanities and Social Sciences at the New Jersey Institute of Technology. He received his Ph.D. from the University of Tennessee. He is currently completing a book on approaching environmental studies from a social science and humanities perspective. He teaches in the graduate programs in professional and technical communication and in environmental policy studies.

M. Jimmie Killingsworth, professor of English at Texas A&M University, has written extensively on American literature and rhetoric. He received his Ph.D. from the University of Tennessee. His many works on environmental rhetoric include *Ecospeak: Rhetoric and Environmental Politics in America,* with Jacqueline S. Palmer. He lives in the Texas hill country with his wife and favorite coauthor and their daughter, Myrth.

Michael E. Kraft is a professor of political science and public affairs and Herbert Fisk Johnson Professor of Environmental Studies at the University of Wisconsin–Green Bay. He holds a Ph.D. from Yale University. He is author of *Environmental Policy and Politics: Toward the Twenty-first Century* and coeditor and contributing author of *Environmental Policy in the 1990s*.

John Opie, Distinguished Professor at the New Jersey Institute of Technology, directs graduate environmental policy studies. He holds a Ph.D. from the University of Chicago. A former Fellow at the National Humanities Center, he has written books and articles on land and water issues, global climate change, technology transfer, landscape aesthetics, and global sustainability. He was founding editor of *Environmental History Review* and founding president of the American Society for Environmental History.

Christine L. Oravec is a professor in the Department of Communication at the University of Utah. She received her Ph.D. from the University of Wisconsin-Madison and is known for her studies of the rhetoric of the early conservation movement, particularly the essays of John Muir. She has published in such venues as the *Quarterly Journal of Speech, Critical Studies in Mass Communication,* and *Philosophy and Rhetoric*. She also contributed to anthologies on the history of rhetoric and the work of Kenneth Burke. She co-coordinated the first Alta Conference on the Discourse of Environmental Advocacy and is active in the development of environmental studies within the communication discipline.

Jacqueline S. Palmer, senior trainer for the Southwest Educational Development Laboratory in Austin, Texas, supports systematic improvement of instruction in mathematics and science education through research, development, and training activities. With ten years' worth of high school teaching experience and a doctorate in education from Memphis State University, she addresses topics in her publications such as science education, alternative assessment, environmental education, and environmental rhetoric. She is currently conducting research on state curriculum frameworks.

Patricia Paystrup is an associate professor at Southern Utah University, where she teaches courses in public relations and advertising. Her research interests focus on environmental public policy advocacy efforts. In her dissertation, she analyzed the controversy surrounding restoring wolves to Yellowstone and won an Outstanding Dissertation Award in 1994 from the International Communication Association (ICA).

Markus J. Peterson is an independent researcher and consultant. He holds a D.V.M. from Washington State University and a Ph.D. in wildlife ecology from Texas A&M University. He has published the results of his studies of wildlife disease ecology, conservation biology, and natural resource policy formation and implementation in journals such as the *Journal of Wildlife Management*, the *Journal of Wildlife Disease, Conservation Biology,* the *Wildlife Society Bulletin,* and *Environmental Values.*

Tarla Rai Peterson is an associate professor in the Department of Speech Communication, Texas A&M University. She holds an interdisciplinary Ph.D. from Washington State University. She is coeditor of *Communication and the Culture of Technology.* She has published analyses of environmental communication, organizational rhetoric, and the rhetoric of science and technology in several book chapters as well as in journals such as the *Quarterly Journal of Speech,* the *Journal of Applied Communication Research, Agriculture and Human Values,* and *Environmental Values.*

David B. Sachsman holds the George R. West Jr. Chair of Excellence in Communication and Public Affairs at the University of Tennessee at Chattanooga. He came to Chattanooga from California State University, Fullerton, where he served as dean and professor of the School of Communications. His research and writing on environmental communication date back to his Stanford University Ph.D. dissertation in the early 1970s.

Harold P. Schlechtweg, who holds a master of science degree in labor studies, is a Ph.D. candidate in the Department of Communication at the University of Massachusetts. He is also the managing editor of the *Plain Dealer,* Wichita, Kansas. He has published articles about organized labor in the popular press and in the University of Massachusetts's *Labor Center Review.* His dissertation addresses the role of culture and ideology in the conflict over the logging of northern California's old-growth redwoods.

Bruce J. Weaver is a professor and chair of the Department of Speech Communication and Theatre at Albion College. He received his Ph.D. from the University of Michigan. His articles have appeared in the *Quarterly Journal of Speech* and *Communication Quarterly.* He is interested in discovering the role that communication played in the establishment of all U.S. national parks.

Diana Wuertz is a legislative program analyst with the Wisconsin Legislative Audit Bureau. She has coauthored several audit reports concentrating on public policy issues, including population growth, regulation enforcement, and program efficiency. She received a master of science degree in environmental science and policy from the University of Wisconsin–Green Bay in 1994.

Index

advocacy
blame-based, 4, 9-10, 90, 143, 214
campaign strategies, 76, 179-80
and culture, 40-41, 46, 53, 55 n
and deep ecology, 131
and emotion, 3, 16, 58, 83
and feminism, 123
and legislation, 98, 100-101
and lobbying, 104-11, 114-15
and persuasion, 33-34, 87, 95, 97-98,
177, 225, 230
and the sublime, 65, 71-73
Alar, 113-14
Amoco, 181, 183-85, 195 n
Arctic National Wildlife Refuge, 109-10,
207
art
conventions of, 62-65
of the wilderness, 59, 65, 66-68
and westward expansion 25-26
autopoiesis, 205-6
awareness of the environment, 76, 82, 126,
219, 233
and attitudes, 4, 73, 80-81, 101, 104,
185-86
and environmental problems, 77-80,
115-16, 206-7, 224, 226, 229, 233-
35, 241, 247, 251-53
and language, 40
and perception, 34, 59, 67, 70, 77, 81,
82, 88, 97, 113, 142, 176, 187, 190-
91, 195 n, 251, 262-63
and place names, 42
and social norms, 25, 80, 84, 85, 88,
131, 219

Bartram, William, 9, 10, 18-20, 22, 33-34
Bhopal, 251
Bierstadt, Albert, 63, 65
biodiversity, 86, 89, 154, 155, 157
Blair, Hugh, 20, 21, 22
Bureau of Land Management, 108
Bureau of Reclamation, 108

Campbell, George, 20, 21, 22
Carson, Rachel, 9, 10, 31-32, 33-35, 72-73,
221, 241, 244
chaos theory, 207

Christianity
as model, 9-10, 67, 136, 138, 229, 232
attacked, 14, 21, 26
influence of, 11-15, 136-40
New Testament, 14
Old Testament, 10, 11, 30
Puritans, 11-13, 14-15, 16, 21, 26, 29,
30, 33
Clean Air Act, 1, 105, 106, 108, 109
Clean Water Act, 105, 106
cluster-agon analysis, 46, 262-64, 272-74,
275 n
cognitive psychology, 81-82, 275 n
common sense realism, 20, 22, 25, 28-29,
32, 129
consciousness
environmental, 3-4, 25, 26, 66, 193,
234-35
and schema, 78-79, 82-83, 88, 257
and self-interest, 33, 80, 81, 86, 87,
129-33
and gender, 125, 144, 162, 237
and political action, 71, 72, 101, 103-
4, 114, 221
perceptual (terministic) screens, 59-61,
70, 199, 200, 210
spiritual, 1, 137, 169, 236
conservation, 215, 219, 233, 241
versus preservation, 25, 26-28, 30, 54,
85, 103, 270
culture
counterculture reaction, 219
influences of, 11, 46-50, 88, 131, 165,
189-90, 215, 232, 260
and empiricism, 4-5, 30, 35
and ideology, 40, 72, 84-86, 95, 124,
220, 223-24, 236, 258
and wilderness, 39, 51-53

Danforth, Samuel, 9, 10, 12, 13-16, 18, 19,
25, 31, 33-35
Darwin, Charles, 23, 28, 34, 133
Dewey, John, 237-38
Dow Chemical, 176, 181, 182, 184, 186-88,
195 n

Earth Day, 1, 85, 187, 224, 243
ecofeminism, 123, 124-28, 130, 219

ecology
   attacked, 4, 26
   as science, 17-20, 23, 30, 32, 33, 155, 248
      and natural history, 10, 18, 19, 34
   web of life, 78, 126, 128, 130-32, 222
education
   drawbacks to, 79, 210
   extent of, 1
   informal, 98, 99-101
   use of, 118, 250
Edwards, Jonathan, 9, 10, 16-17, 21-22, 25,
   28, 29, 30, 31, 33-34
Emerson, Ralph Waldo, 9, 10, 17, 20-22, 23,
   25, 27, 28, 29, 30, 33-35
endangered species, 89, 101, 231
   northern spotted owl, 43, 259, 260, 261
   timber wolf, 43
   whooping cranes, 4, 203-4, 215
Endangered Species Act, 1, 103, 105, 106,
   107, 260
energy conservation, 84, 85, 99, 109-11,
   192, 230, 231, 232-33
Energy Policy Act, 110
environmental concerns, 77-78, 79, 87,
   223, 226
environmental determinism, 23
environmental movement
   categories of, 221-22
   Chipko, 126, 139
   conflict within, 219, 221
   ecofeminism, 125-26, 221
   Earth First!, 128, 129, 135, 137, 143, 235,
      258-62, 264-75, 275 n
   financing, 100, 103-4
   grassroots, 96, 98, 99, 112, 115-16, 118
   Greens, 99, 113, 219, 222
   interest groups, 79, 95-100, 104, 111,
      221-22, 237, 243
   membership in, 101, 102
   Sierra Club, 70, 99, 100, 101, 104, 108,
      109, 113, 119 n, 237, 264
   Wilderness Society, 50, 99, 100, 101,
      119 n
   Wise Use, 103, 111, 118, 233, 260
environmental paradigms, 85, 90, 125
Environmental Protection Agency
   Environmental Impact Statement, 9, 10,
      31-32, 33-34
   mandates, 101, 186
      and political action, 106, 108, 111,
         113-14, 186
ethics
   anthropocentrism, 85, 260
   biocentrism, 134, 141-42, 260

   commitments, 30, 83-84, 87, 88, 116,
      223, 227
   deep ecology, 123, 128-29, 137, 140, 219,
      222, 223
   ecofeminist, 124, 223
   land ethic, the, 29, 30, 132
   pragmatism, 227-28, 231
   social ecology, 123, 127, 143, 129-30,
      222, 223, 225
   wilderness, 237
ethos, 3, 9, 219, 220, 237
extinction, 1, 200
Exxon Valdez, 104, 114, 198-99, 205-9, 211-
   12, 216, 253

feminism, 123, 124, 125-26, 133, 135, 138-
   39, 140, 237
Finland, 46-50, 55 n
Fish and Wildlife Service, United States,
   108
Forest Service, United States, 85, 259, 260
Friedrich, Caspar, 65

Gandhi, 131, 132
Gore, Al, 9, 10, 31, 32, 33-35
green consumerism, 220, 222-23, 227-39
Georgia Pacific, 260, 268, 269, 270, 275 n

Hetch Hetchy valley, 25-29, 34
health communication, 248-49
Huntsman Chemical, 176, 181, 184, 195 n

information processing
   biases, 76, 81, 83, 84, 88-89, 251, 257-58,
      263
   and the self, 79-80, 81, 89, 128, 130-33,
      138, 139-40
   environmental default mechanisms, 81-
      89
   heuristics, 12, 81, 158
   rationalization, 85
instrumental discourse, 219-22
   and identity, 231-33
   and ritual action, 225-29
International Harvester, 241-42
issues management, 179-80

language
   emotional reactions to, 16, 43, 83, 263
   influence on environmentalism, 4, 20,
      34, 68-69, 88, 95, 232
   influence on politics, 115
   symbolic nature of, 1, 45-46, 49, 58, 73,
      125, 159, 190, 199, 227-28, 232, 265

and condensational symbols, 4, 88-89, 101-3, 109, 115-16, 159, 190
and journalism, 250
and landscapes, 42-43
and metaphor, 13, 21, 23, 25, 34-35, 45, 64, 125, 127, 134, 135-36, 141-42, 183
and metonymy, 266, 267
Leopold, Aldo, 9, 10, 29-31, 33-35, 132, 138, 221
Linnaeus, 18, 19, 28
littering, 81, 84
Locke, John, 16
Longinus, 64, 65, 66, 67
Luhmann, Niklas, 11, 200, 205-10, 213, 214

Marxism, 129
mass media
    agenda setting, 88, 179, 252-54
    direct mail, 101, 116, 242
    frames, 257-59, 274-75, 275 n, 276 n
    influence of, 1, 70-71, 78, 86, 89, 191
    lobbying of, 112-14, 245-46
    shortcomings of, 112, 244-45, 257, 258
    use of, 58, 68, 71, 79, 87, 95, 100, 105, 111-12, 157, 177, 182, 191-92, 242, 257, 263
Mather, Cotton, 12, 13
McDonald's, 176-77, 194 n
McKibben, Bill, 9, 10, 31, 32, 33-35
Miller, Perry, 9, 11, 16-17, 26
monkey wrenching, 261-62, 266-67, 269
Montreal Accords, 1
Moran, Thomas, 59, 63, 65, 67
motivation, 76, 80-81, 83, 87, 90, 199, 262-63
Mount Monadnock, 42, 43, 45, 46
Muir, John, 9, 10, 26-29, 30, 33-35, 61, 62-64, 66, 67, 71-73, 74 n, 140, 221

National Energy Act, 1
National Environmental Policy Act, 1, 9, 10, 31-32
national parks
    Grand Canyon, 59, 153, 154, 155, 156, 202
    National Park Service, United States, 71, 103, 108, 156, 165
    Shenandoah, 151, 157, 166
    Smoky Mountains, 151-74
    Yellowstone, 59, 63, 153, 154, 155, 156, 174 n, 202
    Yosemite, 26, 27, 60, 73, 73n, 153, 155, 156, 174 n

New England, 13, 29, 50, 51, 53, 127
New World
    discovery, 11, 22-24, 69, 70, 159
    First Nations, 17, 65, 160, 162, 212, 243
        Aleut, 189-90, 194
        Sioux, 189-90, 194
        Western Apache, 42-42, 45
    use of, 18, 25, 26, 30
    westward expansion, 22, 24, 25
Niagara Falls, 61, 68, 69, 154
North American Free Trade Agreement, 111
nuclear energy, 98, 109, 116, 126

Office of Technology Assessment, 32
ozone, 76, 84, 99

pathos, 10
Peirce, C.S., 227-28
perspective by incongruity, 177-78, 182-84, 188, 190
plastics, 177-96
pesticides, 72, 100, 109, 234
Pinchot, Gifford, 27-29, 32, 34
policy
    104th Congress, 3, 89, 95-96, 97, 101, 103, 111, 118, 119
    agenda setting, 96-98
    change, 97-98, 104, 106, 128, 272
    congressional legislation, 1, 29, 95-96, 109, 152, 178-79, 186-87, 243
    lobbying for, 105-17
    by industry, 106, 111, 114, 193, 195 n, 245
    psychological response to, 78-79, 80, 85, 105, 119, 204, 243
    Regan administration, 97, 100-101, 211, 224-25
population, 1, 5, 84, 99, 124, 143-44
poverty, 125, 129, 139, 165, 174 n
    in opposition to wealth, 163-64, 168, 173 n
    negative images of, 168, 171-72
public relations, 86, 89, 99, 100, 111, 114, 116-17, 178, 179-80, 186, 194, 195 n, 234, 241, 242, 243, 244-248, 253

Quintillian, 12

Ramus, Petrus, 12, 14
ramism, 10, 12-13, 14-15, 16, 18, 21, 22-23, 25, 28, 29, 31, 32
reasoning, 78

recreation
    as valued, 201-3
    and tourism, 54-55, 60, 66, 74n, 151-53,
        156-58, 166
recycling, 81, 176-77, 178, 181-96, 234
redwood forests, 259-62, 265, 270, 274
risk analysis, 81, 99, 241
risk communication, 113, 248-51
rhetoric
    Aristotelian, 10, 12-13, 20, 61, 64, 227
    Burkean, 43, 45, 177, 182-85, 199-200,
        225, 262-63, 164, 165
    instrumental, 226-68
    nature of, 69-70, 72, 88, 153, 219, 231
    Platonic, neo-, 16
    rhetorical situation, the, 178, 188-89,
        210
    The Rhetoric, 10
    training in, 12, 20, 22-23, 27
Rio Accords, 1
Rockefeller, John D., Jr., 152
Rocky Mountains, the, 45, 154, 155, 159,
    162
Roman Empire, the, 61
romanticism, 21-22, 25-26, 62, 67
Roosevelt, Franklin D., 152
Roosevelt, Theodore, 27

Santa Barbara Channel, 244, 252-53
sauna, 47-48, 49
science and technology, 4, 18, 21, 26, 29,
    32, 34-35, 44, 77-78, 123, 140-41, 177,
    208-9, 236, 243
Sierra Nevada Mountains, the, 62-63, 66,
    74 n
solid waste, 177, 179, 180-82, 184-85, 189,
    193, 231, 234
Superfund, 1, 105
sustainability, 118, 119, 219, 222, 229, 231,
    238, 260
symbolic politics, 89

Thoreau, Henry David, 26, 27, 42, 66, 162
timber industry
    in Finland, 49-50, 55 n
    lobbying, 103, 117
    policy, 117, 154

practices, 53-54, 82, 242-43, 259-61,
    270, 274
and deforestation, 1, 129, 154, 163, 228
value of, 201
toxins, 1, 76, 82, 89, 101, 125, 129, 249,
    251
tradeoffs, 52, 87
transcendentalism, 21-22, 27
tree spiking, 261-62, 266-67, 269
Turner, Frederick Jackson, 9, 10, 22-25, 27,
    29, 33-35

values
    aesthetic, 30-31, 58, 59-60, 67, 68, 72,
        89, 165, 238
    conflict, 99, 100-101, 103, 135
    economic, 1, 4, 20, 51, 79, 85, 100-101,
        136, 140, 163, 213-14, 235
    enlightenment, 20, 28
    environmental, 4, 35, 71-72, 224
        and consumption, 71-72, 234
        and deep ecology, 113, 221
        and news reporting, 249-51, 257
        and politics, 105, 220, 221-22
        and pressure groups, 100-101
        and social ecology, 35, 44, 97-98, 133,
            219
    scientific, 35
valuation
    contingent, 202-3
    cost-benefit, 204-5, 209, 213
    total value paradigm, 202-3, 204, 212,
        215
    travel cost, 201-2
Virginia Company, the, 11

wilderness
    as laboratory, 18
    as obstacle, 15-16, 160
    as rejuvenator, 13, 16-17, 21, 26, 67, 138,
        169, 173
    as threat, 11, 14
    value of, 16, 101, 159-60, 200, 202, 214-
        15, 237
    versus development, 13, 20, 26, 28, 52-
        53, 55 n, 70, 89, 103, 160
Wilderness Act, 1